Methods For Monitoring Tiger And Prey Populations

K. Ullas Karanth • James D. Nichols
Editors

Methods For Monitoring Tiger And Prey Populations

Springer

Editors
K. Ullas Karanth
Wildlife Conservation Society (WCS)
New York, NY, USA

Centre for Wildlife Studies
Bengaluru, India

Wildlife Conservation Society
India Program
Bengaluru, India

National Centre for Biological
Sciences-TIFR
Bengaluru, India

James D. Nichols
Crofton, MD, USA

ISBN 978-981-10-5435-8   ISBN 978-981-10-5436-5 (eBook)
DOI 10.1007/978-981-10-5436-5

Library of Congress Control Number: 2017953042

Printed on acid-free paper

This Springer imprint is published by Springer Nature
The registered company is Springer Nature Singapore Pte Ltd.
The registered company address is: 152 Beach Road, #21-01/04 Gateway East, Singapore 189721, Singapore

Foreword

If wild tigers were easy to study, every aspiring field researcher would want to do so. The elegance, ecological role, and aura of this top predator exert a magnetic pull that extends far beyond biologists. Surveys undertaken by conservation groups show that among the general public, tigers consistently come out as the most popular wild species on Earth. The irony, of course, is that catching a glimpse of a tiger outside of zoos and in its natural habitats is often, at best, a once-in-a-lifetime event. It should thus come as no surprise that since George Schaller's pioneering study of wild tigers in India's Kanha National Park over 50 years ago, relatively few long-term field studies have been successfully carried out on this highly territorial, secretive large carnivore. The authors in this volume make up a large subset of those intrepid biologists who have crouched on jungle trails to set camera traps, pick up scats, and look for other signs or monitored the prey of this big cat.

Even though few biologists have had the good fortune to study tigers in the wild, everyone wants to know about their numbers, especially conservationists, for without accurate counts and projections, it's difficult or even impossible to devise adequate conservation strategies. Those who have studied tigers are repeatedly plied with the same four questions: How many tigers are left in the wild? Are their numbers increasing or decreasing? Will they go extinct? What can we do to save them? These are the questions to which government wildlife officials in the tiger range states are hounded to provide answers by concerned citizens and the popular press. Unless we have robust, repeatable methods to answer questions accurately about changes in tiger numbers and forces affecting them, officials must answer by reverting to guessing or crude surveys. The reliance on outdated and less accurate approaches does little to advance science and even less to determine the true fate of wild tigers.

Enter this volume. A gathering of scientists, under the leadership of Ullas Karanth and Jim Nichols, addresses the most important topics related to accurately assessing the status of tigers and their prey and, we can hope, monitoring their recovery. It is a critical task to share information on best ways of analyzing numbers for tiger and tiger prey researchers, for concerned others, and for those who study other species in the region. In the ongoing effort to save and recover wild Asia, the content of this book provides the monitoring bible.

In the 1970s and even today, India has been home to the majority of the world's wild tigers, and it is on that country that the authors focus much of their attention. In the late 1960s, wildlife officials became aware that poaching of tigers and prey and loss of habitat were leading to precipitous drops in the number of tigers in the wild. The prevalent technique at the time to estimate tiger numbers was the well-worn pugmark method, relying on the claim that the footprints of tigers, measured in the appropriate substrate, served as identifying "fingerprints," and adult and subadult tigers could be identified as distinct individuals from tedious field measurements. Ullas Karanth in the 1980s illustrated the inaccuracy of this approach through a clever controlled experiment: allowing a known number of zoo tigers to make tracks and inviting official "tiger pugmark identification experts" to attempt to determine how many tigers had actually been present. The results of this experiment put to rest the idea that "seasoned experts" could eyeball correctly the number of tigers in an area based on an examination of pugmarks. Tracks were useful for establishing the presence of tigers but not to determine their numbers. Karanth and coauthors followed this up with a rigorous overall critique of the pugmark census method (Karanth et al. 2003: "Science deficiency in conservation practice: the monitoring of tiger populations in India," *Animal Conservation* 6:141–146), which eventually led to the method's official demise in 2005.

From the early 1990s, Karanth and his WCS colleagues worked to replace the pugmark method with a new technology—automatically triggered camera traps—that, when used with a systematic sampling design, could begin to improve estimation techniques. Guided by statistical guru Jim Nichols from USGS, the team revolutionized how best to sample wild tigers and, by extension, how populations of many other cryptic but individually identifiable animal species that occur at low densities could be reliably, and noninvasively, estimated.

By 1995, researchers were also using new GIS mapping techniques to determine the tiger's range as a series of landscapes called tiger conservation units (TCUs), formed by the known presence of tigers or their suspected presence and by their ability to disperse. The resulting exercise identified places where tigers could be conserved. It also identified giant polygons as "survey landscapes," large blocks of presumed habitat where no information on tiger status was available. These polygons, mapped in purple, caught the attention of Karanth and his disciples and became the targets of field efforts, many of these far from the boundaries of India. Ten years later, all of the survey TCUs had been thoroughly camera-trapped or at least visited by camera-trapping teams.

These results were collated again in 2005 and published in a series of papers, one of which, "The Fate of Wild Tigers" by biologists from the World Wildlife Fund and the Wildlife Conservation Society, divided the tigers' truncated ranges into 76 tiger conservation landscapes. The critical finding of their analysis could be termed "range collapse": tigers occupied 40% less habitat in 2005 than they had a decade earlier according to a 1995 study. The structural habitat was still present to support tigers, but either the tigers, their large prey, or both had been hunted out. Even accounting for some error in the comparison, the interpretation of the intensive camera-trapping campaign between 1995 and 2005 can be summed up in a depressing, short phrase: learning a lot more about fewer tigers.

Recognizing that a shift in strategy was needed to avoid the extirpation of tiger populations, several tiger biologists hatched the idea of a Global Tiger Initiative (GTI), which culminated in staging a Global Tiger Summit in 2010 in St. Petersburg, Russia. The attending heads of state, hosted by Vladimir Putin, and subsequently other range state leaders not in attendance agreed to the global goal of doubling the wild tiger population from about 3200 in 2010 to upward of 6500 by 2022, the date of the next Year of the Tiger. Suddenly, the work of Karanth and Nichols took on a fresh new urgency: how to provide reliable information to monitor the path to this milestone of recovery.

That work continues in earnest to this day, and the initial chapters here focus on the latest approaches to answering several basic questions: Why are you undertaking the monitoring program? What aspect of tiger or prey biology are you trying to address? How do you go about it in the most robust way? It is an ideal beginning, to state the questions clearly at the outset, to avoid wasting time and money in what could otherwise be glorified fishing expeditions.

A subsequent chapter (Chap. 5) discusses five other questions that biologists and technicians should ponder before proceeding. Another reason this book needs to be widely circulated is that we need to ensure that the estimates of tiger numbers and occupancy are accurate to make sure that the interventions on behalf of tiger conservation are actually working. The GTI helped catalyze a number of range states to allocate more funds for tiger protection. By answering the questions laid out here and adopting the best practices described in this volume, we stand the best chance of these resources being invested wisely.

A further pathbreaking finding of Karanth and his co-workers was that tiger density can be predicted by prey density (Karanth et al. 2004: "Tigers and their prey: Predicting carnivore densities from prey abundance," *Proceedings of the National Academy of Sciences*, USA 101:4854–4858). This insight builds on work of Karanth's mentor, felid biologist Mel Sunquist, who remarked at a conference in 1980 that "to be a tiger biologist, you really need to be a deer biologist." So it is welcome that this volume includes several important chapters on monitoring prey density. If being a tiger biologist means being a deer biologist, one could argue that being a deer biologist requires an understanding of plant ecology, soil science, and geology. Underpinning the correlation of tiger densities with large prey densities is another rarely discussed feature: alluvial soils produce the densest population of tigers because they support the highest concentrations of large herbivores. Soil fertility thus may be one of the ultimate drivers of tiger densities.

By incorporating techniques for monitoring tigers and their prey that can apply to many other large mammals, the authors greatly extend the reach and value of their work. A further benefit of the volume is the penultimate chapter, a treatise on connectivity, a vital aspect of conserving many large terrestrial mammal populations. If tigers are to have a future in Asia, they must be managed as metapopulations—populations linked by dispersal. The fact remains that not a single tiger reserve among the 350 or so in the tiger's range is large enough to maintain a viable population. Establishing corridors connecting tiger conservation areas and developing a healthy matrix through which to disperse are vital.

What will tiger and prey monitoring look like 15 years hence? Advances in camera trapping, remote sensing, and GIS modeling led to a quantum leap in how we have studied tigers and tiger habitats over the past 20 years. Could the emerging technologies of computer vision, machine learning, miniaturization of electronics, smart drones, and long-range (LoRa) radio frequency communication further revolutionize how we count or detect wildlife? Sensors that use cryptic cameras equipped with computer vision algorithms to detect humans (poachers) and relay such information via GSM or LoRa networks will soon be in place in African wildlands. It is only a half-step more to insert new low-cost, computer vision chips into such cameras that can do "onboard" processing to detect tigers, sort and identify them, and transmit photos in real time to a researcher connected to the Internet. When these sensors are mass produced at "cents to the dollar" per unit, it would be only another small step to expanding camera-trapping grids to systematically cover the entire tiger landscape using linked sensors and thus obtain a total census.

These potential developments raise a final question: Will our sophisticated ability to monitor tigers and their prey then be matched by the development of the necessary political will, interest in protecting enough habitat, ability to gain the local buy-in to save tiger landscapes, and then mobilization of sufficient resources to restore thriving populations of tigers and other populations of large Asian mammals? If we only set aside large blocks of habitat and left tigers and their prey alone, there would be much less need for a book like this. But because the development of the tiger range states, among the fastest in the world, has ramped up the pressures, this volume becomes even more vital.

Biodiversity and Wildlife Solutions Program
Washington, DC, USA

Eric Dinerstein

Preface

The tiger is a global conservation icon. Because of its popular appeal, which has religious, cultural, aesthetic, and social roots, an impressive array of leaders in these spheres has also seriously engaged with tiger conservation over the past 50 years. A quick chronological, but nonstatistical, sample includes Indira Gandhi (post-1969), King Birendra of Nepal (1970s), Dalai Lama (1990s), and Vladimir Putin (2000s). Tiger range state officials, tiger scientists, national and international NGOs, and conservation advocates at various levels have all toiled hard to stem the tiger's slide, into what appeared to be inevitable extinction by the mid-1960s. Some of these efforts have succeeded, while many others have failed.

One of us (Karanth) got involved in tiger conservation as an amateur naturalist in the 1970s, around the same time as the other (Nichols) began his professional career as a quantitative ecologist. At that time, in spite of conservation efforts, rigorous science was excluded from the critical task of assessing how tiger populations were faring. Such assessments required a thoughtful synthesis tiger biology and quantitative ecology, relevant to specific conservation contexts. Such a methodological synthesis was simply not available.

In the late 1980s, we met for the first time, somewhat serendipitously, at the University of Florida, Gainesville. Soon after, we decided to collaborate to develop and apply rigorous and defensible methods for monitoring tigers and their principal prey species. With lots of help from various field personnel, wildlife managers, as well as other tiger biologists and quantitative ecologists, we developed a set of monitoring methods that we believed to be useful and representative of the state of the art in inference methods for animal populations.

Application of these methods in India demonstrated their feasibility and utility to the point where we believed that we should recommend them to others involved with tiger conservation monitoring. We organized an international technical workshop, sponsored by the Wildlife Conservation Society (WCS) at Nagarahole Reserve, India, in January 1999. The workshop was conducted in association with Indian government's Project Tiger and some state forestry departments and included several wildlife managers, carnivore biologists, and biostatisticians. Several population monitoring approaches were discussed in detail, with topics ranging from the collection of field data to their final statistical analysis.

As a direct result of this workshop, we produced the edited volume, *Monitoring Tigers and Their Prey: A Manual for Wildlife Researchers, Managers and Conservationists in Tropical Asia* (Karanth and Nichols 2002). The intent of this volume was to provide guidance about field monitoring and statistical methods that we believed to be superior (in the sense of yielding stronger inferences), compared to methods that had been used for decades in India and the rest of tiger range. Recognizing the natural reluctance of practitioners to discard such historical methods in favor of new approaches, we made a special effort to present the logic and rationale underlying our recommended approaches. We tried to describe field methods in sufficient detail that they could be duplicated by others and to provide simple numerical examples to aid understanding of some of the analytic approaches that we were recommending. In 2008, WCS collaborated with Trust for Environmental Education, India, to produce a 47-minute video guide that clearly depicted the various monitoring methods described in Karanth and Nichols (2002). Both PDF version of the book and the video guide were made freely available to all users by WCS on the Internet (http://wcsindia.org/home/media-library/).

As a result, the methods recommended in Karanth and Nichols (2002) have been widely used by carnivore ecologists, conservationists, and wildlife managers. Many users have adapted these methods to monitor not only tigers but also other big cats, such as jaguars, *Panthera onca*; cheetahs, *Acinonyx jubatus*; leopards, *Panthera pardus*; and snow leopards, *Panthera uncia*. We have been surprised and pleased that some of the ideas (e.g., on camera trapping, occupancy modeling of sign survey data) have been applied to conservation problems extending well beyond tigers and prey, to varied species and conservation issues across the globe.

We ended the penultimate paragraph of the preface to Karanth and Nichols (2002:xv) with the following two sentences: "However, we are well aware that, eventually, the methods that we present in this manual will be replaced by better methods generated through the very same process of scientific review that we endorse. In fact, we look forward to such improvements."

This methodological evolution has proceeded even more rapidly than we anticipated, to the point where we believed it necessary to develop this new volume to incorporate these advances. We were fortunate to have collaborated, once again, with an outstanding group of authors who have produced what we believe is an exceptional set of methodological chapters representing the current state of the art in animal monitoring in general and as applied to tigers and their prey specifically.

This new volume differs from the 2002 volume in important ways, and we believe that the relationship between the two volumes merits some discussion here. For the purpose of brevity, in this comparison, we will refer to the earlier manual as *KN (2002)* and the present volume as *KN (2017)*.

First, we note that the methods presented in *KN (2002)* are still foundational and useful. Unlike the earlier tiger census methods that they replaced, these methods have sound conceptual underpinnings and still "work." Rather than scrapping these methods, many of the chapters of *KN (2017)* extend the modeling of *KN (2002)* in exciting and novel ways that increase their value. Even the major "new" method in *KN (2017)*, spatially explicit capture-recapture modeling, can be viewed as a marriage between two classes of methods described in *KN (2002)*, classical capture-recapture models and distance sampling.

An additional distinction between the two volumes is in simplicity of explanations. In both editions, we made special efforts to simplify explanations to the degree possible, hoping that readers can develop an understanding of how the described methods work. However, many of the extensions and new developments in *KN (2017)* were made possible by advances in computing power and associated numerical methods, which are not as amenable to simplified explanations and examples as the initial models described in *KN (2002)*. We have thus come to view the two volumes as complementary. The 2002 volume is probably a better starting place for readers who have never been exposed to these inference methods yet seek to develop an understanding of them. Explanations in the current volume have been simplified to the degree possible but will be more readily understood by those who have some prior familiarity with these general classes of methods. Therefore, we have made the *KN (2002)* volume and its associated video guide to readers of this volume available on the Internet (http://wcsindia.org/home/media-library/). The video guide and related visual material are also available at http://www.conservationindia.org/ and the publisher's online support (www.springer.com). We hope readers will find these features helpful.

An additional feature of this volume is a final chapter describing how the rigorous monitoring approaches explained and recommended here can fit into, and contribute to, adaptive natural resource management (Walters 1986, Nichols et al. 2007) and structured decision-making (Martin et al. 2009), which are being increasingly applied by conservation practitioners and wildlife managers in varied contexts globally. We urge tiger conservationists to seriously consider these suggestions in their own specific situations.

We both continue to strongly believe that the scientific process of peer review and publication in high-quality journals should guide the choice of appropriate methods for monitoring tigers and their prey. Therefore, we are somewhat dismayed that, in spite of availability of superior methods, tiger conservation practitioners are sometimes slow to adopt them or even use demonstrably flawed or obsolete methodologies. We believe this is largely because of intellectual inertia, rather than resource constraints, given the current levels of investment. Unfortunately, we can offer no methodological cure for this problem.

As is the way of all scientific progress, the animal monitoring approaches that authors in this volume recommend too will eventually be superseded by superior ones. As editors, we hope that this volume will inspire cohorts of talented carnivore biologists and quantitative ecologists who will follow, to seriously engage with the innovations necessary to rapidly render our current effort obsolete. The still precarious fate of the wild tiger populations urgently demands such a proactive engagement.

Bengaluru, India K. Ullas Karanth
Crofton, MD, USA James D. Nichols

References

Martin J, Runge MC, Nichols JD, Lubow BC, Kendall WL (2009) Structured decision making as a conceptual framework to identify thresholds for conservation and management. Ecol Appl 19: 1079–1090

Nichols JD, Runge MC, Johnson FA, Williams BK (2007) Adaptive harvest management of North American waterfowl populations: a brief history and future prospects. J Ornithology 148 (Suppl. 2): S343–S349

Walters CJ (1986) Adaptive management of renewable resources. MacMillan, New York

Acknowledgments

Editorial Acknowledgments

We would like to thank all the authors of each chapter for their sterling contributions. The quality of their work speaks for itself and justifies our claim to present to the readers the "state of the art" as far as monitoring of tigers and their prey is concerned.

We also acknowledge the following authors for assisting us, going well beyond the call of duty, by reviewing several chapters as we raced against the submission deadline: Samba Kumar, Devcharan Jathanna, Ravishankar Parameshwaran, and Divya Vasudev. Our editorial assistant Mahi Puri is thanked for her admirable role in "herding the 31 cats" to manuscript submission stage.

We are indebted for the superb assistance from the WCS, India Program staff—H. S. Srikanth Rao, Swapna Lawrence, S. Mythri, Manish Machaiah, and Kiran Yadav—in the final stages of wrapping up the manuscript. The following photographers who contributed images are gratefully acknowledged: Ramki Sreenivasan, Michael Vickers, Jitendra Sankaraiah, Harsh Dhanwatey, Ullas Karanth, Valmik Thapar, Kalyan Varma, Killivalavan Rayar, Eleanor Briggs, and WCS.

Eric Dinerstein wrote an insightful foreword. Cristián Samper, Alan Rabinowitz, David Anderson, David Macdonald, and P. K. have appreciated our initiative. We thank them all.

Representing the publishers of Springer India, Suvira Srivastav, Aakanksha Tyagi, Joseph Daniel, as well as the printers SPi Global were helpful. We thank them for the elegant production of the book.

As we collaborated over the years, we have benefited immensely from the hard work, both on and off the field, of key colleagues associated with the WCS-India Program and Centre for Wildlife Studies. Although the number of such frontline data collection warriors is too large to acknowledge individually, key leaders in the field at various times—Samba Kumar, Arjun Gopalaswamy, Girish D. V., Killivalavan Rayar, and Devcharan Jathanna—deserve our special thanks.

On the analytical front, James Hines, William Link, Andrew Royle, and Robert Dorazio at USGS and Mohan Delampady at the Indian Statistical Institute generously shared their special expertise with our intellectual explorations. We are very grateful to all of them.

Author Acknowledgments

K. Ullas Karanth acknowledges the support from the Wildlife Conservation Society (WCS)-New York, Wildlife Conservation Society, India Program, Centre for Wildlife Studies, and National Centre for Biological Sciences (NCBS), Bengaluru. He is indebted to the Ministry of Environment, Forest and Climate Change, Government of India, and state forest departments of Karnataka, Kerala, Goa, Assam, Maharashtra, Andhra Pradesh, and Telangana for permissions provided for conducting research on tiger and prey species over the years. Karanth also acknowledges the funding support of his work from the US Fish and Wildlife Service-Office of International Affairs, Washington, DC; Liz Claiborne Art Ortenberg Foundation, New York; Satter Foundation, Chicago; and other WCS donors as well as from the Department of Science and Technology (DST) and Department of Biotechnology (DBT), Government of India. He especially acknowledges John Robinson, director of Global Conservation Programs, and Cristián Samper, president and CEO of WCS-New York, for encouraging his work.

James D. Nichols would like to thank the US Geological Survey (USGS) and US Fish and Wildlife Service (USFWS) for the administrative support of his collaboration with the Wildlife Conservation Society, India Program at different times over the years. Judd Howell and the late Greg Smith, former directors of the USGS Patuxent Wildlife Research Center, and Allan O'Connell of USGS are especially thanked for encouraging his work.

John M. Goodrich would like to thank Panthera, the Robertson Foundation, and the Cline Family Foundation, USA, for the support of his work.

G. Viswanatha Reddy thanks the Rajasthan Forest Department, Government of Rajasthan, Wildlife Conservation Society, India Program, and Centre for Wildlife Studies for the support provided to him.

Vinod B. Mathur acknowledges the support of the Wildlife Institute of India and Ministry of Environment, Forest and Climate Change, Government of India.

Hariyo T. Wibisono would like to acknowledge the ongoing support from the Department of Entomology and Wildlife Ecology, University of Delaware, USA, Forum HarimauKita, and Fauna & Flora International-Indonesia Programme.

Sunarto Sunarto would like to thank the field team of WWF-Indonesia, HarimauKita, and the Government of Indonesia for the support provided to him.

Anak Pattanavibool would like to thank the Department of National Parks, Wildlife and Plant Conservation (DNP), Government of Thailand, WCS-India Program, and WCS-New York for all the support. He would also like to thank Saksit Simcharoen and Somphot Duangchantrasiri from DNP for their support.

Melvin T. Gumal thanks the Johor National Parks Corporation, Department of Wildlife and National Parks, Department of Forestry-Peninsular Malaysia, and all the staff of WCS and donors for their support.

Dale G. Miquelle thanks the Wildlife Conservation Society-New York and the many donors who have supported his work for so many years.

Krishnamurthy Ramesh is thankful to the Wildlife Institute of India; National Tiger Conservation Authority, India; and Madhya Pradesh Forest Department for providing an enabling environment and support for his work.

Abishek Harihar would like to thank Panthera; the Robertson Foundation; the Cline Family Foundation, USA; and the Nature Conservation Foundation, India, for the support of his work.

D. Mark Rayan acknowledges WWF-Malaysia and the donors who support his work.

Kanchan Thapa thanks WWF-Nepal for all the support.

Arjun M. Gopalaswamy thanks the Indian Statistical Institute; University of Oxford, UK; and Wildlife Conservation Society-New York and Wildlife Conservation Society, India Program.

Mohan Delampady thanks the Indian Statistical Institute, Bengaluru.

Varun R. Goswami acknowledges the support from the Wildlife Conservation Society-New York, WCS, India Program, and Centre for Wildlife Studies. He also thanks Madan K. Oli, professor at the University of Florida, for encouraging his work and the US Fish and Wildlife Service for the financial support over the years.

Ravishankar Parameshwaran would like to thank the Centre for Wildlife Studies and Wildlife Conservation Society, India Program.

Soumen Dey would like to acknowledge the Indian Statistical Institute for providing support to his work.

N. Samba Kumar acknowledges the long-term support from the Wildlife Conservation Society-New York. He also thanks Mahi Puri, Ravishankar Parameshwaran, Killivalavan Rayar, and Devcharan Jathanna of the WCS, India Program for their assistance.

Mahi Puri acknowledges the support from the Centre for Wildlife Studies, Wildlife Conservation Society, India Program, and Department of Wildlife Ecology and Conservation, University of Florida.

Srinivas Vaidyanathan acknowledges the support from the staff of FERAL, India, and Wildlife Conservation Society, India Program, Bengaluru.

Samantha Strindberg would like to thank the Wildlife Conservation Society-New York.

Len Thomas would like to thank the Centre for Research into Ecological and Environmental Modelling (CREEM), University of St Andrews, UK.

Hannah J. O'Kelly wishes to acknowledge the Wildlife Conservation Society (WCS)-Cambodia Program.

Robert M. Dorazio* is grateful for the institutional support provided by the USGS Wetland and Aquatic Research Center, Gainesville, FL, USA, the Wildlife Conservation Society, India Program, and the Centre for Wildlife Studies, India. He also thanks Marc Kery, Nick Golding, and the editors (Ullas Karanth and Jim Nichols) for review comments that improved earlier drafts of his chapters.

J. Andrew Royle* is grateful for the institutional support provided by the USGS Patuxent Wildlife Research Center, Wildlife Conservation Society, India Program, and Centre for Wildlife Studies, India. He also thanks Nathan Hostetter (USGS, PWRC) and Alec Wong (Cornell University) for reviewing drafts of Chap. 9.

Devcharan Jathanna acknowledges the support of the Centre for Wildlife Studies and Wildlife Conservation Society, India Program. He would also like to thank Arjun Gopalaswamy and Meghana Natesh for discussions and clarifications on the drafts of Chap. 11.

Samrat Mondol acknowledges the Wildlife Institute of India and the DST-INSPIRE Faculty Award, of the Department of Science and Technology, Government of India, for all the support.

Uma Ramakrishnan acknowledges the long-term support from the National Centre for Biological Sciences, Tata Institute of Fundamental Research, India, the Ramanujan Fellowship, and the Department of Atomic Energy-Outstanding Scientist Award. She also acknowledges project funding support from the Wildlife Conservation Society, India Program; Department of Science and Technology, Department of Biotechnology, and National Tiger Conservation Authority, Government of India; Wildlife Conservation Trust; and Microsoft e-Science.

Olutolani Smith wishes to thank the University College London Department of Genetics, Evolution and Environment; the American Museum of Natural History, New York; Panthera; the Robertson Foundation; the Cline Family Foundation; and the Nature Conservation Foundation, India, for the support of her work.

Divya Vasudev acknowledges the financial support from the Department of Science and Technology, Government of India, and Wildlife Conservation Society, New York. She is grateful for the institutional support from the Wildlife Conservation Society-New York, Wildlife Conservation Society, India Program, and Centre for Wildlife Studies. She thanks Mahi Puri for her extensive help, Varun Goswami and the editors for commenting on drafts of Chap. 13, and Robert Fletcher (University of Florida) for his encouragement and academic input.

Bengaluru, India K. Ullas Karanth
Crofton, MD, USA James D. Nichols

*Any use of trade, firm, or product names by the author is for descriptive purposes only and does not imply endorsement by the US Government.

Contents

Contributors

J. Andrew Royle USGS Patuxent Wildlife Research Center, Laurel, MD, USA

Mohan Delampady Statistics and Mathematics Unit, Indian Statistical Institute, Bengaluru, India

Soumen Dey Statistics and Mathematics Unit, Indian Statistical Institute, Bengaluru, India

Robert M. Dorazio Wetland and Aquatic Research Center – United States Geological Survey (USGS), Gainesville, FL, USA

John M. Goodrich Panthera, New York, NY, USA

Arjun M. Gopalaswamy Statistics and Mathematics Unit, Indian Statistical Institute, Bengaluru, India

Department of Zoology, University of Oxford, Oxford, UK

Varun R. Goswami Wildlife Conservation Society, India Program, Bengaluru, India

Centre for Wildlife Studies, Bengaluru, India

Melvin T. Gumal Wildlife Conservation Society (WCS) – Malaysia Program, Kuala Lumpur, Malaysia

Abishek Harihar Tiger Program, Panthera, New York, NY, USA

Nature Conservation Foundation, Mysuru, India

Devcharan Jathanna Wildlife Conservation Society, India Program, Bengaluru, India

Centre for Wildlife Studies, Bengaluru, India

K. Ullas Karanth Wildlife Conservation Society (WCS), New York, NY, USA

Centre for Wildlife Studies, Bengaluru, India

Wildlife Conservation Society, India Program, Bengaluru, India

National Centre for Biological Sciences-TIFR, Bengaluru, India

Vinod B. Mathur Wildlife Institute of India, Dehradun, India

Dale G. Miquelle Wildlife Conservation Society (WCS), New York, NY, USA

Department of Ecology, Far Eastern Federal University, Vladivostok, Russia

Samrat Mondol Wildlife Institute of India, Dehradun, India

James D. Nichols Crofton, MD, USA

Hannah J. O'Kelly Conservation Scientist, Vientiane, Lao People's Democratic Republic

Ravishankar Parameshwaran Wildlife Conservation Society, India Program, Bengaluru, India

Centre for Wildlife Studies, Bengaluru, India

Anak Pattanavibool Wildlife Conservation Society (WCS) – Thailand Program, Bangkok, Thailand

Department of Conservation, Kasetsart University, Bangkok, Thailand

Mahi Puri Wildlife Conservation Society, India Program, Bengaluru, India

Centre for Wildlife Studies, Bengaluru, India

Department of Wildlife Ecology and Conservation, University of Florida, Gainesville, FL, USA

Uma Ramakrishnan National Centre for Biological Sciences, TIFR, Bengaluru, India

Krishnamurthy Ramesh Wildlife Institute of India, Dehradun, India

Department of Forest and Conservation Sciences, University of British Columbia, Vancouver, BC, Canada

D. Mark Rayan World Wide Fund for Nature (WWF) – Malaysia, Selangor, Malaysia

Durrell Institute of Conservation and Ecology (DICE), University of Kent, Kent, UK

G. Viswanatha Reddy Indian Forest Service, Government of Rajasthan, Jaipur, Rajasthan, India

Samantha Strindberg Wildlife Conservation Society (WCS), New York, NY, USA

N. Samba Kumar Wildlife Conservation Society, India Program, Bengaluru, India

Olutolani Smith Tiger Program, Panthera, New York, NY, USA

Department of Genetics, Evolution & Environment, University College London, London, UK

Sunarto Sunarto World Wide Fund for Nature (WWF) – Indonesia, Jakarta, Indonesia

Forum HarimauKita, Bogor, Indonesia

Kanchan Thapa World Wide Fund for Nature (WWF) – Nepal, Kathmandu, Nepal

Len Thomas Centre for Research into Ecological and Environmental Modelling, University of St Andrews, Fife, UK

Srinivas Vaidyanathan Foundation for Ecological Research, Advocacy and Learning (FERAL), Auroville, Tamil Nadu, India

Divya Vasudev Wildlife Conservation Society, India Program, Bengaluru, India

Centre for Wildlife Studies, Bengaluru, India

Hariyo T. Wibisono Department of Entomology and Wildlife Ecology, University of Delaware, Newark, DE, USA

Fauna & Flora International – Indonesia, Jakarta, Indonesia

Forum HarimauKita, Bogor, Indonesia

Blurbs

This book brings together the best minds and decades of experience in tiger conservation. We must continue to improve the way we monitor wildlife to measure the impact of our work. This authoritative volume will be an invaluable tool for us in this important task, for decades to come.

Cristián Samper
President and CEO of the Wildlife
Conservation Society, New York

There are no two people better suited to have produced this important and timely book. A follow-up to their previous authoritative classic *Monitoring Tigers and their Prey: A Manual for Wildlife Researchers, Managers and Conservationists in Tropical Asia*, this incredibly detailed and comprehensive work should be the road map for anyone attempting rigorous science and conservation on tigers and other wildlife. This work is outstanding and stands alone.

Alan Rabinowitz
CEO, Panthera
New York

This book is well written and exciting to read. The focus on tigers allows the reader to understand the deep complexities in estimating demographic parameters in animal populations. The valuable contribution here is to showcase the relevant spatial estimation theory and methods—these new thrusts are very important and useful for tigers and for a wide variety of other animal species.

David R. Anderson
Emeritus Professor of Colorado State University
Fort Collins, USA

Compendious, enlightening, and monumental with relevance far beyond tigers: if you don't know what you've got, you can't know how you are doing or how to do better, which is why counting animals and monitoring trends in their numbers are fundamental to conservation. Counting tigers is difficult, and the answers are policy dynamite, which is why the opening chapter of this benchmark book is right in asserting that its contents are critical and, while technically sophisticated, happily not daunting. In 2002, these editors compiled an earlier book in which they looked forward to their then cutting-edge ideas being superseded by others who would follow. It turns out that others didn't need to do the job; Karanth and Nichols have done it themselves!

David W. Macdonald
CBE, FRSE, Director of the WildCRU
University of Oxford, UK

I was a sponsor of the technical workshop held in January 1999, which led to the popular manual compiled by Karanth and Nichols, titled *Monitoring Tigers and their Prey*, in 2002. I am sure this new book by the same editors titled *Methods for Monitoring Tiger and Prey Populations* will prove to be as popular and useful as its predecessor.

Prashanta Kumar Sen
Former Director of Project Tiger
Government of India, New Delhi

About the Editors

K. Ullas Karanth (b:1948) is the Director for Science, Asia with the Wildlife Conservation Society (WCS)-New York, besides being Emeritus Director at the Centre for Wildlife Studies (CWS), and WCS, India Program based at Bengaluru. Originally trained as a mechanical engineer, he subsequently obtained his graduate education in wildlife biology from the University of Florida, Gainesville, USA, and Mangalore University, India. The ecology of tigers, sympatric carnivores, and their prey species and issues of monitoring of wild animal populations have been his focal areas of research since 1986. Dr. Karanth has published over 125 scientific articles. He has authored or edited 15 technical and popular books. Over the years, Dr. Karanth has served on India's Forest Advisory Committee, the National Tiger Conservation Authority, the Governing Body of the Wildlife Institute of India, and the Indian Board for Wildlife. He is a Fellow of the Indian Academy of Sciences and has received the J. Paul Getty Award from WWF-USA for conservation leadership. Dr. Karanth has been honored with India's presidential award *Padma Shri* and *Rajya Prashasti* by Karnataka State, in recognition of his services to wildlife conservation and science.

James D. Nichols (b: 1949) is now retired after 40+ years of service as a senior scientist at the Patuxent Wildlife Research Center for the US Fish and Wildlife Service and then the US Geological Survey. He received his undergraduate degree in biology from Wake Forest University, his master's degree in wildlife management from Louisiana State University, and his doctorate in wildlife ecology from Michigan State University, in the USA. His research interests are animal population dynamics and wildlife management with a special focus on the estimation of demographic parameters. He has published over 400 scientific articles and is widely recognized for his contributions to wildlife ecology and management. Dr. Nichols is a fellow of the Ecological Society of America and the recipient of various awards including the Outstanding Publication Awards from The Wildlife Society and the American Statistical Association. He is a recipient of the Aldo Leopold Award (The Wildlife Society), Wetland Conservation Achievement Award (Ducks Unlimited), Award of Excellence (Biometrics Working Group of The Wildlife Society), US Presidential Rank Award (Meritorious Senior Professional), and Wings Across the Americas Award (US Forest Service for outstanding contributions to bird conservation).

Role of Monitoring in Global Tiger Conservation

1

K. Ullas Karanth, James D. Nichols, John M. Goodrich, G. Viswanatha Reddy, Vinod B. Mathur, Hariyo T. Wibisono, Sunarto Sunarto, Anak Pattanavibool, and Melvin T. Gumal

1.1 Introduction

This chapter sets the overall context by providing a brief overview of the historical and current status of wild tiger populations and social, cultural, and scientific perspectives on the tiger. It also covers aspects of population biology of tigers, history of conservation efforts, and the need for reliable monitoring for advancing

K.U. Karanth (✉)
Wildlife Conservation Society (WCS), New York, NY, USA

Centre for Wildlife Studies, Bengaluru, India

Wildlife Conservation Society, India Program, Bengaluru, India

National Centre for Biological Sciences-TIFR, Bengaluru, India
e-mail: ukaranth@wcs.org; https://www.wcs.org/; http://cwsindia.org/; http://wcsindia.org/home/; https://www.ncbs.res.in/

J.D. Nichols
Crofton, MD, USA
e-mail: jamesdnichols2@gmail.com

J.M. Goodrich
Panthera, New York, NY, USA
e-mail: jgoodrich@panthera.org; https://www.panthera.org/

G.V. Reddy
Indian Forest Service, Government of Rajasthan, Jaipur, Rajasthan, India
e-mail: gvreddy.rajforests@gmail.com

V.B. Mathur
Wildlife Institute of India, Dehradun, India
e-mail: vbm@wii.gov.in; https://www.wii.gov.in

K.U. Karanth, J.D. Nichols (eds.), *Methods For Monitoring Tiger And Prey Populations*, DOI 10.1007/978-981-10-5436-5_1

tiger science as well as conservation. We also touch upon typical monitoring goals relevant to different ecological and social contexts. Our primary goal here is to convince field practitioners that approaches presented in this volume are not as daunting as they may appear at first glance: they are more sound and practical, compared to many of the methods on which large investments are currently being made.

1.1.1 Status of Wild Tigers

Tigers (*Panthera tigris*) were once widespread across Asia, with a distributional range spanning 30 present-day countries, stretching across a vast region (latitudes 53^0 52′ N to S 8^0 51′ N and longitudes 46^0 42′ E to 134^0 24′ E). After modern humans colonized Asia (~60,000 years BP), forest clearance for shifting cultivation, followed by settled agriculture and livestock raising (~10,000 years BP), squeezed tiger habitats. Expanding human settlement brought tigers under great pressure. In the escalating conflict with people over land and livestock, tigers were systematically hunted out. Finally, with the advent of steel traps and snares—and later firearms, explosives, and chemical poisons—hunters virtually extirpated tigers from most agricultural tracts (Karanth 2001). For example, a molecular genetic estimate (Mondol et al. 2009b), with wide variance, suggests a median estimate of 58,000 tigers for peninsular India ~500 years ago. Just 150 years ago, the extent of tiger habitat across Asia may have exceeded 13 million km^2, an area the size of China and India combined (computation based on Walston et al. 2010, Goodrich et al. 2015).

H.T. Wibisono
Department of Entomology and Wildlife Ecology, University of Delaware, Newark, DE, USA

Fauna & Flora International – Indonesia, Jakarta, Indonesia

Forum HarimauKita, Bogor, Indonesia
e-mail: beebach66@yahoo.com; http://canr.udel.edu/enwc/; https://www.harimaukita.or.id/

S. Sunarto
World Wide Fund for Nature (WWF) – Indonesia, Jakarta, Indonesia

Forum HarimauKita, Bogor, Indonesia
e-mail: macandahan@gmail.com; http://www.wwf.id/en; https://www.harimaukita.or.id/

A. Pattanavibool
Wildlife Conservation Society (WCS) – Thailand Program, Bangkok, Thailand

Department of Conservation, Kasetsart University, Bangkok, Thailand
e-mail: anakp@wcs.org; https://thailand.wcs.org/; http://conservation.forest.ku.ac.th/

M.T. Gumal
Wildlife Conservation Society (WCS) – Malaysia Program, Kuala Lumpur, Malaysia
e-mail: mgumal@wcs.org; http://malaysia.wcs.org/

During the nineteenth and twentieth centuries, the following threats to tigers ratcheted up:

1. Agricultural encroachments through large- and small-scale forest conversions.
2. Major infrastructure and industrial development, such as railways, roads, mines, irrigation, and power projects.
3. Intensive industrial-scale logging by governments and companies as well as severe forest biomass exploitation by rural communities and their livestock.
4. Large-scale sport hunting by social elites and bounty hunting of tigers for conflict mitigation. Depletion of wild ungulate populations by local hunters. For example, in the British Indian Empire, in 50 years between 1875 and 1925, over 80,000 tigers were killed off by sportsmen and bounty hunters (Rangarajan 2001).
5. In the late twentieth century, increased demand for tiger body parts as medicinal curatives and decorative trophies from newly affluent consumers in China and East Asia has emerged as another major threat (Nowell and Ling 2007).

Under such intense pressures, tigers were successively extirpated: from Bali Island by the 1940s, from West Asia by the 1950s, and from Java, Korean Peninsula, Central China, Cambodia, and Vietnam by the late twentieth century (Goodrich et al. 2015). In the last 200 years, the tiger's range has shrunk by >95%, down to ~ 638,000 km^2 spread across 11 countries: India, Nepal, Bangladesh, and Bhutan (all in South Asia) and Myanmar, Thailand, Malaysia, Indonesia, and Lao PDR where they may be virtually extinct (all in Southeast Asia). Tigers also survive in the Russian Far East and adjacent areas of China in Northern Asia (Goodrich et al. 2015; Fig. 1.1).

An educated guess is that ~4000–5000 wild tigers may now survive in Asia, with 75% occurring in Southern Asia, which supports only 30% of remaining habitat. Tiger populations are now secure only in a few protected areas, mostly in India. They are virtually confined to "source populations" occupying about 6% of overall habitat range wide (Walston et al. 2010, Wikramanayake et al. 2011).

1.1.2 Social and Cultural Underpinnings of Tiger Conservation

The tiger is the largest among living felid species. It is an apex predator in ecological communities as diverse as the snow-bound taiga to steaming tropical jungles. Its primary prey are large ungulates. Undoubtedly, the tiger must have made a deep impression on the first humans who colonized Asia ~60,000 years ago. Tigers preyed on human beings and were feared greatly for that reason (McDougal 1987). Both admired and feared, the tiger is deeply embedded in Asian cultures: from the early animistic ones to the Hindu, Taoist, Buddhist, Christian, and Islamic faiths that historically swept across Asia. Tiger iconography is pervasive: in caves and shrines of tribal cultures as deities or spirits and in association with various gods, saints, prophets, warriors, and kings (Boomgaard 2001, Thapar 2011). However, this human fascination has not helped the tiger much in its historical struggle for survival

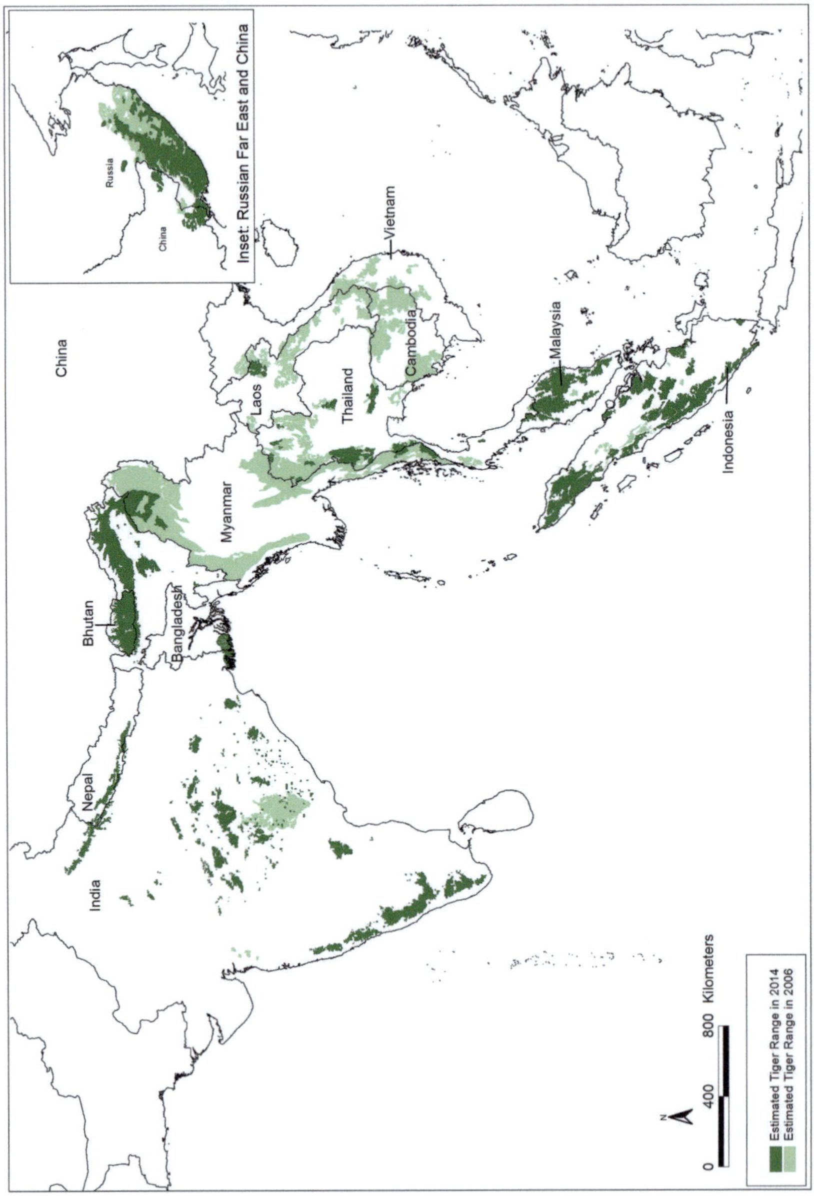

Fig. 1.1 Distribution of wild tiger populations (tiger range) 2006–2014 (Goodrich et al. 2015)

(Karanth 2001). Overhunting of tigers for sport in the past and their continued killing for human consumption are both culturally rooted.

More recently, this fascination for tigers has turned benign, with the cat serving as a mascot of commerce for diverse products such as beer, petrol, breakfast cereals, sports teams, and even hedge funds. The same fascination is now inspiring a rearguard action to save tigers. Because of its wide cultural appeal, the tiger is now a flagship of all global species conservation efforts.

1.1.3 Scientific Underpinnings of Tiger Conservation

Until the 1960s, all knowledge of tiger ecology came from accounts by hunters and amateur naturalists. The first application of methods of modern wildlife biology, such as analyses of tiger scats and kills to understand their food habits or observations at baits to study tiger behavior, was by George Schaller in Central India (Schaller 1967).

Scientific tiger studies advanced greatly with the application of the newly developed radiotelemetry techniques in the early 1970s in Nepal (Sunquist 1981, Smith 1993), in India (Karanth and Sunquist 2000) and Russia (Goodrich and Miquelle 2010) in the 1990s, and in Thailand after 2000 (Simcharoen et al. 2014). These studies have generated fine-grained data on the tiger's spatial ecology, social organization, and behavior.

Application of molecular genetic techniques to tiger DNA obtained from tissue and blood samples has opened up new paths to study tiger taxonomy and evolution (Luo et al. 2004, Mondol et al. 2009b, Wilting et al. 2015). Individual identification using fecal DNA has advanced noninvasive population studies (Mondol et al. 2009a; Chap. 11).

The critical need for practically estimating wild tiger numbers with sufficient statistical rigor was addressed only in the 1990s, with the availability of inexpensive, rugged camera traps activated by tiger movement (Karanth 1995). Such traps could obtain "samples" of tiger photos from wild populations, enabling identification of individuals from their unique stripe patterns. In combination with capture–recapture statistical models (see Chaps. 9 and 10), photographic "capture histories" of individuals were obtained. These enabled rigorous estimation of tiger numbers (Karanth and Nichols 1998).

Many photographic capture–recapture studies have been conducted across the tiger's range, for example, Karanth et al. (2004a, b) and Jhala et al. (2011, 2015) in India, Simcharoen et al. (2007) in Thailand, Rayan and Mohamad (2009) in Malaysia, Sunarto et al. (2013) in Indonesia, Thapa and Kelly (2016) in Nepal, and Xiao et al. (2016) in China. Systematic, long-term camera trap studies using "open-model" analyses in India (Karanth et al. 2006) and Thailand (Duangchantrasiri et al. 2016) have enabled even the estimation of difficult parameters such as rates of survival, recruitment, and transience, providing a comprehensive understanding of dynamics of wild tiger populations reliably and cost-effectively. Because of its practical utility and rigor, the photographic capture–recapture method has now become the predominant method for population surveys of tigers across their range.

1.1.4 History of Tiger Conservation Efforts

Sporadic efforts to preserve tigers, as trophies to be exclusively hunted by social elites, gathered pace during the nineteenth and twentieth centuries. These efforts included exclusive hunting reserves, official tiger hunting licenses, and "bag limits." However, because tiger eradication efforts and habitat encroachment also continued in the wider landscapes outside of hunting reserves, early preservation efforts could not arrest the overall decline of tigers (Boomgaard 2001, Karanth 2001, Rangarajan 2001). Tiger recovery efforts launched under internationally sponsored initiatives after 1969 (Jackson 1999) have been more successful.

However, it is a telling commentary on the difficulties of recovering wild tigers that, after nearly five decades of effort, the needle of tiger numbers probably remains pretty much where it was in 1969. All the key threats to tigers persist: hunting of their prey by local people, direct killing of tigers either for trade or in conflicts, and pressures on habitats from rural as well as urban sectors of Asia's high-growth economies.

Even after preservation efforts began, tigers were extirpated from Iran, the two Koreas, Vietnam, Cambodia, and most parts of China. Tigers appear to have declined in Russia, Lao PDR, Malaysia, Thailand, Bangladesh, and Indonesia. They seem to be just about holding their ground in India, Nepal, and Bhutan. Currently tigers are listed as *endangered* by the World Conservation Union (Goodrich et al. 2015) and included in Appendix I of the Convention on Trade in Endangered Species (CITES).

On the positive side, successful tiger population recoveries that occurred in several reserves in India's Western Ghats, Terai, and Central regions, as well as in the Western Forest Complex of Thailand and Tambling in Indonesia, clearly show that effective management responses can recover depleted tiger populations. There is substantial scope for such recoveries over the 1.5 million km^2 area of potential habitat that still remains. It is clear that recovery efforts must focus on increasing tiger numbers in surviving source populations in order to produce the surpluses necessary to repopulate "sink" landscapes beyond (Karanth et al. 2010, Walston et al. 2010).

It can be argued that past massive investments in tiger recovery have not paid commensurate returns. A contributory factor has been the long history of weak monitoring of tiger numbers used to evaluate the effectiveness of management (Karanth et al. 2003, Karanth 2011, Gopalaswamy et al. 2015). The ability to learn from past successes and failures (see Chap. 14) is hamstrung without reliable data on tiger distribution and numbers. Addressing this lacuna is critical to sustain future tiger recovery.

1.2 Why Monitoring Tiger Populations is a Critical Conservation Need

There are three important reasons why monitoring of wild tiger populations should be a rigorous process, rather than a periodic managerial ritual as is often the case:

1.2.1 Planning, Targeting, and Managing Conservation Interventions

Estimates of numbers and spatial distributions of tigers are critical to plan and implement conservation interventions. Selection of the best sites for tiger recovery, their extent and configuration within wider landscapes, location of protection infrastructure, and targeted interventions should all be based on population data on tigers and their prey. Goals for future tiger population recovery must necessarily rest on rigorous assessments of tiger population dynamics—before, during, and after the anticipated recoveries.

1.2.2 Tracking and Auditing Conservation Successes or Failures

Tiger conservation efforts may succeed, or fail, due to internal management factors or extraneous social ones. Typically, a number of surrogate metrics for conservation success are used on an ad hoc basis. Such surrogates may include amount of money invested, local employment or income generated, human welfare measures undertaken, or the number of staff deployed. Assessments of threats to tigers or habitats are sometimes made either in the field or with remotely sensed data. However, it is not possible to be certain about the utility of such surrogate metrics, without *directly* measuring the response of tiger and prey populations (Karanth and Nichols 2002). For example, in Lao PDR, conservation interventions appeared to reduce poaching and increase prey populations, yet tigers were still lost because poaching intensity was not reduced sufficiently (Johnson et al. 2016). In an even more extreme case, after showing healthy numbers (using an unreliable counting method), tigers rapidly plunged to extinction in Sariska and Panna reserves of India during 2003–2004 (Karanth 2011). Tiger extirpations, which have initially gone undetected because of poor survey methods, added up to an overall range loss of 42% between 2006 and 2010 (Goodrich et al. 2015).

1.2.3 Adaptively Learning and Making Management Predictions

Experience from other well-studied examples of conservation programs, such as those involving management of North American duck hunting or horseshoe crab harvests in Chesapeake Bay (Williams et al. 2002), shows that conservation funds and efforts can be deployed optimally and social friction among stakeholders minimized, when management is demonstrably adaptive rather than merely "seat-of-the-pants." Adaptive management is designed to learn about population responses to management actions, thus improving management decisions as time goes on. Such adaptive management would require learning from results of interventions on a continual basis (see Chap. 14). It would certainly involve statistically rigorous measurement of demographic parameters (Williams et al. 2002, Karanth et al. 2006, Gopalaswamy et al. 2015, Chap. 9).

Although it may not be easy to apply a formal adaptive management structure to tiger conservation, given the current pervasiveness of subjective and ad hoc managerial decision-making in tiger conservation, the need for better evidence-based approaches is clear (Chap. 14).

1.3 Demographic Parameters to be Monitored in Tiger and Prey Populations

Tiger populations may occur as single "source" populations or as interconnected meta-populations (Chap. 13). In some cases, they may have the potential to become connected through suitable interventions. Clusters of breeding female territories (Chap. 2) are typically embedded in much wider "sink" landscapes. The abundance of ungulate prey species will be the single most important determinant of potential size for most tiger populations (Karanth and Stith 1999, 2004a, b, Miquelle et al. 1999).

Two broad questions are asked while trying to measure tiger population parameters: (1) where are the tigers found (the spatial distribution question) and (2) how many tigers are there at each location (population density/abundance questions). The former involves measurement of tiger-occupied space, and the latter involves estimating tiger numbers.

Investigations of both of these questions are frequently focused on characteristics of a tiger population at a given single point in time. However, conservationists are often more interested in measuring changes in tiger habitat occupancy, or in tiger numbers, over multiple years. Indeed, conservation entails implementation of management actions designed to bring about specified changes to focal populations. Assessment of such changes is thus a key to useful conservation.

Sometimes, conservationists are interested in assessing connectivity between two or more source populations. Therefore, they might want to assess movement rates of tigers among protected areas (Chap. 13).

In some cases, it might be practical only to get some idea of "relative abundance" of tigers (or prey species) rather than absolute numbers or densities. However, the formidable methodological challenges of achieving even this seemingly easier goal are often not understood by managers (Chaps. 3 and 10, Williams et al. 2002, Gopalaswamy et al. 2015). Furthermore, effort and expense involved in getting such "indices" (Jhala et al. 2011, 2015) are often not significantly lower than reliably estimating densities. As an example, one can compare the protocols labeled

Phase IV section 2 and 3 recommended by India's National Tiger Estimation effort (NTCA 2012): while both protocols are officially approved, the approach in section 2 that yields only "minimum counts" and weaker indices involves as much effort and as many resources as the section 3 protocol which can yield a comprehensive picture of tiger and prey population dynamics.

Sometimes conservationists may want to know how intensively tigers (or prey species) may use particular patches of habitat within a reserve. In this case, the goal is not to measure animal distribution or "habitat occupancy." It is only to measure intensity of the animal's use of habitat in a smaller area within its wider distributional range. Often the differences between these two definitions of "occupancy" are not appreciated by practitioners, leading to poor survey designs (Chaps. 4 and 5).

1.3.1 Measuring Spatial Distribution and Habitat Use Intensity

Several kinds of questions relevant to conservation can be addressed by investigating the spatial distribution of animal populations. We list below some kinds of estimation efforts that could prove useful:

1. Estimating overall tiger/prey population distribution across wider regions or landscapes (true habitat occupancy) and identifying ecological and management covariates (e.g., prey presence, human disturbance) that might drive these occupancy rates. Assessing temporal changes in habitat occupancy across large landscapes and likely covariates influencing them (Chaps. 4 and 5).
2. Assessing intensity of habitat use by tigers or prey across smaller areas, such as single reserves, to study habitat selection behaviors and covariates influencing them (Chaps. 4 and 5). These covariates can be ecological- (e.g., prey density) or management-related factors (e.g., human disturbance).
3. Assessing connectivity and movement among source populations is necessary to maintain long-term demographic and genetic viability. Such assessments of connectivity can indirectly measure structural habitat features such as type of land cover or, more usefully, functional responses of animals directly (Chaps. 4, 5, and 13).

1.3.2 Measurement of Potential Tiger Numbers and Actual Population Dynamics

Inferences about tiger populations and their dynamics are relevant to a variety of conservation issues, as exemplified by the following kinds of investigations:

1. Assessing the potential carrying capacity of tiger source populations in key conservation sites by estimating prey density and abundance (Chaps. 6, 7, and 8) to know "how many tigers" can be there.

2. In critical tiger habitats and key reserves, estimating tiger densities and abundances, as well as annual changes in these parameters. Over the years, estimating rates of survival, recruitment, and movement parameters that drive changes in tiger abundance (Chaps. 9 and 10).
3. Investigating ecological and management covariates likely to influence demographic factors such as survival, recruitment, and movement in tiger source populations (Chaps. 9 and 10).

1.4 Challenges and Opportunities in Monitoring Tiger and Prey Populations

It is important to realize that a tiger monitoring program is not just a routine set of activities to be carried out each year in isolation. Instead, monitoring should be an integral part of an overall program of science or management and thus should be driven by the need to answer specific scientific or management questions (Chap. 14). The initial step in establishing a monitoring program is thus to identify the scientific questions or conservation decisions that the program is intended to inform. The monitoring is then tailored to these identified needs.

Here we briefly list the typical challenges that tiger conservationists or managers face, when they establish any monitoring program:

1. Ecological traits of tigers (including low population density) and behavioral traits (e.g., large home range size, elusive behavior).
2. Features of the physical environment, such as terrain (steep cliffs, wide rivers), ease of access (roads, waterways), and substrate and seasonal conditions (presence of snow for tracking, torrential rains).
3. Administrative and social factors, such as availability of funds, man power, technical or field skills, and local work ethic.
4. Often there are "political" constraints, such as governmental or donor pressure to report "better results" from tiger monitoring (Karanth et al. 2003, Karanth 2011). Building up the capacity of wildlife managers and reserve staff for carrying out routine monitoring is desirable. However, the technical nature of the task, continuing scientific advances, and the desirability of isolating monitoring activity from political pressures also need to be considered. We believe a collaborative approach that also involves adequately qualified scientists and capable naturalist volunteers is the best way to bring in rigor, transparency and efficiency to the monitoring of tigers.

1.5 Conclusion

The tiger—in the broadest sense including its prey species and habitats—is a powerful flagship for global conservation. In the past 50 years, major efforts to recover wild tiger populations have been made. Governments of range states have played the most important role, often in collaboration with local, national, and international conservation groups. There is no denying that great success has been achieved at specific sites, particularly in India and to a lesser extent in other regions. Generally, success has been achieved wherever availability of necessary resources as well as social and political support for tiger conservation has been high.

In spite of these sporadic successes, the status of the tiger remains precarious. While conservation efforts have prevented extinctions in many places, vast areas of potential habitat are now devoid of tigers (Fig. 1.1).

More attention and investments are needed to effect significant improvements in the status of wild tigers. The necessary interventions, such as improved law enforcement, creation of more tiger-friendly habitats, or depressing the demand for tiger body parts, will require major societal commitments. Rapid economic and cultural changes now sweeping Asia present both challenges and opportunities for conservation initiatives.

Adverse human impacts on tigers are driven by politics serving populism as well as powerful economic interests. Market-driven local forest encroachment, exploitation, and infrastructure development projects pose severe threats to tigers. The reality is that wildlife managers—or even the tiger conservation community as a whole—have little influence on underlying factors such as increased incomes, rising aspirations, and changing cultures. However, managers are still mandated to recover tiger populations across a potential habitat spanning 1.5 million km^2.

Tiger conservationists must address the above challenges by being innovative while learning from past successes and even more importantly from past failures. In this context, monitoring programs should be rigorous ecological audits of conservation efforts. As we shall demonstrate in each chapter, there is much room for "creative destruction" by replacing many of the obsolete, inefficient monitoring methods with better ones. Authors here have tried hard to meet this critical need.

The proposed methods have sound statistical and ecological foundations. They have been applied by the authors and their collaborators to generate reliable results published in high-quality scientific literature. We hope this factor will motivate at least the more progressive minds among researchers, conservationists, and wildlife managers to apply these methods.

References

Boomgaard P (2001) Frontiers of fear: tigers and people in the Malay world, 1600–1950. Yale University Press, New Haven

Duangchantrasiri S, Umponjan M, Pattanavibool A, Chaiwattana S, Maneerat S, Kumar NS, Jathanna D, Srivathsa A, Karanth KU (2016) Dynamics of a low-density source population of tigers in Southeast Asia in the context of improved law enforcement. Conserv Biol 30:639–648

Goodrich JM, Miquelle DG (2010) Tiger telemetry. In: Tilson R, Nyphus PJ (eds) Tigers of the world: the science, politics and conservation of *Panthera tigris*, 2nd edn. Elsevier, New York, pp 263–276

Goodrich JM, Miquelle DG, Smirnov EN, Kerley LL, Quigley HB, Hornocker MB (2010) Spatial structure of Amur (Siberian) tigers (Panthera tigris altaica) on Sikhote- Alin Biosphere Zapovednik, Russia. J Mammal 91:737–748

Goodrich JM, Lynam A, Miquelle D, Wibisono H, Kawanishi K et al (2015) *Panthera tigris,* Tiger. The IUCN Red List of Threatened Species 2015:eT15955A50659951

Gopalaswamy AM, Delampady M, Karanth KU, Kumar NS, MacDonald DW (2015) An examination of index-calibration experiments: counting tigers at macroecological scales. Methods Ecol Evol 6(9):1055–1066

Jackson P (1999) The tiger in human consciousness and its significance in crafting solutions for tiger conservation. In: Seidensticker J, Christie S, Jackson P (eds) Riding the tiger: tiger conservation in human-dominated landscapes. Cambridge University Press, Cambridge, pp 50–54

Jhala YV, Qureshi Q, Gopal R (2011) Can the abundance of tigers be assessed from their signs? J Appl Ecol 48:14–24

Jhala YV, Qureshi Q, Gopal R (eds) (2015) The status of tigers, co-predators and prey in India 2014. National Tiger Conservation Authority, New Delhi and Wildlife Institute of India, Dehradun TR2015/021

Johnson A, Goodrich JM, Hansel T, Rasphone A, Saypanya S, Vongkhamheng C, Venevongphet SS (2016) To protect or neglect? Design, monitoring and evaluation of a law enforcement strategy to recover small populations of wild tigers and their prey. Biol Conserv 202:99–109

Karanth KU (1995) Estimating tiger *Panthera tigris* populations from camera-trap data using capture-recapture models. Biol Conserv 71:333–338

Karanth KU (2001) The way of the tiger: natural history and conservation of the endangered big cat. Voyageur Press, Inc, Stillwater

Karanth KU (2011) India's tiger counts: the long march to reliable science. Econ Polit Week 46:22–25

Karanth KU, Nichols JD (1998) Estimation of tiger densities in India using photographic captures and recaptures. Ecology 79:2852–2862

Karanth KU, Nichols JD (eds) (2002) Monitoring tigers and their prey: a manual for researchers, managers and conservationists in tropical Asia. Centre for Wildlife Studies, Bangalore

Karanth KU, Stith BM (1999) Prey depletion as a critical determinant of tiger population viability. In: Seidensticker J, Christie S, Jackson P (eds) Riding the tiger: tiger conservation in human-dominated landscapes. Cambridge University Press, Cambridge, pp 100–113

Karanth KU, Sunquist ME (2000) Behavioural correlates of predation by tiger *Panthera tigris*, leopard *Panthera pardus* and dhole *Cuon alpinus* in Nagarahole, India. J Zool (Lond) 250:255–265

Karanth KU, Nichols JD, Seidensticker J, Dinerstein E, Smith JLD, McDougal C, Johnsingh AJT, Chundawat RS, Thapar V (2003) Science deficiency in conservation practice: the monitoring of tiger populations in India. Anim Conserv 6:141–146

Karanth KU, Chundawat RS, Nichols JD, Kumar SN (2004a) Estimation of tiger densities in the tropical dry forests of Panna, Central India, using photographic capture–recapture sampling. Anim Conserv 7:285–290

Karanth KU, Nichols JD, Kumar NS, Link WA, Hines JE (2004b) Tigers and their prey: predicting carnivore densities from prey abundance. Proc Natl Acad Sci USA 101:4854–4858

Karanth KU, Nichols JD, Kumar NS, Hines JE (2006) Assessing tiger population dynamics using photographic capture-recapture sampling. Ecology 87:2925–2937

Karanth KU, Goodrich JM, Vaidyanathan S, Reddy GV (2010) Landscape scale, ecology-based management of wild tiger populations. Global Tiger Initiative, World Bank, and Wildlife Conservation Society, Washington, DC

Luo SJ, Kim JH, Johnson WE, Walt V d, Martenson J, Yuhki N et al (2004) Phylogeography and genetic ancestry of tigers (*Panthera tigris).* PLoS Biol 2:2275–2293

McDougal C (1987) The man-eating tiger in geographical and historical perspective. In: Tilson RL, Seal US (eds) Tigers of the world: the biology, biopolitics, management and conservation of an endangered species. Noyes Publications, New Jersey, pp 435–448

Miquelle DG, Smirnov EN, Merrill TW, Myslenkov AE, Quigley HB, Hornocker MG, Schleyer B (1999) Hierarchical spatial analysis of Amur tiger relationships to habitat and prey. In: Seidensticker J, Christie S, Jackson P (eds) Riding the tiger: tiger conservation in human-dominated landscapes. Cambridge University Press, Cambridge, pp 71–99

Mondol S, Karanth KU, Kumar NS, Gopalaswamy AM, Andheria A, Ramakrishnan U (2009a) Evaluation of non-invasive genetic sampling methods for estimating tiger population size. Biol Conserv 142:2350–2360

Mondol S, Karanth KU, Ramakrishnan U (2009b) Why India holds the genetic key to tiger recovery. PLoS Genet 5:1–9

Nowell K, Ling X (2007) Taming the tiger trade: China's markets for wild and captive tiger products since the 1993 domestic trade ban. TRAFFIC East Asia

NTCA (2012) A protocol on phase IV monitoring. In: Technical document No.1/2011. National Tiger Conservation Authority. Available via http://projecttiger.nic.in/WriteReadData/CMS/Protocol_Phase_IV_Monitoring_r.pdf. Accessed 4 Jan 2017

Rangarajan M (2001) India's wildlife history. Permanent Black, New Delhi

Rayan DM, Mohamad SW (2009) The importance of selectively logged forests for tiger *Panthera tigris* conservation: a population density estimate in peninsular Malaysia. Oryx 43:48–51

Schaller G (1967) The deer and the tiger. University of Chicago Press, Chicago

Simcharoen S, Pattanavibool A, Karanth KU, Nichols JD, Kumar NS (2007) How many tigers are there in Huai Kha Khaeng Wildlife Sanctuary Thailand? An estimate using photographic capture-recapture sampling. Oryx 41:447–453

Simcharoen A, Savini T, Gale GA, and Smith JD (2014) Female tiger Panthera tigris home range size and prey abundance: Importance metrics for management. Oryx 48:370–377

Smith JLD (1993) The role of dispersal in structuring the Chitwan tiger population. Behaviour 124:165–195

Sunarto S, Kelly MJ, Klenzendorf S, Vaughan MR, Zulfahmi HMB, Parakassi K (2013) Threatened predator on the equator: multi-point abundance estimates of the tiger *Panthera tigris* in central Sumatra. Oryx 47:211–220

Sunquist ME (1981) Social organization of tigers *Panthera tigris* in Royal Chitawan National Park, Nepal. Smithson Contrib Zool 336:1–98

Thapa K, Kelly MJ (2016) Density and carrying capacity in the forgotten tigerland: tigers in understudied Nepalese Churia. Integr Zool. doi:10.1111/1749-4877.12240

Thapar V (2011) The tiger – soul of India. Oxford University Press, New Delhi

Walston J, Robinson JG, Bennett EL, Breitenmoser U, da Fonseca GAB et al (2010) Bringing the tiger back from the brink - the six percent solution. PLoS Biol 8:e1000485

Wikramanayake E, Dinerstein E, Seidensticker J, Lumpkin S, Pandav B et al (2011) A landscape-based conservation strategy to double the wild tiger population. Conserv Lett 4:219–227

Williams BK, Nichols JD, Conroy MJ (2002) Analysis and management of animal populations. Academic Press, San Diego

Wilting A, Courtiol A, Christiansen P, Niedballa SAK, Orlando L, Balkenhol N, Hofer H, Schadt SK, Fickel J, Kitchener A (2015) Planning tiger recovery: understanding intraspecific variation for effective conservation. Sci Adv 1:e1400175

Xiao W, Feng L, Mou P, Miquelle DG, Hebblewhite M, Goldberg JF, Robinson HS, Zhao X, Zhou B, Wang T, Ge J (2016) Estimating abundance and density of Amur tigers along the Sino-Russian border. Integr Zool 11:322–332

Tiger Ecology in Relation to Monitoring Issues 2

K. Ullas Karanth, John M. Goodrich, Dale G. Miquelle, Krishnamurthy Ramesh, Abishek Harihar, Anak Pattanavibool, Sunarto Sunarto, D. Mark Rayan, and Kanchan Thapa

2.1 Introduction

Early naturalists recorded descriptive accounts of tigers in tropical Asia during the past two to three centuries (Karanth 2001). For example, during the twentieth century, important anecdotal accounts of tiger behavior were provided by Brander (1923), Corbett (1944), McDougal (1977), and Singh (1984) for India and Nepal; Baikov (1927) in Russia and China; Locke (1954) in Malaya; Baze (1957) in

K.U. Karanth (✉)
Wildlife Conservation Society (WCS), New York, NY, USA

Centre for Wildlife Studies, Bengaluru, India

Wildlife Conservation Society, India Program, Bengaluru, India

National Centre for Biological Sciences-TIFR, Bengaluru, India
e-mail: ukaranth@wcs.org; https://www.wcs.org/; http://cwsindia.org/; http://wcsindia.org/home/; https://www.ncbs.res.in/

J.M. Goodrich
Panthera, New York, NY, USA
e-mail: jgoodrich@panthera.org; https://www.panthera.org/

D.G. Miquelle
Wildlife Conservation Society (WCS), New York, NY, USA

Department of Ecology, Far Eastern Federal University, Vladivostok, Russia
e-mail: dmiquelle@wcs.org

K. Ramesh
Wildlife Institute of India, Dehradun, India

Department of Forest and Conservation Sciences, University of British Columbia, Vancouver, BC, Canada
e-mail: ramesh@wii.gov.in; http://www.wii.gov.in; http://www.forestry.ubc.ca/

K.U. Karanth, J.D. Nichols (eds.), *Methods For Monitoring Tiger And Prey Populations*, DOI 10.1007/978-981-10-5436-5_2

Indochina; and Hoogerwerf (1970) in Java. Although qualitative, these accounts laid the foundation for subsequent scientific studies.

Scientific studies of tigers within the framework of modern wildlife biology were pioneered by George Schaller (1967). Further scientific advances were made in the 1973–1985 period, with the first radio-telemetry studies of tigers at Chitwan, Nepal, by the Smithsonian Institution (Sunquist 1981; Smith 1993). During and after the 1990s, long-term ecological studies of tigers have been conducted in India (Karanth and Sunquist 1995, 2000; Karanth and Nichols 1998, Karanth et al. 1999, 2004, 2006, 2011) and in Russia (Miquelle et al. 1999, 2010; Goodrich et al. 2008; Goodrich et al. 2010; Kerley et al. 2002, 2003; Miller et al. 2013, 2014; Petrunenko et al. 2015; Robinson et al. 2015). These studies involved radio-telemetry, camera trapping, diet analysis, and assessments of prey and habitats. They included the first efforts to monitor population parameters in wild tiger populations using rigorous methods.

All the challenges to tiger population monitoring (Chap. 1) must consider three key factors: biology of the tiger, nature of the environment, and how we actually conduct the monitoring. In this chapter, we provide a brief background on the general biology, behavior, and population ecology of wild tigers. Thereafter, we examine how tiger biology and environmental features interact with resource availability while trying to establish reasonable monitoring goals for specific contexts.

A. Harihar
Tiger Program, Panthera, New York, NY, USA

Nature Conservation Foundation, Mysuru, India
e-mail aharihar@panthera.org; aharihar@ncf-india.org; https://www.panthera.org/; http://ncf-india.org/

A. Pattanavibool
Wildlife Conservation Society (WCS) – Thailand Program, Bangkok, Thailand

Department of Conservation, Kasetsart University, Bangkok, Thailand
e-mail: anakp@wcs.org; https://thailand.wcs.org/; http://conservation.forest.ku.ac.th/

S. Sunarto
World Wide Fund for Nature (WWF) – Indonesia, Jakarta, Indonesia

Forum HarimauKita, Bogor, Indonesia
e-mail: macandahan@gmail.com; http://www.wwf.id/en; https://www.harimaukita.or.id/

D.M. Rayan
World Wide Fund for Nature (WWF) – Malaysia, Selangor, Malaysia

Durrell Institute of Conservation and Ecology (DICE), University of Kent, Kent, UK
e-mail: mdarmaraj@wwf.org.my; markrayan78@gmail.com; http://www.wwf.org.my/; https://www.kent.ac.uk/dice/

K. Thapa
World Wide Fund for Nature (WWF) – Nepal, Kathmandu, Nepal
e-mail: kanchan.thapa@wwfnepal.org; https://www.wwfnepal.org

2.2 General Biology, Behavior, and Ecology of the Tiger

2.2.1 Morphology and Physiological Adaptations

Tigers are quite variable in body size across their extensive range, with records of the largest specimens (adult males weighing 175–250 kg) coming from India, Nepal, and the Russian Far East. Tigers in Southeast Asian mainland tend to be smaller (adult males 150–210 kg). Specimens from Sundarbans of India and Bangladesh and from Indonesian islands are the smallest (adult males 100–140 kg). Tigers are strongly sexually dimorphic, with females weighing 20%–40% less than males. Typical range of body size measures reported for tigers are: head and body length, 270–310 cm; tail length 85–110 cm; and shoulder height of 90–110 cm.

Tigers possess 30 teeth, represented by the dental formula: *incisors* 3/3; *canines* 1/1; *premolars* 3/2; and *molars* 1/1. The upper and lower canines (lengths, 50–60 mm and 40–50 mm, respectively) are the tiger's primary weapons for killing prey. The tiger's lower jaw is anchored to the skull by powerful muscles that enable powerful and precise killing bites. The shear-like carnassial teeth can rapidly slice flesh, while the molars can grind meat and smaller bones, with the incisors helping in plucking hair, gripping, and peeling.

The forequarters of the tiger are especially powerful, with strong skeletal structures and powerful muscles that enable it to grasp and pull down prey animals three to five times its own size (Plate 2.1). The long hind legs enable speedy movement and rearing up to pull down taller prey animals. Tigers have five curved, retractile claws on their forepaws and four on the hind, which assist in pinning down prey. At other times these claws remain sheathed, enabling noiseless movement, which is also facilitated by the soft underfoot pads. Several scent glands are located in the tiger's digital pads, orbits, and cheeks. The anal gland produces a signature scent, when mixed with the cat's urine. Spraying such scent deposits continuously on vegetation or substrate is the tiger's most important communication tool. Tigers also employ a range of calls—purrs, grunts, growls, and deep roars which carry 3–5 km—for communication in real time.

The tiger's pelage can vary from pale yellow to dark ochre, with white underside and orbital patches. The stripe patterns vary considerably and are unique to each individual. This rather striking pattern, in reality, helps to break the tiger's outline, facilitating stealthy ambuscade of prey animals. Rarely, genetic mutations involving an excess of melanin pigment can result in partially black tigers. In contrast, a single gene mutation resulting in the absence of a specific melanin (pheomelanin) results in the famous "white tigers," which are not true albinos (Xu et al. 2013), nor a distinct subspecies, as some believe.

Plate 2.1 Tigers are powerfully built predators capable of killing prey animals 3–5 times their own size (Image Copyright©: Ramki Sreenivasan)

2.2.2 Taxonomy, Evolution, and Biogeography

Within the mammalian order *Carnivora*, the tiger is a member of the family *Felidae* together with 37 other cat species. It belongs to the genus *Panthera* or "roaring big cats," which also includes the lion (*P. leo*), jaguar (*P.onca*), leopard (*P. pardus*), and the snow leopard (*P. uncia*).

Further taxonomic subdivisions are debated, with disagreements among molecular geneticists, paleontologists, and biogeographers. The disagreements arise from differing interpretations from sparse data available across the species range: one view is that there are eight to nine extant and extinct subspecies, whereas another posits only two. The former recognizes nine subspecies: Indian or Bengal *tigris*; northern Indo-Chinese *corbetti*; Malayan *jacksonii*; Sumatran *sumatrae*; Amur *altaica*, all of which still survive in wild populations; the South China subspecies *amoyensis*, which survives only in captivity; and the extinct Javan *sondaica*; Bali *balica*; and Caspian *virgata* (Luo et al. 2004). Some argue that the Caspian and Amur tigers are of one subspecies (Driscoll et al. 2009).

However, others argue based on zoological, morphological, and biogeographic evidence that variation among tigers is clinal and can be explained by biogeography and ecological rules. They suggest all tigers, extant or extinct, on mainland Asia belong to one subspecies, with those on islands forming a second subspecies (Kitchener and Dugmore 2000; Wilting et al. 2015).

The broad consensus appears to be that a smaller ancestral species evolved from the *Panthera* lineage in China and Southeast Asia about two million years ago. Thereafter, the modern tiger evolved 72,000–108,000 years ago (Luo et al.

2004). The tiger's distribution expanded thereafter, reaching its limits in India and West Asia during the last glacial maximum 20,000 years ago. Biogeographers attribute variations observed in the tiger's range to the timing of the radiation of tigers in relation to major climatic events in the late Pleistocene. During the cold-dry periods (stadials), forests in which tigers could survive shrank, but the lower sea level connected the mainland to islands and archipelagos, thereby permitting tiger range expansion. In contrast, the expansion of forests in the warm, moist periods (interstadials) increased the extent of tiger habitat, but as sea levels rose, connectivity between landmasses was severed, preventing further range expansion (Kitchener and Dugmore 2000).

Although tigers are an adaptable species, they cannot tolerate arid bio-climates the way cheetahs, leopards, and lions can. Tigers tolerate temperatures from -50^0 C in the snow-bound Russian Far East to $+50^0$ C in Western India, where they overlapped the lion's former range. They are adapted to a wide range of topographies from sea level to about 2500 m altitude and are known to transit Himalayan passes at 4000 m. As a result of their adaptability, tigers were distributed across a diversity of forests: the freezing Taiga in the Russian Far East; mangrove swamps in the delta of Sundarbans; montane mixed conifer forests in the Himalayas and China; reedbeds and riverine forests in West and Central Asia, and in evergreen, semi-evergreen, and deciduous forests as well as alluvial moist grasslands of tropical Asia. However, shade and water are critical habitat elements wherever they occur.

Despite the tiger's general adaptability, a key determinant of their persistence is the presence of ungulate prey *at adequate densities*. Of critical importance is the presence of preferred prey of 50–250 kg body mass, essential for females to successfully raise young (Karanth and Sunquist 1995, Miller et al. 2014). If prey densities are depressed by human impacts, tiger populations cannot persist even in suitable undisturbed forests. Tiger persistence must be examined primarily through the lens of available prey density (Karanth et al. 1999, 2004, Miquelle et al. 1999), a factor that conservationists sometimes ignore.

2.2.3 Social Organization, Spacing, and Land Tenure

Tigers are solitary felids, with associations between adults lasting just 4–6 days for mating, when females come into estrus (typically once in 17–21 days) or when related animals share kills. A tigress and her cubs associate for 18–24 months (Smith 1993, Kerley et al. 2003), with the cubs being dependent on their mother through the first year. (Plate 2.2) Thereafter, they gradually acquire survival skills to become independent.

Female home range size is governed by prey density and hence varies from ~20 km^2 in the Indian subcontinent, through ~80–400 km^2 in Thailand, to over 400 km^2 in the Russian Far East. Male home range size is dependent on female density and only indirectly on prey, and hence varies much more. Male ranges can be as small as 50–300 km^2 (Simcharoen et al. 2014) in tropical habitats to over 2000 km^2 in temperate forests in Russia (review in Goodrich et al. 2010). Tigers are generally territorial, i.e., they exclude conspecifics of the same sex from their

Plate 2.2 Tiger cubs are dependent on their mother for nearly 2 years (Image Copyright©: Michael Vickers)

home ranges. However, territories tend to overlap on average about 10% (Smith et al. 1987; Goodrich et al. 2010). Male territories tend to completely overlap 1–4 female territories. There may be exceptions (e.g., Hernandez-Blanco et al. 2015) as a result of human interference.

Tigers are active mostly from dusk to dawn, resting ~18 h or more during the daily cycle. When active, they may travel 5–30 km/day, to hunt, seek mates, or, in the case of residents, to secure their home ranges against intruding conspecifics.

The keystone of tiger social organization is the resident breeding female, who becomes reproductively active at 3.5–4 years. Breeding will usually depend on her first acquiring a territory that she defends while raising cubs. By defending their home ranges, which overlap territories of 1–4 adjacent females, male tigers also defend their own offspring from infanticide by challenging males. After a gestation of 102–108 days, a tigress gives birth to 1–5 cubs, which she carefully raises for the next 18–20 months. The cubs are nursed and fed regurgitated meat for the first 2–3 months, before being taken to feed off kills. In the second year the juveniles learn to hunt wild prey and acquire survival skills such as avoiding humans or dominant conspecifics. At 18–20 months, subadults gradually disperse away from natal ranges. Males disperse long distances (50–300 km or more), whereas females try to carve out territories inside or close to their natal ranges (Smith 1993, Goodrich et al. 2010).

Throughout their lives, tigers always face risks associated with intraspecific aggression: fights over kills, territories, and for mating access as well as from infanticide of cubs by encroaching males. Tigers often die from starvation following injuries suffered during fights or while hunting large, dangerous prey. They are particularly at risk at all times from humans, for a variety of reasons (Chap. 1).

During the post-dispersal years (1.5–5 years), tigers move extensively across larger landscapes seeking to establish their own territories. During such "transience," they may crisscross boundaries of resident territories, human-dominated landscapes, or even countries (Smith 1993). Consequently, transients suffer high mortality rates from intraspecific aggression and human persecution (Karanth et al. 2006, Robinson et al. 2015, Duangchantrasiri et al. 2016). As a cumulative consequence of these natural dynamics—high rates of births, deaths, emigration, immigration, and transience—tiger populations have high turnover of individuals. This fact is often ignored in monitoring schemes, resulting in inflated tiger numbers and flawed estimates of demographic parameters.

2.2.4 Ecological Determinants of Tiger Population Dynamics

Tigers depend on large ungulate prey: several species of deer, wild pigs, and wild cattle that inhabit Asian forests (Plate 2.3). Occasionally, they may kill tapirs and calves of elephants and rhinoceros.

Tigers are solitary "stalk-and-ambush" hunters that stealthily charge at their unwary prey from 10 to 50 m distance. If the prey animal is large, the tiger strangulates it with a throat bite, while trying to pull it down to avoid its dangerous hooves and horns (Plate 2.4). If the animal is smaller, it is brought down from the impact itself and killed swiftly with a bite through its skull or neck.

The tiger stays with its kill for 3–4 days in the tropics consuming about 65% of the kill, but in colder climates they normally stay with a kill until over 90% of the meat is consumed, which may take over 7 days for larger prey (Miller et al. 2013). Typically, an adult tiger kills about 3200–3400 kg of live prey animals per year, which translates to about 50–60 kills/year (Miller et al. 2013). Females raising cubs need to acquire approximately twice as much biomass as a lone female to successfully raise an average-sized litter of cubs over the course of a year (Miller et al. 2013). Studies show that tigers may crop about 10% of available prey biomass or numbers, depending on prey size. Thus, there appears to be an approximate ratio of 500 prey animals per tiger (Karanth et al. 2004). This ratio of numbers of available prey to tigers influences resident female home range size, overall tiger density, and, consequently, potential "carrying capacity" of a given habitat for tigers.

2.2.5 Tiger Population Dynamic Parameters: Abundance, Survival, Recruitment, and Movement

Most wild tiger populations now occur in small isolated clusters. Although such tiger "source populations" occupy just 6% of the remaining habitat, they support 70% of the remaining population, highlighting their critical importance for future recovery of the species (Walston et al. 2010).

Plate 2.3 The tiger's principal prey are several species of deer, wild cattle and wild pigs found in Asian forests (Image Copyright©: Clockwise from *top left*: Gaur – Kalyan Varma; Wild pig – Ramki Sreenivasan; Banteng – Jitendra Shankaraiah; Nilgai – Harsh Dhanwatey; Barasingha – Ullas Karanth; Muntjac, Sambar, Chital, Tiger – Ramki Sreenivasan)

Population densities of tigers can vary greatly, because of ecological as well as anthropogenic factors. Even under adequate protection, tiger densities are low (often less than 1.0 tiger/100 km^2) in the Russian Far East where prey is naturally scarce. A density of 1–4 tigers/100 km^2 is typical of tropical evergreen and mangrove forests that support somewhat higher prey abundance. Higher densities of 8–18 tigers/100 km^2) are attained in alluvial grasslands and tropical deciduous forests that support very high prey densities of 15–70 ungulates/ km^2 (Karanth et al. 2004). Thus, tiger habitats in the Russian Far East that spread over 200,000 km^2 currently support ~400 tigers, whereas just five tiger reserves covering 2000 km^2 in Southwestern India can support as many tigers.

Long-term studies in Nagarahole, India, show that even under strict protection, tiger densities naturally fluctuate widely from 8 to 15 tigers/100 km^2. Densities vary depending on numbers of resident females that successfully bred in the previous year. Because of its high productivity, over a 10-year span, this tiger population remained stable, although ~23% of its members were lost annually from mortalities and emigration (Karanth et al. 2006).

Theoretical population models (Kenney et al. 1995, Karanth and Stith 1999) generally use the following estimates of annual survival rates: cubs = 60%;

Plate 2.4 Tigers kill large dangerous prey through strangulation, trying to avoid their dangerous horns and hooves (Image Copyright©: Valmik Thapar)

juveniles = 90%; female transients = 70%; male transients = 65%; breeding females = 90%; and breeding males = 80%. These rates will be much lower in sub-optimal habitats. Actual estimates of survival rates combined for tigers over 1 year age, estimated from photographic capture-recapture studies, were 77% in India (Karanth et al. 2006) and 82% in Thailand (Duangchantrasiri et al. 2016). In Russia, annual survival of adult females estimated from radio-telemerty data (85%) was greater than that of adult males (62%) and subadults (55%) (Goodrich et al. 2008, Robinson et al. 2015). The data set included periods of high poaching.

It should be noted that these high natural mortality rates are not necessarily a cause for alarm because they are offset by high reproductive potential and recruitment rates. With sufficient prey, a tigress can have 12–15 cubs in her lifetime, producing large "surpluses" of tigers to more than make up for the heavy losses. However, when direct tiger poaching is intense and impacts the resident female segment of the population, it can drive local extinctions (Johnson et al. 2016, O'Kelly et al. 2012, Gopal et al. 2010, Sankar et al. 2010). Potential growth rates of a tiger population will be largely determined by which segment of the population is suffering high mortality rates. High mortality rates of subadults will have a much lower impact on growth rates than when adult reproductive females are suffering high mortality rates. Theoretical models suggest that when adult female mortality rates exceed 15%, population growth may stall (Chapron et al. 2008). Therefore, protection of the core breeding population is the key to recovery of tiger populations. Monitoring the fate of these breeders can assess potential trajectory of a population.

2.3 Challenges to Reliable Monitoring: Tiger Ecology and Environmental Factors

2.3.1 Tiger Ecology in Relation to Monitoring

Most aspects of tiger ecology pose serious challenges to monitoring populations. Tigers avoid human contact. Their secretive nature, preference for dense cover, and largely nocturnal movements render them difficult to observe in field surveys. Even tiger signs such as tracks and scats are dispersed across wide areas and degrade rapidly making them difficult to find.

Normal daily ranging patterns of tigers also pose challenges. Even resident breeders can move over 10–15 km per day, covering wide areas in a week or less. The problem is worse with transients that may traverse tens or even hundreds of kilometers in a few days. Tigers deposit signs as they move back and forth across boundaries of census blocks or reserves that monitoring teams are typically expected to cover. Furthermore, tigers are widely distributed, with potentially suitable forests extending over 1.2 million km^2 across Asia (Walston et al. 2010).

Perhaps the most difficult challenge to tiger monitoring is that, being large-bodied carnivores, tigers live at naturally low densities. Even in the best habitats, densities are as low as 10–20 tigers/100 km^2. Across 90% of tiger range, tiger densities are just 0.5–1.0 tigers/100 km^2 (Karanth and Nichols 1998, 2000). Tiger population surveys are like searching for the proverbial needle in a haystack!

The combination of all the above factors—low densities, secretiveness, wide-ranging behavior, high rates of population turnover, extensive distribution, and the low probabilities of detecting tiger signs—poses serious methodological problems for monitoring. Collection of necessary quantitative data becomes problematic. These problems are compounded by additional managerial constraints imposed by the environments in which tigers live.

On the other hand, the fact that tigers have complex stripe patterns that make each individual uniquely identifiable and that they commonly use the same trails as travel routes present substantial advantages to sampling their populations in the face of all the challenges (Chaps. 9 and 10).

2.3.2 Environmental Variables that Govern Tiger Monitoring

In practical terms, tiger habitats in Asia fall under two major categories: the temperate zone areas in the Russian Far East and China, and the tropical forests in South and Southeast Asia. In the Russian Far East, extremely low tiger densities and the huge geographic scale at which tigers are monitored clearly set the monitoring challenges apart from the monitoring problems in most of tropical Asia However, similar challenges are faced in the vast, extremely rugged forested mountains of Sumatra and in Myanmar.

On the positive side, in the Russian Far East and adjacent tiger habitats in China, fresh snowfall provides an ideal substrate that retains tiger tracks. Similarly, muddy intertidal zones in the mangrove forests of Sundarbans of India and Bangladesh can retain track prints for several hours between tides. Although light rain can improve conditions for finding tiger signs during surveys, heavy rainfall can wash signs away.

In contrast, almost everywhere tiger tracks are seen along trails or streambeds only if suitable substrate occurs naturally or is created by vehicular movement. Although tigers prefer natural trails as travel routes (Karanth and Suquist 1995, 2000), if such trails are not readily identifiable, even counting signs becomes a real challenge.

2.3.3 Issues of Access for Survey Personnel

Because tigers are so wide-ranging and occur at such low densities, surveys must necessarily cover extensive areas to detect their travel routes. In this context, access to the entire area to be sampled is a critical challenge. Poor access may require greater investment of money, human power and time, to attain sampling intensities essential for reliable monitoring, yet even when resources are available to cover such areas, access may be so difficult as to make sampling in a short enough time frame, to avoid violating the assumption of population closure, extremely difficult.

Inadequate access can result from difficult topography (high altitude, cliffs, rivers) or seasonal factors (extreme heat or cold, deep snow, incessant rain, insect infestations, forest fires, dense vegetation, etc.) and is usually a combination of these factors. Lack of roads or trails to quickly reach remote areas on vehicles, boats, or draft animals can also be impediments at times. Although not common, high probabilities of encounters with dangerous animals such as elephants, rhinoceros, or even individual tigers with a tendency to prey on humans could be factors that limit access.

In many countries specific social factors such as barriers in the form of political or administrative boundaries or the presence of violent insurgencies or criminals may hinder access to survey areas.

2.4 Basic Steps Toward Establishing Reliable Tiger Monitoring Programs

Although monitoring challenges described are often recognized, there is another challenge that is not always explicitly recognized. Because of the almost totemic conservation significance of tigers, massive attention and resources have already been expended for monitoring efforts across tiger range over several decades. However, many of these monitoring schemes are based on ad hoc methodologies developed without adequate attention to critical assumptions, details of design, or statistical rigor. They consequently suffer from serious statistical flaws and practical deficiencies (Karanth et al. 2003; Gopalaswamy et al. 2015; Harihar et al. 2017) that hinder proper estimation of critical population parameters (Chap. 1).

However, we would like to emphasize here that progress toward modifying flawed monitoring approaches has been slow and reluctant. This hesitancy has several causes: misguided loyalty to "management traditions;" lack of familiarity or comfort-level with technically superior alternatives; lack of adequate understanding of tiger ecology and of basic statistical issues.

A fundamental antidote to such mindsets is to develop a deeper appreciation of the universal approach of all modern science: reliance on constant refinement through critical reviews. Such an appreciation rests on the foundation of "peer review and publication" of the monitoring methods in scientific literature. Continuous criticism and suggestions for improvement are at the heart of this process. This process should not be viewed as an attack on authority or persons responsible for tiger monitoring, but as a continuous attempt to improve the reliability and accuracy of survey methods.

We submit that many of the current controversies about tiger numbers, their rates of increase or decrease, and future prospects of wild tigers, result from impractical monitoring goals and subsequent application of weak methodologies to attain them. This we believe is a consequence of the widespread lack of appreciation of the scientific process of constant review and refinement, among tiger conservationists and wildlife managers.

In the following section, we examine the key features of a sound approach for establishing reliable, science-based tiger monitoring programs across the range of the species.

2.4.1 Adopting a Unified, Sampling-Based Approach to Tiger Monitoring

We do not want to "reinvent the wheel" by trying to review all established scientific methods for animal population estimation in general: there are several excellent manuals that address this need. These works cover conceptual issues (Thompson et al. 1998; Buckland et al. 1993, 2004; Williams et al. 2002; MacKenzie et al. 2006; Royle and Dorazio 2008; Link and Barker 2010; Royle et al. 2014; Kéry and Royle 2016), as well as more focused works directed at monitoring of rare, elusive species (Thompson 2004; Long et al. 2008; Boitani and Powell 2012). Some publications even cover specific field techniques we advocate here, such as camera-trap surveys (O'Connell et al. 2011).

We expect users of this book to be aware of such standard literature on animal population estimation. However, for the sake of completeness, brief reviews of key sampling-based population estimation methods are provided in Chaps. 3, 4, 5, 6, 7, 8, 9, 10, 11, 12, and 13. It is important to note that these seemingly disparate approaches are, in fact, conceptually unified elegantly as shown in Chap. 3.

Specifically, the methods we recommend in this manual will consider the manner in which monitoring approaches deal with the core problems of *spatial sampling* and *observability* or *imperfect detection*. Spatial sampling concerns the frequent inability to use animal survey methods over an entire area of interest. In such cases, we need to survey some subset of the entire area of interest and then use these results to draw inference about the entire area. Observability concerns the typical inability to detect and count all animals present in an area that is selected for survey.

Regardless of the particular survey method, comparisons of resulting count statistics over time or space require consideration of the associated detection probabilities (probability that an animal appears in the count statistic). Some approaches to animal monitoring permit direct estimation of these detection probabilities (these methods tend to be those with the greatest requirements for effort and resources), whereas others rely on strong assumptions about the absence of variation in these probabilities over time and/or space.

Focus on the key features of animal monitoring below will be useful for monitoring tigers and their prey.

2.4.2 Importance of Defining and Setting Clear, Practical Monitoring Objectives

The very first step in monitoring any tiger or prey population is to define the objectives of the exercise. These specific objectives are linked to one or more of the following specific goals.

1. Mapping the distribution of tigers and principal prey species on a regional or countrywide basis.

Tigers and their principal prey are landscape animals: as habitats become degraded and fragmented, their local populations may get extirpated. The distributional range of tigers may contract and shrink. On the other hand, if conservation measures are successful, habitat fragments may get reconnected, enlarging populations. Dispersal into newly improved habitats can establish new populations. Tracking changes in distributions of tigers and prey at large spatial scales once every 5 years or so is thus a useful objective. Methods to attain this objective are covered in Chaps. 4 and 5.

2. Estimating population trajectory or "trend" using quantitative indices of relative abundances of tigers and prey species (e.g. number of tiger track sets seen per 100 km walked by surveyor, number of sambar pellets found in a 100 m^2 plot), in selected reserves.

At the level of specific individual reserves, managers, researchers, or conservationists lacking adequate resources may only be capable of monitoring population trends of tigers and prey. Although desirable, it may not always be possible to obtain estimates of population size or densities of tigers at a desired precision. In such cases sometimes management needs may be met by some reasonably reliable metric of population size. Such metrics or "indices" may allow comparison of population densities at the same site across the years at relatively low cost. The issues concerning the use of indices are covered in Chap. 3. We note that various methods covered in Chaps. 4, 5, 6, 7, 8, 9, and 10, because they address the issue of detection probability squarely, can be helpful for generating better indices also.

3. Estimating potential 'carrying capacity' for tigers in an area based on estimates of absolute densities of principal prey species.

With trained personnel and simple equipment such as compasses and range finders, it is often possible to estimate densities and abundances of principal prey species of tigers with reasonable accuracy. Thereafter, using the approximate 1 tiger to 500 prey ratio, it is possible to estimate (at least roughly) how many tigers could potentially live in the surveyed area. Methods to assess this potential carrying capacity are explained in Chaps. 6 and 7.

4. Estimating densities, numbers, rates of annual survival, recruitment and movement rates in tiger populations.

At priority tiger conservation sites, where the primary goal is to understand tiger ecology at a fine-grained level, parameters such as tiger population size, absolute density, rates of survival, mortality, immigration, and dispersal may have to be measured. In this case, there is no escape from deploying the most advanced equipment, statistical techniques and skilled personnel. Such fine-grained data cannot be obtained by simpler techniques used to attain objectives 2.4.2 (1, 2 and 3). However, estimating carrying capacity may permit you to make reasonable guesses at results expected when rigorous methods for estimating tiger numbers covered in Chaps. 9 and 10 can be employed.

2.4.3 Assessing Resources Available for Monitoring

Which of the monitoring objectives outlined above can actually be achieved depends on ecological and environmental factors outlined in section 3.0. It also depends on the time available, material resources, manpower, and technical skills at the disposal of the wildlife manager or tiger conservationist. Therefore, assessing the resources available is the second important step of tiger monitoring.

Usually monitoring of tigers and prey is carried out by either government agencies, such as forestry or wildlife personnel, or by wildlife researchers. Survey personnel may vary greatly in terms of their technical skills and field abilities. Their numbers may range from a handful of highly trained naturalists to dozens, or even hundreds, of field personnel without great scientific skills.

The types of skills required for carrying out field surveys of tigers and prey also vary. *Field skills*, such as the ability to observe carefully, recognize and record animals or their signs accurately, and do hard field work, are all of prime importance. Persons with such *field skills* may be wildlife biologists, wildlife staff or local hunters, and naturalists.

Very different skills are necessary for designing the field surveys and, subsequently, for analyzing the resulting data. We will call these *analytical skills*. These skills include knowledge of population sampling methods and field conditions and an ability to organize field survey data collection as well as interpreting their results correctly. It is critically important for a wildlife manager or conservationist to assess the kinds of skills that are available (or not available) in a particular monitoring situation.

Similarly, the material resources available for tiger monitoring vary. In most cases, particularly where wildlife or forestry departments carry out the tiger monitoring, only basic tools exist, such as rudimentary maps of the area being surveyed, polythene bags, identification tags, etc. Sometimes, additional "special tools" such as compasses, range finders, and global positioning systems (GPS) may be available. Increasingly, even "advanced" tools such as camera traps and computers may be on hand.

2.4.4 Matching Resources to Objectives

A successful tiger monitoring program depends on the manager being able to come up with a realistic, robust survey scheme that defines achievable objectives after carefully considering available manpower, technical skills, and material resources. Setting up grand goals based on wishful thinking is not useful, but is not uncommon in many monitoring schemes we have seen. Tiger monitoring goals must be realistic in a specific context: if not, the failure of the monitoring program is almost guaranteed.

For a wildlife manager or a biologist, the following guidelines may be helpful to assess monitoring methods that can be employed reliably in a given situation.

1. Recognize that it is simply impossible to achieve some of the objectives when dealing with *large regions*, *states*, or *countries*. Therefore, if you have access to only basic tools and untrained field personnel, you can only do surveys of presence of tigers and prey species (2.4.2(1)), applying concepts and methods described in Chaps. 4 and 5. Most likely this is the situation that wildlife managers will encounter over much of the tiger's range in Asia.

2. If there are well-trained wildlife biologists or survey personnel, you can try to get relative density estimates of tigers from track or scat-based indices and relative density estimates of prey species using dung counts (objective 2.4.2(2)). It is likely that you can do such index-based surveys only in *individual reserves* or *study areas* for making comparisons across the years. Some concepts and methods used for attaining these objectives are described in Chaps. 3 and 8.
3. If you have access to special tools, enough trained personnel, and skilled biologists, then, in addition to objectives 2.4.2(1 and 2), you can estimate absolute densities of prey species (Objective 2.4.2(3)), using the concepts and methods detailed in Chaps. 6 and 7. Such surveys are likely to be feasible only in *individual reserves* or *study areas*, but not across wider landscapes.
4. If you have access to all the above resources, plus adequate number of camera traps and persons with technical skills to deploy them properly, photographic capture-recapture sample surveys of tigers using methods described in Chaps. 9, 10, and 11 can estimate absolute densities and abundances of tigers. This type of survey may be feasible only in a few sites. If repeated across years, under rigorous open-population model scenarios, the same surveys can also additionally yield reliable estimates of population change (rates of growth) and other demographic parameters such as rates of survival and recruitment (Chaps. 9 and 10). When deployment of camera traps is not feasible and reliable facilities are available for analyses of fecal DNA extracted from tiger scat samples, methods elaborated in Chap. 11 can be helpful.

Further studies of tiger movements and connectivity among populations across large spatial scales (Chap. 12) and of meta-population structure and dynamics (Chap. 13) are possible over the longer term by assessment of these data collected at various spatial scales.

However, it is likely that over much of tiger range, the sample sizes attained from field counts of tiger tracks, camera trap photos, or prey dung counts are likely to be too small for deriving density estimates or even reliable metrics of relative numbers. Therefore, managers may have to be realistically content with objective 2.4.2(1) (mapping the presence of tigers and principal prey) diligently over the years. There is absolutely nothing wrong with this goal: monitoring of tiger distribution is a critically important first step in implementing any landscape level conservation program. Therefore, survey efforts can be initiated with objective 2.4.2(1) and over the years gradually build necessary capacity and resources to try to reach objectives 2.4.2(3 and 4).

We emphasize that if one really wants to estimate tiger population and meta-population parameters, such as abundance, densities, survival, recruitment, movement, and connectivity listed under objective 2.4.2(4), there is no escape from employing the advanced methods described in Chaps. 9, 10, 11, 12, and 13.

However, we note that all sorts of *ad hoc* methods are sometimes used in conjunction with deployment of camera traps or fecal DNA kits without clarity as to how such data should be collected and analyzed. These advanced methods are relevant for monitoring tigers over a very small fraction of the animal's vast range and work only if they are correctly used. On the other hand, the three other critical needs of monitoring, mapping spatial distributions over large regions (2.4.2(1)), tracking tiger population trends in specific reserves (2.4.2(2)), and establishing carrying capacities for tigers in reserves (2.4.2(3)) are attainable goals using relatively simple tools and methods.

2.5 Conclusion

Considering the different possible monitoring objectives, the variety of field situations encountered, and the differences in available manpower, equipment, and resources, it is clear that an array of different monitoring methods—rather than any single approach—is needed. While attempts to successfully deploy these tiger monitoring methods under different ecological, social, and technological contexts are sometimes made, it is not often recognized that they should also nest within a sound scientific framework (Karanth et al. 2003, Karanth 2011).

The four monitoring goals identified above are distinct and often pursued by different sets of managers, researchers, or conservationists. However, they are mutually contributory. It is unwise for these practitioners to pursue their objectives in isolation from each other or from the scientific or management programs to which they contribute (Karanth 2011).

The underpinning for all tiger population monitoring must be provided by sound science. While methods applied must match the resources and skills available, lack of resources or training should not become an excuse to practice substandard science. In other words, having worthless or misleading data is worse than an honest acknowledgment of data deficiency.

It is clear that the methods employed to meet data collection needs at different levels of refinement are likely to be quite different in terms of the sampling design, effort, equipment used, and the kinds of personnel skills required for the survey. However, regardless of the variables involved in a particular situation, the monitoring methods need to be statistically robust. This means they should have a reasonable theoretical basis, a history of empirical validation, and a record of being refined through the universally accepted process of scientific peer review.

Tiger biologists, managers, and conservationists cannot claim exemption from these criteria, if they are to retain credibility. Therefore, the editors of this manual have deliberately chosen to present only those approaches to monitoring that are adequately justified in the peer-reviewed scientific literature. A conceptually unified, rigorous array of valid methods for monitoring tigers in an adaptive management framework is provided in Chaps. 3, 4, 5, 6, 7, 8, 9, 10, 11, 12, 13, and 14 that follow.

References

Baikov NA (1927) Tigers in the Far East. Okhotnik:23–25

Baze W (1957) Tiger! Tiger! Elek Books, London

Boitani L, Powell RA (eds) (2012) Carnivore ecology and conservation: a handbook of techniques. Oxford University Press, New York

Brander AAD (1923) Wild animals in Central India. Edward Arnold & Co, London

Buckland ST, Anderson DR, Burnham KP, Laake JL (1993) Distance sampling: estimating abundance of biological populations. Chapman & Hall, London

Buckland ST, Anderson DR, Burnham KP, Laake JL, Borchers DL, Thomas L (2004) Advanced distance sampling: estimating abundance of biological populations. Oxford University Press, Oxford

Chapron G, Miquelle DG, Lambert A, Goodrich JM, Legendre S, Clobert J (2008) The impact on tigers of poaching versus prey depletion. J Appl Ecol 45:1667–1674

Corbett J (1944) Man eaters of Kumaon. Oxford University Press, New Delhi

Driscoll CA, Yamaguchi N, Bar-Gal GK, Roca AL et al (2009) Mitochondrial phylogeography illuminates the origin of the extinct Caspian Tiger and its relationship to the Amur Tiger. PLoS One 4:e4125

Duangchantrasiri S, Umponjan M, Pattanavibool A, Chaiwattana S, Maneerat S, Kumar NS, Jathanna D, Srivathsa A, Karanth KU (2016) Dynamics of a low-density source population of tigers in Southeast Asia in the context of improved law enforcement. Conserv Biol 30:639–648

Goodrich JM, Kerley LL, Smirnov EN, Miquelle DG, McDonald L et al (2008) Survival rates and causes of mortality of Amur tigers on and near the Sikhote-Alin Biosphere Zapovednik. J Zool 276:323–329

Goodrich JM, Miquelle DG, Smirnov EN, Kerley LL, Quigley HG, Hornocker MG (2010) Spatial structure of Amur (Siberian) tigers (*Panthera tigris altaica*) on Sikhote- Alin Biosphere Zapovednik, Russia. J Mammol 91:737–748

Gopal R, Qureshi Q, Bhardwaj M, Singh JRK, Jhala YV (2010) Evaluating the status of the endangered tiger Panthera tigris and its prey in Panna Tiger Reserve, Madhya Pradesh, India. Oryx 44:383–393

Gopalaswamy AM, Delampady M, Karanth KU, Kumar NS, Macdonald DW (2015) An examination of index-calibration experiments: counting tigers at macroecological scales. Methods Ecol Evol 6:1055–1066

Harihar A, Chanchani P, Pariwakam M, Noon BR, Goodrich J (2017) Defensible inference: questioning global trends in Tiger populations. Conserv Lett. https://doi.org/10.1111/conl.12406

Hernandez-Blanco JA, Naidenko SV, Chistopolova MD, Lukarevskiy VS, Kostyrya A, Rybin A et al (2015) Social structure and space use of Amur tigers (*Panthera tigris altaica*) in Southern Russian Far East based on GPS telemetry data. Integr Zool 10:365–375

Hoogerwerf A (1970) Udjung Kulon. The land of the last Javan Rhinoceros. E J Brill, Leiden

Johnson A, Goodrich J, Hansel T, Rasphone A, Saypanya S, Vongkhamheng C et al (2016) To protect or neglect? Design, monitoring and evaluation of a law enforcement strategy to recover small populations of wild tigers and their prey. Biol Conserv 202:99–109

Karanth KU (2001) The way of the tiger: natural history and conservation of the endangered big cat. Voyageur Press, Inc, Stillwater

Karanth KU (2011) India's tiger counts: the long march to reliable science. Econ Polit Week 46:22–25

Karanth KU, Nichols JD (1998) Estimation of tiger densities in India using photographic captures and recaptures. Ecology 79:2852–2862

Karanth KU, Nichols JD (2000) Ecological status and conservation of tigers in India. Final Technical Report to the US Fish and Wildlife Service and Wildlife Conservation Society, New York. Centre for Wildlife Studies, Bangalore

Karanth KU, Stith BM (1999) Prey depletion as a critical determinant of tiger population viability. In: Seidensticker J, Christie S, Jackson P (eds) Riding the tiger: tiger conservation in human-dominated landscapes. Cambridge University Press, Cambridge, pp 114–122

Karanth KU, Sunquist ME (1995) Prey selection by tiger, leopard and dhole in tropical forests. J Anim Ecol 64:439–450

Karanth KU, Sunquist M (2000) Behavioural correlates of predation by tiger (Panthera tigris), leopard (Panthera pardus) and dhole (Cuon alpinus) in Nagarahole, India. J Zool 250:255–265

Karanth KU, Sunquist ME, Chinappa M (1999) Long-term monitoring of tigers: lessons from Nagarahole. In: Seidensticker J, Christie S, Jackson P (eds) Riding the tiger: tiger conservation in human-dominated landscapes. Cambridge University Press, Cambridge, pp 114–122

Karanth KU, Nichols JD, Seidensticker J, Dinerstein E, Smith JLD, McDougal C, Johnsingh AJT, Chundawat RS, Thapar V (2003) Science deficiency in conservation practice: the monitoring of tiger populations in India. Anim Conserv 6:141–146

Karanth KU, Nichols JD, Kumar NS, Link WA, Hines JE (2004) Tigers and their prey: predicting carnivore densities from prey abundance. Proc Natl Acad Sci U S A 101:4854–4858

Karanth KU, Nichols JD, Kumar NS, Hines JE (2006) Assessing tiger population dynamics using photographic capture-recapture sampling. Ecology 87:2925–2937

Karanth KU, Gopalaswamy AM, Kumar NS, Vaidyanathan S, Nichols JD, MacKenzie DI (2011) Monitoring carnivore populations at the landscape scale: occupancy modelling of tigers from sign surveys. J Appl Ecol 48:1048–1056

Kenney JS, Smith JLD, Starfield AM, McDougal C (1995) The long-term effects of tiger poaching on population viability. Conserv Biol 9:1127–1133

Kerley LL, Goodrich JM, Miquelle DG, Smirnov EN, Nikolaev IG, Quigley HB, Hornocker MG (2002) Effects of roads and human disturbance on Amur tigers. Conserv Biol 16:1–12

Kerley LL, Goodrich JM, Miquelle DG, Smirnov EN, Nikolaev IG, Quigley HB, Hornocker MG (2003) Reproductive parameters of wild female Amur (Siberian) tigers (*Panthera tigris altaica*). J Mammal 84:288–298

Kéry M, Royle JA (2016) Applied hierarchical modeling in ecology: analysis of distribution, abundance and species richness in R and BUGS. Academic Press, San Diego

Kitchener AC, Dugmore AJ (2000) Biogeographical change in the tiger, *Panthera tigris*. Anim Conserv 3:113–124

Link WA, Barker RJ (2010) Bayesian inference: with ecological applications. Academic Press, San Diego

Locke A (1954) The tigers of Terengganu. Museum Press Ltd, London

Long RA, MacKay P, Zielinski WJ, Ray JC (eds) (2008) Noninvasive survey methods for carnivores. Island Press, Washington, DC

Luo SJ, Kim JH, Johnson WE, Walt V d, Martenson J, Yuhki N et al (2004) Phylogeography and genetic ancestry of tigers (Panthera tigris). PLoS Biol 2:2275–2293

MacKenzie DI, Nichols JD, Royle JA, Pollock KH, Bailey LL, Hines JE (2006) Occupancy estimation and modelling: inferring patterns and dynamics of species occurrence. Academic Press, San Diego

McDougal C (1977) The face of the tiger. Rivington Books, London

Miller CS, Hebblewhite M, Petrunenko YK, Seryodkin IV, DeCesare NJ, Goodrich JM, Miquelle DG (2013) Estimating Amur tiger (*Panthera tigris altaica*) kill rates and potential consumption rates using GPS collars. J Mammal 94:845–855

Miller CS, Hebblewhite M, Petrunenko YK, Seryodkin IV, Goodrich JM, Miquelle DG (2014) Amur tiger (*Panthera tigris altaica*) energetic requirements: implications for conserving wild tigers. Biol Conserv 170:120–129

Miquelle DG, Smirnov EN, Merrill TW, Myslenkov AE, Quigley HB et al (1999) Hierarchical spatial analysis of Amur tiger relationships to habitat and prey. In: Seidensticker J, Christie S, Jackson P (eds) Riding the tiger: tiger conservation in human-dominated landscapes. Cambridge University Press, Cambridge, pp 71–99

Miquelle DG, Goodrich JM, Kerley LL, Pikunov DG, Dunishenko YM, Aramiliev VV et al (2010) Science-based conservation of Amur tigers in Russian Far East and northeast China. In: Tilson

R, Nyphus PJ (eds) Tigers of the world: the science, politics and conservation of Panthera tigris, 2nd edn. Elsevier, New York, pp 403–424
O'Connell AF, Nichols JD, Karanth KU (eds) (2011) Camera traps in ecology. Springer, Tokyo
O'Kelly HJ, Evans TD, Stokes EJ, Clements TJ, Dara A, Gately M et al (2012) Identifying conservation successes, failures and future opportunities; assessing recovery potential of wild ungulates and tigers in Eastern Cambodia. PLoS One 7:e40482
Petrunenko YK, Montgomery RA, Seryodkin IV, Zaumyslova OY, Miquelle DG, Macdonald DW (2015) Spatial variation in the density and vulnerability of preferred prey in the landscape shape patterns of Amur tiger habitat use. Oikos 125:55–75
Robinson HS, Goodrich JM, Miquelle DG, Miller CS, Seryodkin IV (2015) Mortality of Amur tigers: the more things change, the more they stay the same. Integr Zool 10:344–353
Royle JA, Dorazio RM (2008) Hierarchical modeling and inference in ecology: the analysis of data from populations, metapopulations and communities. Academic Press, San Diego
Royle JA, Chandler RB, Sollmann R, Gardner B (2014) Spatial capture-recapture. Academic Press, San Diego
Sankar K, Qureshi Q, Nigam P, Malik PK, Sinha PR et al (2010) Monitoring of reintroduced tigers in Sariska Tiger Reserve, Western India: preliminary findings on home range, prey selection and food habits. Trop Conserv Sci 3:301–318
Schaller G (1967) The deer and the tiger. University of Chicago Press, Chicago
Simcharoen A, Savini T, Gale GA, Simcharoen S, Duangchantrasiri S, Pakpien S, Smith JLD (2014) Female tiger Panthera tigris home range size and prey abundance: Importance metrics for management. Oryx 48:370–377
Singh A (1984) Tiger! Tiger. Jonathan Cape, London
Smith JLD (1993) The role of dispersal in structuring the Chitwan tiger population. Behaviour 124:165–195
Smith JLD, McDougal CW, Sunquist ME (1987) Female land tenure system in tigers. In: Tilson RL, Seal US (eds) Tigers of the world: the biology, biopolitics, management, and conservation of an endangered species. Noyes Publication, New Jersey, pp 97–109
Sunquist ME (1981) Social organization of tigers *Panthera tigris* in Royal Chitawan National Park, Nepal. Smithson Contrib Zool 336:1–98
Thompson WL (2004) Sampling rare or elusive species: concepts, designs and techniques for estimating population parameters. Island Press, Washington, DC
Thompson WL, White GC, Gowan C (1998) Monitoring vertebrate populations. Academic Press, San Diego
Walston J, Robinson JG, Bennett EL, Breitenmoser U, da Fonseca GAB et al (2010) Bringing the tiger back from the brink - the six percent solution. PLoS Biol 8:e1000485
Williams BK, Nichols JD, Conroy MJ (2002) Analysis and management of animal populations. Academic Press, San Diego
Wilting A, Courtiol A, Christiansen P, Niedballa J, Scharf AK, Orlando L et al (2015) Planning tiger recovery: understanding intraspecific variation for effective conservation. Sci Adv 1:e140017
Xu X, Dong GX, XS H, Miao L, Zhang XL, Zhang DL et al (2013) The genetic basis of white tigers. Curr Biol 23:1031–1035

Animal Population Monitoring: A Unified Conceptual Framework

3

James D. Nichols, K. Ullas Karanth, Arjun M. Gopalaswamy, and Mohan Delampady

3.1 Introduction

As emphasized in the previous two chapters, monitoring of tiger and prey populations must necessarily rest on a solid foundation of established statistical theory and associated methodological practice. In this chapter, we describe a unified framework for monitoring populations of tigers as well as their prey species. For generality and convenience, we refer to the species populations that are targeted for monitoring as "animal populations." A graphic visualization that introduces this topic is available in the online support material and in the companion video guide Monitoring Tigers & Their Prey: The Right Way (Trust for Environmental Education 2008).

J.D. Nichols (✉)
Crofton, MD, USA
e-mail: jamesdnichols2@gmail.com

K.U. Karanth
Wildlife Conservation Society (WCS), New York, NY, USA

Centre for Wildlife Studies, Bengaluru, India

Wildlife Conservation Society, India Program, Bengaluru, India

National Centre for Biological Sciences-TIFR, Bengaluru, India
e-mail: ukaranth@wcs.org; https://www.wcs.org/; http://cwsindia.org/; http://wcsindia.org/home/; https://www.ncbs.res.in/

A.M. Gopalaswamy
Statistics and Mathematics Unit, Indian Statistical Institute, Bengaluru, India

Department of Zoology, University of Oxford, Oxford, UK
e-mail: arjungswamy@gmail.com; https://www.isibang.ac.in; https://www.zoo.ox.ac.uk/

M. Delampady
Statistics and Mathematics Unit, Indian Statistical Institute, Bengaluru, India
e-mail: mohan@isibang.ac.in; https://www.isibang.ac.in

K.U. Karanth, J.D. Nichols (eds.), *Methods For Monitoring Tiger And Prey Populations*, DOI 10.1007/978-981-10-5436-5_3

"Monitoring" of animal populations can be defined as the estimation of some state variable (e.g., absolute or relative abundance, occupancy) or vital rate (e.g., survival or reproductive rate) relevant to populations, for the purpose of drawing inferences about variation in this quantity over space and/or time. For example, we might focus on a single time period and ask whether animal abundance varies among different locations at that time in order to assess the distribution patterns of animals or to address questions about the relationship between animal abundance and such factors as habitat or management actions. We might also focus on a single location and ask whether animal abundance varies over time. Rates of temporal change are sometimes referred to as "trends," and some workers view monitoring as restricted to such assessments of temporal change. We make no such restriction.

As noted in Chap. 1, development of an animal monitoring program requires answers to three basic questions (Yoccoz et al. 2001, Nichols and Williams 2006): why, what, and how? For scientific endeavors, the basic answer to the "why" question is to provide estimates of key quantities to be compared with model-based predictions for use in discriminating among the competing hypotheses represented by the models. In the case of tigers and their prey, scientific hypotheses could involve questions about distribution of tigers as a function of prey (e.g., Karanth et al. 2004) and distribution of both tigers and prey with respect to habitat (Karanth et al. 2011) and human density (Karanth et al. 2009), for example.

When monitoring is to be a component of a conservation program, it provides estimates for use in making state-dependent decisions, assessing progress toward conservation objectives, discriminating among competing hypotheses about effects of conservation actions, and populating models used to predict effects of conservation. For example, decisions in tiger conservation will depend on current estimates of tiger densities, as well as on inferences about the effects of different management actions on tiger population dynamics. We cannot overemphasize the importance of having a clear objective (answer to the "why" question) in mind when developing a monitoring program. Indeed, the answers to the other two fundamental questions about monitoring, "what" and "how," are basically inherited from the answer to "why."

The answer to the question of what to monitor is thus determined by the objective(s) of the monitoring effort. When monitoring is used to inform conservation, then it is typically focused on the state variable of primary conservation interest (e.g., tiger abundance or occupancy), state variables relevant to those of primary interest (e.g., abundance of important prey species), and perhaps vital rates useful in models that project dynamics of state variables and predict management effects.

Answers to the why and what questions then provide guidance about how we monitor. Methods used to estimate state variables and vital rates of animal populations fill many books (e.g., Seber 1982, Borchers et al. 2002, Williams et al. 2002). However, in this volume we focus on specific methods that are potentially useful to tiger conservation. Although these methods are seemingly diverse, we note that a single conceptual framework underlies virtually all inference methods for state variables and vital rates of animal populations. Our framework clarifies the relationship among these seemingly disparate methods and provides a basis for considering and developing modifications and new methods. In particular, this framework clarifies the (often misunderstood) relationship between *indices* and *abundance estimates*. It should thus facilitate informed decisions about which of these approaches should be most appropriate and useful for different situations.

3.2 Statistical Framework

3.2.1 Basic Issues in Counting Animals

Virtually all inferences about animal populations are based on count statistics. In many cases, count statistics are provided by direct counts of animals themselves. For example, we might count the number of chital deer observed while walking on a line transect or the number of tigers identified (caught) by camera trapping. In other situations, count statistics are based on animal sign such as ungulate tracks or dung, or pugmarks (footprints) or scats of tigers.

Two basic problems confront biologists and managers who would like to use such count statistics to estimate and draw inferences about animal populations: *observability* or detectability and *spatial sampling* (see Nichols 1992; Thompson 2002; Lancia et al. 1994, 2005; Skalski 1994; Nichols and Conroy 1996; Williams et al. 2002). With respect to direct counts of animals, observability refers to the usual inability to detect and enumerate all animals, regardless of the sampling or survey method being used. With respect to indirect evidence of animal presence such as tracks, pugmarks, scat, and dung, observability refers to the fact that sign may go undetected as well, such that sites occupied by animals may not be identified. Even when sign is detected, sign counts will seldom equal the true number of animals. Spatial sampling refers to the fact that we are frequently interested in areas so large that we are unable to obtain count statistics over the entire area. Instead, we must select locations thought to be representative of the entire area and then try to use counts on these sampled areas to draw inferences about the number of animals in the entire area.

3.2.2 Observability/Detectability

We can state more precisely what we mean by observability and detectability by first defining the following quantities for a sample area (or for the entire area, if we assume there is no need to subsample areas). Here we assume that the quantity or state variable of interest is animal abundance, N:

C = count statistic or number of animals (or indirect sign) counted
N = abundance or true number of animals
p = proportionality constant relating the count statistic and abundance

In the case of direct counts of animals, p reflects the probability that an animal in the sampled area is counted. This can be restated as the probability that a member of N appears in the count statistic; p can also be viewed as the expected proportion of animals appearing in the count statistic. The count, C, is known to the investigator and is a result of his or her sampling efforts, whereas the other two quantities, N and p, are not known and must be estimated. The following expression shows the relationship between the count statistic and abundance:

$$E(C) = Np \tag{3.1}$$

where $E(C)$ denotes the expected value (or expectation) of C. C is a random variable and can assume different values each time a count is made. $E(C)$ can be viewed as the average value of C that would be obtained if the count statistic could be collected a large number of times under the same exact sampling conditions with the same N. In cases where C represents some count of animal sign (or of something other than the animals themselves), Eq. 3.1 still provides a reasonable model. However, in this case, p should not be thought of as a detection probability but simply as a coefficient relating N and C.

One consequence of Eq. 3.1 is that it is not possible to draw strong inferences about N based only on our counts, C. Lower counts in one place than another may result from lower abundance or simply lower detection probability (see Chap. 7). Thus, we must know something about detection probability, p, in order to draw inferences about abundance. If we are able to estimate the p associated with a particular count statistic (denote this estimate as $\widehat{p}$ where the "hat" denotes an estimate), then abundance can be estimated as

$$\widehat{N} = \frac{C}{\widehat{p}} \tag{3.2}$$

The estimator in Eq. 3.2 is very general, as virtually all population estimation methods (Seber 1982; Lancia et al. 1994, 2005; Borchers et al. 2002, Williams et al. 2002) for a single location can be written in this basic form. For example, the count statistic under distance sampling (Buckland et al. 1993, 2001) is the number of animals observed and counted (e.g., along a line transect), and the perpendicular distances of these observations to the transect line are used to estimate the detectability function and, hence, p. The count statistic under classical capture-recapture sampling (Otis et al. 1978; Seber 1982; White et al. 1982; Pollock et al. 1990, Williams et al. 2002) is the number of different animals caught, and the patterns of capture and recapture for individual animals are modeled in order to estimate p. The number of different animals captured is also the count statistic for spatially explicit capture-recapture models (e.g., Royle et al. 2014), although in this case the extra information provided by the location of each individual capture provides information that is used to estimate capture probability.

As a numerical example to help illustrate the rationale and intuition underlying Eq. 3.2, assume that we count 20 deer in an area and estimate a corresponding detection probability of $\widehat{p} = 0.25$; that is, we estimate that we detected about

25% of the animals when we conducted our counts. Our abundance estimate is then obtained as $\widehat{N} = 20/0.25 = 80$. This estimate is intuitively reasonable in that we estimate that we detect approximately one of every four deer, so our estimated abundance is four times the number of animals counted. Perhaps the most important consideration resulting from Eqs. 3.1 and 3.2 is that the count statistic itself does not permit unambiguous inference about abundance (see subsequent section, *Counts as Indices*). Instead, such inference requires information about the detection probability associated with the count statistic.

The quantity of primary biological interest in Eqs. 3.1 and 3.2 is abundance, N, total number of animals present in the surveyed area. However, these equations are also fundamental to drawing inferences about other quantities of interest. For example, in Chap. 4 we focus on a different state variable, occupancy, used in the assessment of spatial distribution patterns of animals. As explained in Chap. 4, the count statistic for inference about occupancy is the number of surveyed sites at which a focal species is detected. Chapter 8 focuses on dynamics of animal populations and on the vital rates responsible for those dynamics. Survival rate is a key vital rate usually obtained by asking how many of a set of marked animals detected at one sampling occasion, t, are still alive and in the population at the next sampling occasion, $t + 1$. The relevant count statistic is the number of marked animals detected at occasion $t + 1$. In both cases, occupancy and survival rate, Eq. 3.2 provides the conceptual basis for translating count statistics into estimates.

3.2.3 Spatial Sampling

Often, we cannot survey an entire area of interest, so we must select sample locations that are representative of the entire area. One way to obtain a representative sample is to select locations using simple random sampling (Thompson 2002, Williams et al. 2002). These sample locations will represent some fraction, α, of the total area of interest. Unlike the situation where the fraction of animals present in a sampled area that is counted (p) must be estimated, the spatial sampling fraction (α) is often known (but see below) and requires no estimation. Define N' to be the true abundance of animals in sampled areas representing fraction α of the total area of interest. Then if this abundance on sample sites can be estimated, abundance for the entire area of interest can be estimated as

$$\widehat{N} = \frac{\widehat{N'}}{\alpha} \tag{3.3}$$

i.e., we simply divide the estimated abundance for the sampled locations by the fraction of the entire area represented by those locations.

As a numerical example, assume that we have randomly, or at least representatively, sampled several locations representing 10% of the entire area of interest ($\alpha = 0.10$) and that we have obtained an estimate of 80 deer on these locations ($\widehat{N'} = 80$). Then the population estimate for the entire area is computed as $\widehat{N} = 80 / 0.10 = 800$ deer.

Equation 3.3 is based on the simple case in which the fraction of the total area that is surveyed is known. However, this simple case does not apply to all animal survey methods. For example, in standard capture-recapture sampling, the area over which we deploy our traps is typically known with certainty. But this does not mean that we know the area to which the resulting capture-recapture abundance estimate applies. Animals with home ranges that lie largely outside the sampled area, but that overlap it, may be caught, such that the abundance estimate actually pertains to a larger area than that covered by traps. Spatially explicit capture-recapture models deal with this issue explicitly, but the point is that α is not always known and sometimes must be estimated.

3.2.4 Canonical Estimator

If we seek to estimate abundance for some large area, based on survey counts from representative sample locations within this area, we must deal with both observability/detectability and spatial sampling. We cannot survey all locations within the area of interest, and we cannot detect all animals even at the locations that we do survey. Thus, we must combine Eqs. 3.2 and 3.3 into a general estimator that deals with both issues. We can readily accomplish this in a general manner that deals with variation from one sample location to the next in either true abundance or detection probability or both (see Skalski 1994, Thompson 2002, Williams et al. 2002). However, for ease of understanding, we present the canonical estimator for the simpler case in which expected abundances and detection probabilities are identical across all sample locations.

Define C' as the summed counts over all sampled locations that comprise area α. If detection probability is similar for the different sample locations, then combining 3.2 and 3.3 yields the following canonical estimator for abundance of the entire area:

$$\widehat{N} = \frac{C'}{\widehat{p}\alpha} \tag{3.4}$$

We thus estimate population size for the entire area of interest by dividing the count statistic for the sampled locations by both the estimated fraction of the animals on the sampled area(s) that was detected ($\widehat{p}$) and the proportion of the total area from which the count statistic was taken (α). The two terms in the denominator reflect the two sources of incompleteness of the counts, as the counts include neither the entire area of interest nor all animals present at the sampled locations.

In order to illustrate the canonical estimator, assume that we count 20 deer ($C' = 20$) at sample locations representing 10% of the total area of interest, so $\alpha = 0.10$. Further assume a detection probability for these counts of 0.25 ($\widehat{p} = 0.25$). Then we estimate abundance for the entire area as

$$\widehat{N} = \frac{C'}{\alpha\widehat{p}} = \frac{20}{(0.10)(0.25)} = 800 \text{ deer}$$

We would thus view 800 deer as our best estimate of the number of deer in the area of interest. But this estimate is characterized by uncertainty, as we did not obtain counts over the entire area, and the counts on sample locations were incomplete. The uncertainty is characterized by a variance, $\mathrm{var}\left(\widehat{N}\right)$, which will be influenced by the details of the spatial sampling design; the degree of variation among counts at sampled locations; the proportion of total area sampled, α; the magnitude of detection probability; and the manner in which it was estimated. Nevertheless, we can make some general statements about the factors that influence the magnitude of the abundance estimator's variance.

The variance of the counts across sample locations, $\mathrm{var}(C')$, is smaller when counts are similar across locations and larger when animals are clumped, with large counts at some locations and small counts at others. A variance will also be associated with the estimate of detection probability, $\mathrm{var}\left(\widehat{p}\right)$, and will depend on the method used to estimate detection and the associated sample size. Smaller variances of abundance, $\mathrm{var}\left(\widehat{N}\right)$, are desirable and result from smaller values of $\mathrm{var}(C')$ and $\mathrm{var}\left(\widehat{p}\right)$ and larger values of p and α. In the case of methods in which sign counts are used to estimate abundance, then an additional variance component will be needed to deal with the relationship between number of sign and number of individual animals.

The above discussion of area surveyed is very simplistic. The ability to draw inferences about locations not surveyed based on data collected at a set of surveyed locations is usually dependent on probabilistic sampling. Under probabilistic sampling designs, each potential location in an area is sampled with some probability that is known *a priori*. A variety of such designs is possible (simple random, stratified random, adaptive random, dual frame, etc.), and Thompson (2002) is an excellent source of information on them (also see Cochran 1977, Skalski 1994, Williams et al. 2002). The basic discussion above, and indeed much of the discussion of spatial sampling throughout the book, is based on simple random sampling but is relevant to other probabilistic designs (stratified random, adaptive random, etc.) as well.

3.2.5 Indices

The canonical estimator thus provides the generic answer to the question "how should we monitor." Unfortunately, this advice is often not followed in tiger monitoring efforts. Frequently, detection probability issues are ignored, and inferences about state variables and vital rates are based directly on the raw count data themselves. The counts are referred to as indices and typically viewed as reflecting a constant proportion of the true state variable. Because the proportionality constant is not known or estimated, index proponents do not typically have good estimates of state variables or vital rates. Instead they claim that indices can be used to draw inferences about *relative* changes or differences in state variables (Johnson 2008), so here we examine the basis for this claim.

In Chaps. 6 and 7, we discuss the use of distance sampling methods for estimating abundance of ungulates based on counts of animals obtained while the investigator walks along a line transect. Information about the distance of each detected animal from the transect line is used to estimate detection probability, assumed to be a decreasing function of distance between observer and animal. Index proponents would not advocate expending the effort to obtain these distance data, but rather would assume that about the same proportion of animals was being seen in each year or at each sampled location. If this assumption is true, then we can estimate the rate of population change between 2 years using the ratio of the counts themselves.

Consider the quantities used in Eq. 3.1; except now add a subscript denoting time or location and a new parameter reflecting the ratio of abundances at two times or locations:

C_i = count statistic or number of animals (or indirect sign) counted at time/place i
N_i = abundance or true number of animals, at time/place i
p_i = proportionality constant relating the count statistic and abundance at time/place i
$\lambda_{ij} = N_i/N_j$= ratio of abundances at two times or places, i and j

So if i and j are successive years, then λ_{ij} is a population trend or rate of change. If i and j are different locations, then λ_{ij} is often referred to as relative abundance. Index proponents typically estimate λ_{ij} as the ratio of the counts themselves. We can explore the possibility of bias in this estimator by writing out its approximate expected value:

$$E\left(\frac{C_i}{C_j}\right) \approx \frac{E(C_i)}{E(C_j)} = \frac{N_i p_i}{N_j p_j} = \lambda_{ij}\left(\frac{p_i}{p_j}\right) \tag{3.5}$$

So when $p_i = p_j$, then $p_i/p_j = 1$, and the ratio of counts provides an approximately unbiased estimator for λ_{ij}, where bias is defined as the difference between the expected value of an estimator, $\widehat{\theta}$, and the true quantity being estimated, θ (i.e., bias $= \widehat{\theta} - \theta$). But when $p_i \neq p_j$, the index-based estimator of Eq. 3.5 will be biased, with the direction and magnitude of bias dictated by the difference between the detection probabilities of the two times or locations being compared.

As an example, assume that we are twice as likely to see an ungulate such as sambar deer in drier open forest habitat (location i) than in more humid forest with dense understory (location j), say $p_i = 0.6$ and $p_j = 0.3$. If we count 50 sambar along transects in each of the two habitats, then our index-based estimate of relative abundance is 1, indicating similar abundances in the two habitats. However, if we estimate the detection probabilities well and apply the estimator of 3.2, we obtain estimated abundances of about 83 in habitat i and 167 in habitat j, yielding an estimate of relative abundance of 0.5. So in this case, the index-based approach led to substantial bias in our estimate of relative abundance.

Note that the above discussion and example focused on detection probability and omitted the α parameter indicating proportional area surveyed. Usually index proponents recognize the need to compare areas of similar size, in which case the α terms drop out and leave us with the above expression (3.5). But if abundance comparisons are based on areas of different sizes, then of course this influences estimates as well. Finally, we note that we based the above discussion on the state variable of abundance, but the very same arguments apply to estimates of other state variables such as density or occupancy, and to vital rates (e.g., survival and reproduction) as well. Detection probability should be a pervasive concern in the estimation of animal population parameters and deserves our serious attention.

3.3 Discussion

The establishment of monitoring programs for tiger and associated prey populations has become a popular endeavor of conservation biologists and wildlife managers across the species' range over the last three decades. In a general sense, this is certainly a good thing, as any information about endangered species is potentially useful in recovery and conservation. However, in times of limited availability of funds and human effort for conservation, we believe that efficient use of these resources is important. In particular, we are concerned that many current tiger monitoring programs are established without adequate thought devoted to exactly how resulting data are to be used for either scientific or conservation endeavors (Karanth et al. 2003, Gopalaswamy et al. 2015). Hence, we began this chapter with the recommendation that developers of monitoring programs consider three basic questions: why, what, and how?

Addressing the question "why monitor" is extremely important, as all other aspects of a monitoring program are inherited from the answer. Our perspective is that monitoring should not be viewed as a stand-alone activity, but rather is most useful when viewed as a component of a larger process, typically either of science or conservation. We elaborate on the specific roles of monitoring in conservation programs in Chap. 14. For the question of "what to monitor," programs designed for the purpose of informing tiger conservation will likely be focused generally on the dynamics of tiger and prey populations and specifically on the responses of such populations to management actions. This focus will typically lead to the monitoring of state variables such as species abundance and/or occupancy and the vital rates responsible for their dynamics (e.g., reproduction, survival, movement, local extinction, local colonization).

The question of "how to monitor" was addressed broadly in this chapter, but the remainder of the book is devoted mostly to such specifics. Two critical issues for the sampling of animal populations were identified: (1) observability/detectability and (2) spatial sampling. With respect to (1), animal survey methods entail counts, and these counts represent some unknown fraction of the animals actually present in a sampled location. Estimation of that unknown fraction (i.e., estimation of detection probability) then becomes a key to inference about animal numbers and related quantities (e.g., occupancy, species richness, survival rate, reproductive rate, local extinction rate, local colonization rate).

With respect to (2), investigators are often interested in drawing inferences about animal populations inhabiting areas so large that they cannot possibly be completely surveyed. In such cases investigators must sample space; that is, they must use their survey methods at a subset of locations within the entire area of interest. Inferences about locations not surveyed must then be based on survey data from the locations that are surveyed. Such inferences are generally made possible by probabilistic sampling approaches such as simple random, stratified random, etc.

A canonical estimator was presented that essentially corrects raw count data for these two sources of incompleteness: not all locations are surveyed and not all animals are seen at surveyed locations. This estimator forms the basis for inferences about state variables, such as animal abundance, occupancy and species richness, and vital rates, such as survival and reproductive rates, rates of local extinction and colonization, etc. The various specific estimation methods and associated estimators described in this volume can all be viewed as special cases of this canonical approach to estimation.

Despite the canonical nature of this basic approach to inference, the specific sampling and estimation methods described in this volume are very diverse in detailed application. For example, approaches to inference about detection probability may be based on detection histories (detection or not, and location for detected animals) of marked individuals over multiple sampling occasions; on the distance between the observer and each detected, unmarked animal; or on detection/nondetection of a species over replicate surveys at multiple locations. Similarly, approaches for dealing with spatial variation include random sampling; random sampling within strata; sampling many locations with an inexpensive survey method and then sampling a subset of these with a more expensive, but also more informative, method; sampling initially with a random approach and subsequently with an approach that is conditional on results of the first survey; etc. These very different approaches to dealing with detectability and spatial variation lead to very different ways to select locations to survey and to very different field survey methods to be used at these locations.

The detailed methods of selecting locations to be surveyed and collecting field data at these locations are the major components of monitoring program design. Indeed, design is critical to the ultimate ability to draw inferences. The chapters of this volume that deal with specific methods will emphasize those features of study design that are critically important to program success, with the hope that practitioners will carefully follow recommendations in order to maximize chances of obtaining data that are useful for the intended purposes. The statistical methods and algorithms used to analyze data are tailored to, and dependent on, the monitoring program design. Although practitioners need not understand all of the details of the statistical methods to which their data will be subjected, we believe it is important that they understand the conceptual basis of each methodology and the resultant close pairing of survey design features and analytic method.

In summary, the framework presented in this chapter focuses on the conceptual thread that runs through all of the specific methods described in this volume. It is our hope that this general framework will facilitate a basic understanding of these methods and how they work. Subsequent chapters then focus on specific methods and their associated designs, hopefully providing the practitioners with the information needed to develop field surveys of tigers and prey based on these approaches. We hope that this overall approach not only facilitates understanding but also leads to development of new field and analytic methods tailored to the logistical and ecological specifics of new estimation problems.

References

Borchers DL, Buckland ST, Zucchini W (2002) Estimating animal abundance: closed populations. Springer, New York

Buckland ST, Anderson DR, Burnham KP, Laake JL (1993) Distance sampling: estimating abundance of biological populations. Chapman and Hall, New York

Buckland ST, Anderson DR, Burnham KP, Laake JL, Borchers DL, Thomas L (2001) Introduction to distance sampling: estimating abundance of biological populations. Oxford University Press, Oxford

Cochran WG (1977) Sampling techniques, 2nd ed. Wiley, NewYork

Gopalaswamy AM, Delampady M, Karanth KU, Kumar NS, Macdonald DW (2015) An examination of index-calibration experiments: counting tigers at macroecological scales. Methods Ecol Evol 6:1055–1066

Johnson DH (2008) In defense of indices: the case of bird surveys. J Wildl Manag 72:857–868

Karanth KK, Nichols JD, Hines JE, Karanth KU, Christensen NL (2009) Patterns and determinants of mammal species occurrence in India. J Appl Ecol 46:1189–1200

Karanth KU, Nichols JD, Seidensticker J, Dinerstein E, Smith JLD, McDougal C, Johnsingh AJT, Chundawat RS, Thapar V (2003) Science deficiency in conservation practice: monitoring tiger populations in India. Anim Conserv 6:141–146

Karanth KU, Nichols JD, Kumar NS, Link WA, Hines JE (2004) Tigers and their prey: predicting carnivore densities from prey abundance. Proc Natl Acad Sci U S A 101:4854–4858

Karanth KU, Gopalaswamy AM, Kumar NS, Vaidyanathan S, Nichols JD, MacKenzie DI (2011) Monitoring carnivore populations at landscape scales: occupancy modeling of tigers from sign surveys. J Appl Ecol 48:1048–1056

Lancia RA, Nichols JD, Pollock KH (1994) Estimating the number of animals in wildlife populations. In: Bookhout T (ed) Research and management techniques for wildlife and habitats. The Wildlife Society, Bethesda, pp 215–253

Lancia RA, Kendall WL, Pollock KH, Nichols JD (2005) Estimating the number of animals in wildlife populations. In: Braun CE (ed) Research and management techniques for wildlife and habitats. The Wildlife Society, Bethesda, pp 106–153

Nichols JD (1992) Capture-recapture models: using marked animals to study population dynamics. Bioscience 42:94–102

Nichols JD, Conroy MJ (1996) Estimation of mammal abundance: introduction. In: Wilson D, Cole FR, Nichols JD, Rudran R, Foster M (eds) Measuring and monitoring biological diversity: standard methods for mammals. Smithsonian Institution Press, Washington, DC, pp 177–179

Nichols JD, Williams BK (2006) Monitoring for conservation. Trends Ecol Evol 21:668–673

Otis DL, Burnham KP, White GC, Anderson DR (1978) Statistical inference from capture data on closed animal populations. Wildlife Monographs 62

Pollock KH, Nichols JD, Brownie C, Hines JE (1990) Statistical inference for capture-recapture experiments. Wildlife Monographs 107

Royle JA, Chandler RB, Sollman R, Gardner B (2014) Spatial capture-recapture. Academic Press, Amsterdam

Seber, G.A.F. 1982. The estimation of animal abundance and related parameters. 2nd ed. Macmillan, New York.

Skalski JR (1994) Estimating wildlife populations based on incomplete area surveys. Wildl Soc Bull 22:192–203

Thompson SK (2002) Sampling. 2nd ed. Wiley, New York

Trust for Environmental Education (2008) Monitoring tigers and their prey the right way. Wildlife Conservation Society – India Program, and Enduro India. (Video)

White GC, Anderson DR, Burnham KP, Otis DL (1982) Capture-recapture and removal methods for sampling closed populations. Los Alamos National Laboratory LA-8787-NERP, Los Alamos

Williams BK, Nichols JD, Conroy MJ (2002) Analysis and management of animal populations. Academic Press, San Diego

Yoccoz NG, Nichols JD, Boulinier T (2001) Monitoring of biological diversity in space and time. Trends Ecol Evol 16:446–453

Concepts: Assessing Tiger Habitat Occupancy Dynamics

4

James D. Nichols, Varun R. Goswami, Ravishankar Parameshwaran, Soumen Dey, and K. Ullas Karanth

4.1 Introduction

Conservationists are often interested in making inferences about *spatial distributions* of tigers, potential competitor species, and prey species (Chap. 2). For this purpose, we do not require precise estimates of abundance or density over many locations across large spatial landscapes. Instead, we can use detection-nondetection survey data for animals, or even their sign, to draw such inferences in a cost-effective manner. We can investigate key spatial relationships (e.g., between tigers and habitat variables, tigers and prey, etc.), as well as occupancy dynamics and the factors influencing the vital rates (local probabilities of extinction and colonization) driving

J.D. Nichols (✉)
Crofton, MD, USA
e-mail: jamesdnichols2@gmail.com

V.R. Goswami • R. Parameshwaran
Wildlife Conservation Society, India Program, Bengaluru, India

Centre for Wildlife Studies, Bengaluru, India
e-mail: varunr.goswami@gmail.com; ravishankar.cws@gmail.com

S. Dey
Statistics and Mathematics Unit, Indian Statistical Institute, Bengaluru, India
e-mail: soumenstat89@gmail.com

K.U. Karanth
Wildlife Conservation Society (WCS), New York, NY, USA

Centre for Wildlife Studies, Bengaluru, India

Wildlife Conservation Society, India Program, Bengaluru, India

National Centre for Biological Sciences-TIFR, Bengaluru, India
e-mail: ukaranth@wcs.org; https://www.wcs.org/; http://cwsindia.org/; http://wcsindia.org/home/; https://www.ncbs.res.in/

K.U. Karanth, J.D. Nichols (eds.), *Methods For Monitoring Tiger And Prey Populations*, DOI 10.1007/978-981-10-5436-5_4

these dynamics. These detection-nondetection occupancy surveys are especially useful for large-scale inferences about geographic range and range changes in tigers and other species.

Just as counts of animals in a sample unit are likely to miss animals (detection probability <1), so are surveys of a sample unit likely to miss species that are actually present there (see Chap. 3). Karanth and Nichols (2002) presented an initial approach to dealing with nondetection in such surveys. This approach used closed capture-recapture models with detection-nondetection data from replicate temporal sampling at multiple sites. The sites were treated as "individual animals," and the detection history data were used to first estimate the number of occupied sample units, $\widehat{N}$. If s sample units are surveyed, then the proportion of these that is actually occupied by the species can be estimated as:

$$\widehat{\psi} = \frac{\widehat{N}}{s} \tag{4.1}$$

The estimator (4.1) thus entails two steps, estimation of N and then ψ. Estimation of N can be based on any of the closed capture-recapture models to be described in Chap. 8.

A key difference between the inference problems of closed-population capture-recapture and occupancy is that inference about N typically is unconstrained in capture-recapture modeling, in the sense that no upper bound on abundance is imposed. Stated differently, the number of individuals exhibiting detection histories of all 0's (never detected) is not constrained and can be very large. However, in occupancy modeling, we know the exact number of all-0 detection histories, imposing the logical constraint that $N \leq s$; that is, the number of sample units occupied by the focal species cannot exceed the total number of units. This knowledge of s led J.A. Royle to conceive a single-step approach to occupancy modeling that imposes this constraint, and this is the approach that we have developed for occupancy modeling (MacKenzie et al. 2002).

4.2 Detection-Nondetection Surveys

Detection-nondetection surveys for the purpose of collecting data to study occupancy can be carried out in a variety of ways. Here, we provide survey design requisites and outline some basic approaches. Detailed discussion of field methods that can be used for tigers and their prey is provided in Chap. 5. Data and corresponding modeling can be used to make inferences about occupancy during a single season or extended to deal with occupancy dynamics across multiple years and seasons. We begin by focusing on sampling for single-season models.

There are few restrictions on the kind of sampling that is conducted, as the primary objective is to simply select a survey method that provides a reasonably high probability of detecting the focal species, given its presence in the sample unit. Investigators often visit a sample unit and then move through it recording individuals of the focal species. For species that are very difficult to detect in this manner, sign surveys may be used, with focal species identified as present based on tracks, scats, dung, or other signs. In other situations, stationary sampling devices such as traps, camera traps, or audio recording devices may be used to survey an area (see Chap. 5).

The key requisite for classic, single-season occupancy modeling is replication. The investigator is interested in S sites or sample units and in the proportion of these sites that is occupied by the focal species. A total of s sites would be selected for survey, with $s \leq S$. A typical single-season sampling design would entail multiple samples at each site, where the sampling would involve implementation of any specific survey method(s) capable of detecting the species of interest or its sign. For a study design with $K = 3$ sampling occasions at each site, raw detection history data for site i would be written as a row vector of 1's and 0's indicating detection or nondetection, respectively, at each sampling occasion. For example, $h_i = 1\ 0\ 1$ would indicate detection of the focal species at site i on occasions 1 and 3, but not on occasion 2. Each site has a corresponding detection history, and the histories for all surveyed sites are the data from which inferences about occupancy are derived.

The key assumption underlying single-season occupancy models is that the sampled sites are closed to changes in occupancy for all surveys during the season. This simply means that a site is either occupied for all K sampling occasions or not. The total period over which temporal replicate visits are made is typically fairly short in order to try to satisfy this closure assumption. If the study system is very large, it may be necessary to conduct all K surveys for one group of sites during one period and for another group of sites during a different period. For this type of sampling design, the closure assumption applies to the replicate sampling within each period. Approaches exist for testing the closure assumption (MacKenzie et al. 2003; Rota et al. 2009), and models have been developed to relax this assumption in specific ways (e.g., Kendall et al. 2013). This and other requisite assumptions are discussed in Sect. 4.3.3.

There are several ways to achieve this sort of replicate sampling over a relatively short period of time (see Chap. 5). In addition to the standard approach of one or more investigators sampling the same sites at multiple times (e.g., multiple days), logistics (e.g., safety reasons) may dictate that multiple individuals visit each site simultaneously. In such cases, it may be possible for different individuals to survey independently, in which case the individual survey results can be used as replicates.

Sometimes logistics dictate so-called removal designs. When sites require substantial effort and travel time to reach, or are otherwise expensive to sample, then investigators may visit all sites on an initial occasion. Sites at which the focal species is detected on the initial survey are not revisited. Instead, second surveys only occur on sites with no detections initially. Similarly, sites with no detections at surveys 1 or 2 are surveyed a third time, etc. This design approach constrains modeling to some degree, but can still be wise when sampling is very expensive.

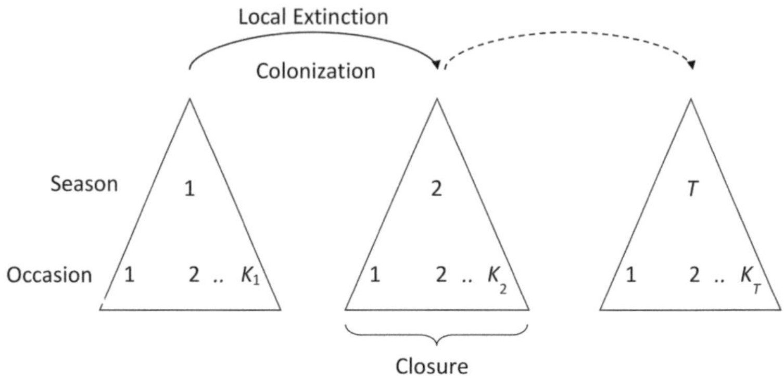

Fig. 4.1 Multiseason occupancy studies of animal populations are conducted at two temporal scales: (i) over multiple (K_t) occasions within each season t, across which sites are assumed to be "closed" to changes in occupancy state (occupied sites remain occupied; unoccupied sites remain unoccupied) and (ii) across multiple (T) seasons, during which unoccupied sites may be colonized and occupied sites may go locally extinct

It is sometimes possible to substitute space for time and to survey small replicate spatial units, such as segments along a trail, within each sample unit. Whenever possible, such selection of spatial units for subsampling should be conducted randomly *and* with replacement (Kendall and White 2009). However, when the focal species is mobile and moves about extensively within sample units, sampling randomly and with replacement becomes less important. In addition, special models have been developed for certain types of dependent spatial sampling (Hines et al. 2010, Chap. 5). In fact these models were initially developed specifically to incorporate tiger movement ecology, by allowing dependence of local tiger occupancy on adjacent trail segments.

Studies focusing on changes in occupancy over time typically follow a so-called robust design (Pollock 1982) consisting of replicate sampling over a short period within a season followed by a longer period with no sampling over which change is expected, and then another short period of replicate sampling (Fig. 4.1). The basic requisites for sampling under single-season models apply as well to the within-season sampling components of multiseason designs.

4.3 Basics of Single-Season Occupancy Modeling

4.3.1 Basic Model

The basic ideas underlying occupancy modeling were summarily noted in Sect. 4.1. The detection history data for surveyed sites contain information about nondetection. For example, consider the detection history 101 for a site. The failure to detect the species at occasion 2 clearly reflects nondetection, as the focal species was known to use the site based on detections at occasions 1 and 3. So occupancy models use the site replicates to draw inferences about detection probability and thus the number of occupied sites. We can estimate the number of occupied sites using the canonical estimator (3.2):

$$\widehat{N} = \frac{C}{\widehat{p}} \tag{4.2}$$

where C is now the number of sites at which the focal species was detected on at least one occasion, and $\widehat{p}$ is the detection probability estimated from the replicate surveys. More explicitly, p in expression (4.2) is the probability that the focal species is detected *on at least one sampling occasion*, given that the sample unit is occupied by the species. Expression (4.1) is used to translate the estimated number of occupied sites into the estimated proportion of sites occupied, $\widehat{\psi}$.

It is useful to model the parameter, ψ, directly as a function of covariates, as species-habitat relationships are an important area of interest in species distribution modeling, for example. J.A. Royle thus developed an approach to occupancy modeling that incorporates the ψ parameter directly into the likelihood (MacKenzie et al. 2002). Under this model, a probability can be assigned to every distinct detection history, and that probability is based on two kinds of parameters, occupancy (ψ) and occasion-specific detection probability, p_j, where j denotes sampling occasion. Specifically, p_j is the probability of detecting the focal species on sampling occasion j, given that the sample unit is occupied by the focal species. Consider the probability corresponding to detection history 101:

$$\Pr\left(101|\psi, p_j\right) = \psi p_1 \left(1 - p_2\right) p_3$$

We know the species was present at the site, because we detected it there, and the probability associated with presence/occupancy is ψ. The species was detected at sampling occasions 1 and 3, with corresponding probabilities, p_1 and p_3, and was not detected at occasion 2, with probability $(1 - p_2)$.

The probability associated with detection history 000 can be written as follows:

$$\Pr\left(000|\psi, p_j\right) = (1 - \psi) + \psi \prod_{j=1}^{3} \left(1 - p_j\right)$$

There are two possible events that could give rise to this detection history, and the underlying probability must include both possibilities. The first is that the species was not present at the site, with corresponding probability $(1 - \psi)$. The second possibility is that the species was present and simply went undetected at all three sampling occasions, represented by the second additive term in the above expression.

Estimation is based on the model likelihood for the observed data which is simply the product of the probabilities associated with the observed detection histories:

$$L\left(\psi, \{p_j\} \mid h_1, h_2, \ldots, h_s\right) = \prod_{i=1}^{s} \Pr\left(h_i\right) \tag{4.3}$$

The method of maximum likelihood determines the values of occupancy and detection probabilities for which the above expression is a maximum, essentially asking "what are the most likely values for occupancy and detection parameters given the model and the observed data?" (see MacKenzie et al. 2006). Programs PRESENCE (Hines 2006), MARK (White and Burnham 1999), and unmarked (Fiske and Chandler 2011) can all be used to compute estimates using detection-nondetection data.

4.3.2 Covariate Relationships

The parameters of expression (4.3) were written without subscripts for specific sample units, essentially assuming that parameter values are the same for all sample units. Estimates of parameters such as occupancy under this assumption can be useful, but we are frequently interested in relationships between these parameters and site-specific covariates. Such relationships can be investigated using the so-called logit link, which can be used to model covariate relationships for parameters (such as occupancy) that are constrained to the interval [0, 1]. For example, we can write a relationship between occupancy of site *i*, ψ_i, and two covariates associated with site *i*, x_{i1} and x_{i2}, as:

$$\log\text{it}\left(\psi_i\right) = \ln\left(\frac{\psi_i}{1-\psi_i}\right) = \beta_0 + \beta_1 x_{i1} + \beta_2 x_{i2} \tag{4.4}$$

Expression (4.4) can be solved for ψ_i to obtain the following expression for occupancy as a function of the covariates:

$$\psi_i = \frac{\exp\left(\beta_0 + \beta_1 x_{i1} + \beta_2 x_{i2}\right)}{1 + \exp\left(\beta_0 + \beta_1 x_{i1} + \beta_2 x_{i2}\right)} \tag{4.5}$$

Such covariate modeling can be used to test hypotheses about ecological relationships, for example, between tiger occupancy and habitat, prey abundance, distance from roads or villages, etc. (see Karanth et al. 2011).

Detection probabilities, p_{ij}, can also be modeled as functions of covariates. In this case, covariates can be associated with either site, or sampling occasion, or both, x_{ijm}, where i denotes sample unit, j denotes sampling occasion, and m indicates the specific covariate. Selection of potential covariates for detection depends on the nature of the sampling. For example, detection probabilities for camera trap sampling might include site-specific covariates such as number of cameras in a sample unit and occasion-specific covariates such as any weather variable that might influence activity of the focal species (e.g., tigers, ungulates). Covariates for detection probabilities based on sign surveys might include site-specific covariates such as substrate (e.g., pugmarks may be more easily detected along a dusty trail than one covered by grass) and occasion-specific covariates such as rainfall (influencing both animal activity and distinctness of tracks). Finally, we note that other link functions can be used for such modeling as well (e.g., Williams et al. 2002), although the logit link will be selected for most occupancy modeling.

Inference about relationships can be obtained by substituting the right-hand side of (4.5) for the occupancy parameter in the likelihood, and estimates of the β_i parameters obtained directly (e.g., MacKenzie et al. 2006). Note that this sort of direct inference precludes our need to use 2-step analyses (doing statistics on statistics) that correlate estimates of occupancy with covariates. Covariate modeling based on Eqs. (4.4) and (4.5) can be accomplished using the abovementioned software. The β_i parameters are estimated directly, and the modeled parameters (e.g., ψ_i and p_{ij}) are estimated as derived parameters.

4.3.3 Assumptions

Several assumptions underlie these basic occupancy models. When assumption violations are anticipated, they can frequently be dealt with via small changes in sampling design. In addition, the models have been extended in many cases to accommodate specific assumption violations. Here we discuss assumptions briefly and provide citations to more detailed discussions and model extensions.

The first assumption is referred to as "closure," that is, surveyed sites do not change occupancy status over the entire period of survey. The sites are either occupied or not occupied for the full survey period. Violations occur if, for example, surveys of migratory animals begin before the focal species has migrated to the breeding grounds or end after all members of the species have departed for nonbreeding areas. The simplest approach to dealing with this issue is to reduce the time between successive sampling occasions, such that violation of closure becomes less and less likely. Kendall et al. (2013) developed a single-season occupancy model to deal with such violations and then extended it for multiple seasons (Chambert et al. 2015b). The model permits entry of the focal species to the surveyed sites after the initiation of surveys and departure before the end of surveys.

A more common form of violation of the closure assumption concerns animals that may move in and out of sample units during the course of their daily movement. This can result in a focal species being present in a sample unit during some survey occasions and not during others. If this movement is completely random, then estimates of occupancy now reflect the proportion of sites that is *used* by the focal species, with *use* reflecting some non-negligible probability of the species being in the sample unit at any time. This probability is <1, reflecting the fact that the species is not confined to the sample unit. Similarly, detection probability for the case of random movement in and out of sampling units is now defined as the product: Pr(at least one member of the focal species is present in the sample unit during the survey

| sample unit is used by the species) × Pr(focal species is detected | at least one member of the focal species is present in the sample unit during the survey).

Closure tests can be developed by combining consecutive groups (e.g., pairs) of sampling occasions within a season and fitting multiseason models (MacKenzie et al. 2003; Rota et al. 2009). For example, a season with six sampling occasions could be treated as being comprised of three "seasons," each of which contains two sampling occasions. Multiseason occupancy models in which local extinction and colonization parameters occurring between occasions are constrained to equal 0 can be viewed as null hypotheses that reflect closure and compared against models that permit these parameters to attain values >0. It should be noted that this test for absence of closure should detect random movement in and out of sample units, a "violation" that does not really cause inference problems (see above). Finally, the single-entry, single-exit model of Kendall et al. (2013) can also be used as an alternative hypothesis against the standard single-season occupancy model that reflects closure.

A second assumption of single-season occupancy models is that there are no differences in occupancy probabilities across sites, or else that variation in occupancy is explained by modeled covariates. However, if unmodeled heterogeneity in occupancy occurs within a surveyed system, random selection of sample units to survey (or survey of all sample units) should produce an approximately unbiased estimate of average occupancy across the units (MacKenzie et al. 2006:106–107).

One specific source of heterogeneity in occupancy across sample units comes from the local neighborhood of a sample unit. Units surrounded by occupied sites are more likely to be occupied than isolated sample units, because of the proximity of sources of colonists (e.g., Hanski 1998). Royle and Dorazio (2008) recommended the use of so-called autologistic models for dealing with this situation. Such models express focal site occupancy as some function (e.g., logistic, Eq. 4.4) of the occupancy of neighboring sites (e.g., mean occupancy, $\overline{\psi}_i^{N_i}$, where N_i denotes the neighborhood of sample unit i). If x_i is a standard covariate (e.g., habitat at sample unit i), then we could write an autologistic model of occupancy as a function of both habitat and neighborhood occupancy as follows:

$$\psi_i = \frac{\exp\left(\beta_0 + \beta_1 x_i + \beta_2 \overline{\psi}_i^{N_i}\right)}{1 + \exp\left(\beta_0 + \beta_1 x_i + \beta_2 \overline{\psi}_i^{N_i}\right)}$$

However, neighborhood occupancy is not a standard covariate, as the same uncertainty that characterizes inference about focal patch occupancy characterizes neighborhood occupancy as well. For this reason, inference in such single-season autologistic models is best carried out using the expectation-maximization (EM) algorithm (Dempster et al. 1977) or Markov Chain Monte Carlo (MCMC; Royle and Dorazio 2008; Link and Barker 2010) approaches. Again, we stress that even when autologistic processes actually generate occupancy data, random sampling yields reasonable estimates of average occupancy, even with standard occupancy models. However, if mapping is an objective of the investigation, then site-specific occupancy is of more interest, and autologistic models can be very useful.

A third assumption of single-season occupancy models is that there are no differences in detection probabilities across sites, or else that variation in detection is explained by modeled covariates. Heterogeneity of detection probabilities typically leads to negatively biased estimates of occupancy. Estimates of detection probability tend to be positively biased, as most of the information about detection is provided by sites with at least one detection, and sites with high detection probabilities are overrepresented in this group. Positively biased estimates of detection probability then produce negatively biased estimates of the number and proportion of occupied sites (expressions 4.1 and 4.2).

The best way to deal with heterogeneous detection probabilities is via selection of site-specific covariates that can explain much of the existing variation. Finite and continuous mixture models are frequently useful (MacKenzie et al. 2006: 137–140) when substantial variation in detection probability remains after covariate modeling. These mixture models assume distributional forms for heterogeneity and then estimate the parameters of these distributions. One specific source of heterogeneous detection probabilities is produced by site-specific variation in animal abundance. The focal species is more likely to be detected at sample units with large numbers of animals and less likely to be detected when only 1 or 2 animals are present. Royle and Nichols (2003) provide one approach to this source of heterogeneity using assumptions about the distribution of animals across sample units.

A fourth assumption is that the species detections are independent both across sample units and at replicates within a sample unit. Lack of independence of detection probabilities across sample units may not be a common issue, but could be dealt with using autologistic modeling (see above) of detection probabilities.

Dependence of detection probabilities across replicates within a sample unit can occur in a number of ways. Consider the situation in which initial detection of the focal species at a sample unit influences detection probability for all subsequent sampling occasions. An example of this situation occurs in tiger sign surveys when an investigator discovers a pugmark impression in hardened mud at a particular location in the sample unit. If the same observer conducts all the sampling, then the observer can revisit that pugmark impression on each subsequent sampling occasion, increasing the detection probability to approach 1 and clearly violating the independence assumption. Such dependence can be addressed by either modifying sampling design or using specific types of models. For example, a useful approach to sampling design is to specify that different observers visit a given site on different sampling occasions. Another way to deal with such dependence is via a removal modeling approach in which there are different detection probability parameters before and following initial detection of the species at a sample unit. Under the removal design described in Sect. 4.2 (investigator stops visiting the sample unit after initial detection), the detection probability following initial detection is fixed at 0. In the case of more general dependence, detection probabilities following initial detection can differ from those prior to detection but are not fixed to any value. In fact, models intermediate between removal models and models assuming independence can include a relationship between detection probabilities before and after initial detection.

Another form of sampling dependence across replicates occurs with spatial sampling. Consider a popular design for tigers and other large felids in which long trails are surveyed for sign, with segments of specified length treated as spatial replicates. It is likely that a single animal can move along the trail for distances that cover multiple segments, creating dependence in detection probability between adjacent segments. This kind of dependence can produce biased estimates of occupancy. Hines et al. (2010) showed that the bias can be substantial and thus developed a model that can be described as a spatial Markov process in which the probability of local presence of the focal species on a segment differs depending on whether the previous segment was locally occupied. This model includes not only the standard occupancy parameter, ψ_i, that corresponds to a sample unit (the area sampled by the entire trail) but also parameters for probability of local occupancy for segments that are, and are not, preceded by locally occupied segments. Detection probability for this model is then conditional on local occupancy, being 0 for segments that are not locally occupied. This kind of model may be useful for other sampling designs in which spatial sample units are linearly arrayed (e.g., Hines et al. 2014) and even for designs based on temporal replication (dependence of detection probabilities for adjacent sampling periods). Guillera-Arroita et al. (2011) provide an alternative approach to modeling data along trails in the face of correlated segments.

A final assumption is that all detections of the focal species are certain. That is, all of the uncertainty associated with detection-nondetection data is with nondetections. However, for certain types of data, so-called false positives are not only possible, but likely. Surveys based on animal sign can lead to false positives. For example, pugmarks and scats of small tigers may be misidentified as leopards, whereas pugmarks and scats of large leopards may be identified as tigers sometimes. False positives can lead to positively biased estimates of occupancy, with substantial problems associated with even small rates of misidentification (McClintock et al. 2010b).

In some cases, false positives can be dealt with via changes in survey design. For example, rather than classifying felid scats based on visual cues, samples of all scats may be subjected to molecular classification based on DNA. Sometimes, additional training of survey personnel would seem to be a useful design approach to the elimination of false positives. However, limited experience with this approach indicates that training may reduce misclassification, but not necessarily eliminate it (Miller et al. 2012a).

Royle and Link (2006) developed a general occupancy model permitting false positives (misidentification) and negatives (nondetection). Miller et al. (2011) built on this model, showing that the existence of subsets of detection observations for which truth is known can greatly improve inferences under such models. Chambert et al. (2015a) then generalized the work of Miller et al. (2011) to include situations in which misclassification probabilities are estimated, not as a part of the occupancy survey, but experimentally in the laboratory, for example. Thus, there exist several

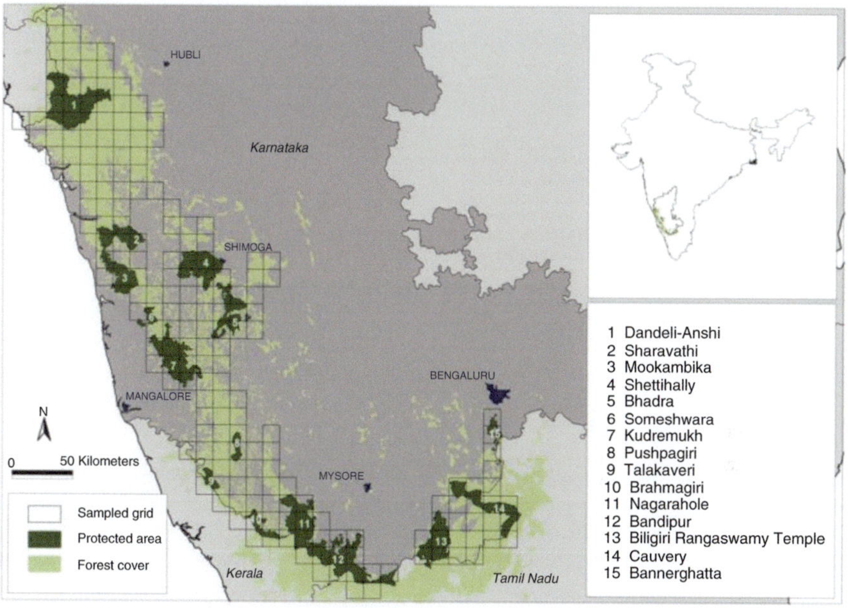

Fig. 4.2 Design of the WCS surveys (2006–2007) to assess tiger occupancy in the Malenad Tiger Landscape (MTL), Karnataka, India. The figure shows grid cells selected for sampling as potential tiger habitat, overall forest cover in the landscape (*light green*), and wildlife reserves (*dark green*). Inset shows the geographic location of the Malenad landscape in India

modeling approaches to dealing with false positives. They are all effective, and selection of a specific method depends on the details of the sampling situation and the manner in which false positives arise (Chambert et al. 2015a).

4.3.4 Example

Karanth et al. (2011) assessed habitat occupancy for a tiger *Panthera tigris* metapopulation across the 38,000-km^2 Malenad Tiger Landscape (MTL) in southwestern India. Survey design used spatial replication, with investigators detecting sign along trail segments within 205 sample units (grid cells, Fig. 4.2), each 188 km^2 in size. Model selection provided strong evidence of the inadequacy of the basic model assuming independent replicates and of the need for the spatial dependence model developed by Hines et al. (2010).

Tiger occupancy was positively related to the proportion of tiger habitat in a sample unit and density of ungulate prey and was negatively related to levels of human disturbance (Karanth et al. 2011). Model-averaged probabilities of local occupancy were substantially greater for trail segments preceded by a locally occupied segment. The model-averaged estimate of replicate-level detection prob-

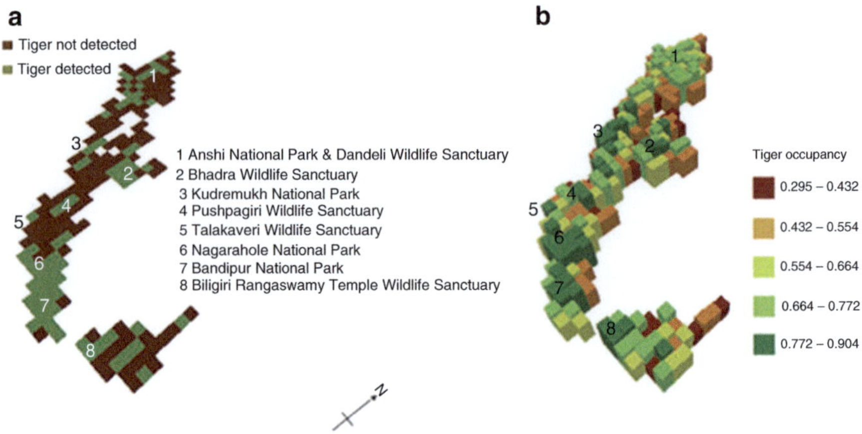

Fig. 4.3 Results of tiger occupancy survey by WCS in Malenad (2006–07): **(a)** Naïve tiger occupancy assuming perfect detectability and **(b)** cell/site-specific estimates of tiger occupancy derived using the Hines et al. (2010) model that allows for imperfect and spatially varying detection rates for tiger signs, as well as spatial dependence in replicate-level occupancy (Source: Karanth et al. 2011)

ability (conditional on local occupancy) was $\widehat{\overline{p}} = 0.17$, $\widehat{SE}\left(\widehat{\overline{p}}\right) = 0.17$. Tiger occupancy for the entire system was estimated to be $\widehat{\overline{\psi}} = 0.67$, $\widehat{SE}\left(\widehat{\overline{\psi}}\right) = 0.086$, or 14,076 km^2 of the 21,167 km^2 of potential habitat. In contrast, a traditional presence-absence approach, assuming tiger absence from all sample units with no detections, underestimated occupancy by 47%. Mapped probabilities of local site occupancy clearly identified areas viewed as tiger source populations and corresponded well with observed tiger density variations across locations at which density had been estimated (Fig. 4.3).

4.4 Dynamic Occupancy Models

4.4.1 Motivation

Occupancy models described in the above sections are useful for estimating static occupancy patterns associated with specific short intervals of time. They are especially useful in identifying relationships between focal species occurrence and spatial covariates. There is a long history in ecology of investigating such static patterns with the objective of drawing inferences about the underlying processes that generated them. However, such inferences about process based on estimated patterns are typically weak. Much stronger inferences can be obtained by directly studying dynamics and the vital rates that generate them (MacKenzie et al. 2006; Yackulic et al. 2015). Multiseason occupancy models (Barbraud et al. 2003; MacKenzie et al. 2003) were developed for this purpose.

4.4.2 Survey Design

Data needed for multiseason or dynamic occupancy models are similar to those collected for single-season modeling. In the first season of a multiseason study, a set of sample units is surveyed multiple times or using spatial replication. The sample units are assumed to be closed during the sampling within the season. The next season, a set of sample units within the system of interest is again sampled. These units need not be exactly the same units as surveyed in season 1, although the strongest inferences about process come from repeat sampling of the same set of sample units. Sample units are not assumed to be closed between seasons, but rather are permitted to be possibly open to changes in occupancy state (occupied or not). However, within season 2, sample units are again assumed to be closed to occupancy changes. This robust sampling design (Pollock 1982) permits inference not only about season-specific occupancy but also about the vital rates that govern changes in occupancy, local probabilities of extinction and colonization (Fig. 4.1).

Data from this design are extended detection histories for each sample unit (Fig. 4.4). For example, consider the following history for sample unit i, $h_i = 110\ 000$. This history reflects a 2-season study with three sampling replicates within each season. The space between the first and second 0's separates the two seasons. So this particular history indicates a sample unit at which the focal species

	Season			
Unit	**1**	**2**	**...**	***T***
1	101	001	...	011
2	000	100	...	110
3	100	000	...	000
.	.	.	...	.
.	.	.	...	.
.	.	.	...	.
.	.	.	...	.
S	000	000	...	000

Fig. 4.4 Matrix of detection versus non-detection data structure for multiseason occupancy studies. This example shows surveys at s sites, across T seasons, and over $K = 3$ occasions (replicates) within each season. Spaces separate surveys in different seasons

was detected on occasions 1 and 2, but not 3, of season 1 and was not detected on any of the three occasions in season 2.

4.4.3 Basic Model

Data from multiseason surveys can be modeled using either *implicit* or *explicit* approaches (MacKenzie et al. 2006). The implicit approach essentially models the data as a sequence of single-season studies, with no necessary dependence between site occupancy in successive seasons. Sometimes, implicit modeling can incorporate a parameter reflecting rate of change in occupancy between seasons. Implicit modeling may be recommended in cases where few sample units are sampled in multiple seasons or where data are generally sparse.

The explicit modeling approach accounts for heterogeneity in occupancy probabilities among sample units attributable to their prior histories. For example, at any season *t*, sample units can be grouped according to whether or not they were occupied in the previous season, *t*-1. Under many ecological hypotheses (e.g., metapopulation models, Hanski 1998), we would expect greater probabilities of being occupied in season *t* for sample units that were also occupied in season *t*-1. This kind of thinking leads us to an explicit Markov model for community dynamics.

Probabilities of local extinction and colonization, respectively, are defined as follows:

$$\varepsilon_t = \Pr\ (\text{not occupied at } t+1 \mid \text{occupied at } t)\,;$$

$$\gamma_t = \Pr\,(\text{occupied at } t+1 \mid \text{not occupied at } t)\,.$$

A site or sample unit at any season, *t*, can be in one of two occupancy states, occupied or not occupied. The transition probability of changing state for an occupied site is the local extinction probability and that for an unoccupied state is the colonization probability (Fig. 4.5). Explicit modeling of detection history data for multiseason models requires these extinction and colonization parameters in addition to probabilities of occupancy (ψ_t, occupancy in season *t*) and detection ($p_{t,j}$, detection on occasion *j*, season *t*).

Consider the multiseason detection history 110 000. The probability associated with this detection history can be written as:

$$\Pr\ \left(110\ 000|\psi_1, p_{t,j}, \varepsilon_t, \gamma_t\right) = \psi_1 p_{1,1} p_{1,2}\left(1-p_{1,3}\right)\left[\varepsilon_1 + (1-\varepsilon_1)\prod_{j=1}^{3}\left(1-p_{2,j}\right)\right]$$

We are certain that the site was occupied in season 1, as the focal species was detected then. The terms in brackets reflect the two possible outcomes for season

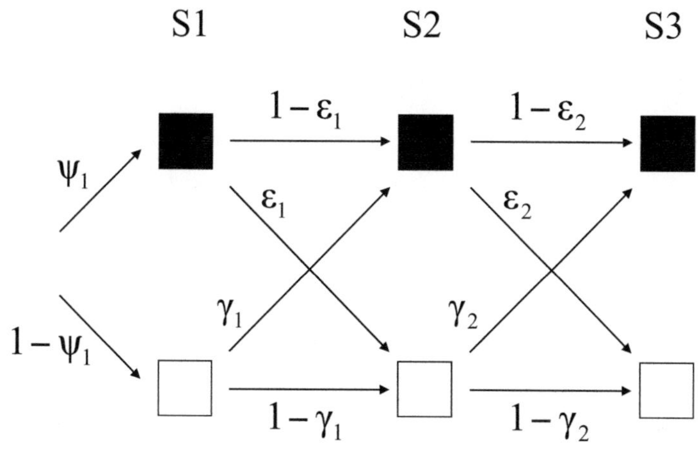

Fig. 4.5 Markov model for occupancy dynamics, where the probability of occupancy at time $t+1$ depends on occupancy state at t. Changes in occupancy state occur through colonization (γ_t) of unoccupied sites or local extinction (ε_t) of occupied sites from season t to $t+1$. Black squares denote occupied sites and white squares denote unoccupied sites

2. The first possibility is that the site went locally extinct. The second is that the species persisted until season 2 but was not detected at any of the three surveys in that season.

As for single-season models, every possible detection history has a corresponding probability statement about the processes that could have generated the history. These probability statements now include not only initial occupancy and occasion- and season-specific detection probabilities but also time-specific probabilities of local extinction and colonization. Note that we would refer to this general multiseason model as model ($\psi_1, p_{t,j}, \varepsilon_t, \gamma_t$). Estimates of occupancy probabilities following season 1 can be obtained as derived parameters using the fundamental equation of occupancy dynamics:

$$\psi_{t+1} = \psi_t (1 - \varepsilon_t) + (1 - \psi_t)\,\gamma_t \tag{4.6}$$

The relationship expressed in Eq. (4.6) shows the redundancies among parameters.

We believe that model ($\psi_1, p_{t,j}, \varepsilon_t, \gamma_t$) will frequently be most useful in permitting direct modeling of local probabilities of extinction and colonization as functions of covariates, as described in Sect. 4.3.2. Relevant covariates may be environmental or may include management actions, depending on the questions being addressed. Some questions may be better addressed by modeling time-specific occupancy directly, in which case we can reparameterize the basic model to include time-specific occupancy and estimate either colonization ($\psi_t, p_{t,j}, \varepsilon_t$) or extinction ($\psi_t, p_{t,j}, \gamma_t$) as derived parameters (MacKenzie et al. 2003, 2006; Hines 2006).

As with single-season models, the likelihoods for the multiseason models can be written as the product of the probabilities for the detection histories of the sampled sites. For example, the likelihood for model ($\psi_1, p_{t,j}, \varepsilon_t, \gamma_t$) can be written as:

$$L\left(\psi_1, \left\{p_{t,j}\right\}, \left\{\varepsilon_t\right\}, \left\{\gamma_t\right\}\right) \mid h_1, h_2, \ldots, h_s\Big) = \prod_{i=1}^{s} \Pr\left(h_i\right) \tag{4.7}$$

This is a multinomial likelihood, and parameters can be estimated via maximum likelihood using software packages such as PRESENCE (Hines 2006), MARK (White and Burnham 1999), and unmarked (Fiske and Chandler 2011). Note that we dropped the subscript i from this description of multiseason models primarily in order to simplify notation, but parameters will be site specific when modeling with site-specific covariates.

4.4.4 Assumptions

The robust design (Pollock 1982) sampling used for multiseason modeling basically combines models for open and closed populations. The single-season models represent the closed model components, and all of the assumptions required for single-season models (see Sect. 4.3.3) thus apply to the within-season sampling conducted each season.

In multiseason models, the "closure" assumption still applies within each season, but not between seasons. Thus, sites may change occupancy state between seasons, but not within them. The potential for closure violations within seasons can be addressed via study design, for example, reducing the time intervals between replicate surveys. The single-season model of Kendall et al. (2013) permits the focal species to enter the surveyed sites after the initiation of surveys and depart before the end of surveys. This model was recently extended to multiple seasons (Chambert et al. 2015b). Another form of violation of the closure assumption concerns animals that may move in and out of sample units during the course of their daily movement. As discussed for single-season models, random movement of this sort simply requires a reinterpretation of the model parameters, such that occupancy now reflects *use* of a sample unit (as opposed to full-time residency) during the season.

As noted in Sect. 4.3.3, tests for closure in single-season models can be developed by combining consecutive groups (e.g., pairs) of sampling occasions within a season and fitting multiseason models (MacKenzie et al. 2003; Rota et al. 2009). This kind of thinking can be readily extended to multiseason models. Intervals between sampling occasions that extend from one season to the next (i.e., between the last sample occasion in season t and the first occasion in $t + 1$) will be parameterized with the usual transition probabilities of local extinction and colonization. Intervals between sampling occasions within a season will be parameterized with extinction and colonization parameters that are constrained to equal 0 in the null model reflecting closure, but permitted to be non-0 in the more general model of an open system. The single-entry, single-exit model of Chambert et al. (2015b) can be also used as a specific alternative hypothesis to the null model of within-season closure.

As is the case for single-season models, multiseason models assume that there are no differences in occupancy probabilities across sites, or else that variation in occupancy is explained by modeled covariates. However, random selection of sample units should produce approximately unbiased estimates of average occupancy across the units (MacKenzie et al. 2006:106–107) even in the face of unmodeled heterogeneity.

In multiseason models, the assumption of no heterogeneity across sites is extended to the rate parameters of local extinction and colonization. Modeling such variation using site-specific covariates not only reduces heterogeneity but is also a focus of important ecological questions. The occupancy state of neighboring sites is likely to be an important determinant of both local colonization (e.g., Hanski 1998) and extinction (Brown and Kodric-Brown 1977) probabilities. Autologistic models express focal site colonization and extinction as some function (e.g., logistic, Eq. 4.4) of the occupancy of neighboring sites (e.g., mean occupancy). However, neighborhood occupancy is not a standard covariate, but rather a quantity that is also estimated. Multiseason models have thus been developed to implement autologistic models using both MCMC (Bled et al. 2011, 2013) and likelihood (Yackulic et al. 2012, 2014; Eaton et al. 2014) approaches to inference. Program PRESENCE (Hines 2006) implements a general likelihood approach to multiseason, autologistic modeling that allows neighborhood occupancy to be defined in a variety of ways (see discussion in Eaton et al. 2014).

The assumption of no differences in detection probabilities across sites, or else that variation in detection is explained by modeled covariates, is also required by multiseason models. As discussed in Sect. 4.3.3, heterogeneity of detection probabilities typically leads to negatively biased estimates of occupancy. Heterogeneity in detection probabilities is also expected to produce bias in estimates of probabilities of local extinction and colonization, but such effects have not been explored to our knowledge. The Royle-Nichols (2003) model for abundance-induced heterogeneity of detection probabilities could be extended to multiseason models but would require the modeling of site-specific abundance dynamics.

Multiseason models require the assumption that species detections are independent both across sample units and over replicates within a sample unit. One form of dependence of detection probabilities across replicates occurs when initial detection of the focal species at a sample unit influences detection probability for all subsequent sampling occasions. Multiseason models can be based on a removal model approach to within-season data, in which detection probabilities within a site differ before and after initial detection of the focal species.

As noted in Sect. 4.3.3, a popular occupancy survey design for tigers and other large felids is to survey long trails for sign, with segments of specified length treated as spatial replicates. The spatial Markov process model of Hines et al. (2010) was developed to model local presence of the focal species on a segment as a function of whether the previous segment was locally occupied. This model has been extended to multiseason studies (Hines et al. 2014).

A final assumption is that all detections of the focal species are certain; i.e., that "false positives" are not possible. The general single-season modeling approaches of Royle and Link (2006), Miller et al. (2011), and Chambert et al. (2015) provide several effective ways to deal with false positives, depending on sampling design and the existence and type of data for which "truth" is known. In particular, the approach of Miller et al. (2011) has been extended to multiseason models (Miller et al. 2013). The latter models have been incorporated into PRESENCE (Hines 2006).

4.5 Extensions

The discussion of occupancy modeling in a previous edition of this manual (Nichols and Karanth 2002) was very simple, requiring only a few pages. Together with MacKenzie et al. (2002), we were introducing a new approach to inference about species occurrence, and most of our discussion centered on its potential for large-scale studies. Subsequent interest in these models was substantial and accompanied by a great deal of work, requiring a synthetic book (MacKenzie et al. 2006). New developments have continued, resulting in synthetic reviews (Bailey et al. 2014; Guillera-Arroita 2016) and work on a second occupancy book. We cannot thoroughly review all developments and extensions here, so will simply provide pointers to literature that we believe to have some potential for use in tiger-prey studies.

A major advance in occupancy modeling was the introduction of multistate occupancy models (MacKenzie et al. 2004; Royle 2004; Royle and Link 2005; Nichols et al. 2007; MacKenzie et al. 2009). The occupancy models discussed thus far have assumed each site or sample unit to be in one of two possible states, occupied or unoccupied. However, "occupied" may not be an adequate descriptor for some kinds of inferences. MacKenzie et al. (2004) considered 2-species models in which sites could be characterized by any of 4 states: occupied by neither species, just one species, just the other species, or both species. Royle (2004) and Royle and Link (2005) considered occupancy modeling of data from anuran call surveys and wanted to distinguish among different levels of calling intensity, denoted 1–3, with higher levels reflecting more animals calling. Nichols et al. (2007) considered states of not occupied, occupied with no reproduction, and occupied with reproduction. Epidemiological studies can focus on sites occupied by a focal species at which a species pathogen was and was not present (McClintock et al. 2010c; Lachish et al. 2012; Miller et al. 2012b). Goswami et al. (2014) evaluated elephant habitat use considering states of not used, used with low intensity, and used with high intensity, where intensity of use was defined based on density of elephant signs encountered during the surveys.

The multistate occupancy modeling described above was initially developed for single-season occupancy studies. Extension of this approach to multiple seasons provides a direct focus on system dynamics, but carries the cost of more complex modeling (MacKenzie et al. 2009; Miller et al. 2012c). Uses of dynamic multistate models include territory transitions between reproductive and nonreproductive states (MacKenzie et al. 2009), dynamics of relative abundance (MacKenzie et al. 2009), joint habitat-occupancy dynamics (Martin et al. 2010; MacKenzie et al. 2011; Miller et al. 2012c), predator-prey dynamics (Miller et al. 2012c), and dynamics of purported competitors (Yackulic et al. 2014). The simultaneous modeling of dynamics of multiple species or habitat-occupancy should be a much more powerful approach to modeling system dynamics than the more usual approach of modeling focal-species occupancy dynamics as functions of covariates. Joint modeling is especially well-suited to making projections of future dynamics (e.g., Yackulic et al. 2014).

Sometimes sampling may be carried out using multiple detection methods or devices. Models have been developed to fully use such data and to deal with lack of independence of detections of two devices at the same location (Nichols et al. 2008). These models also permit inferences about method-specific detection probabilities. When tigers are the focal species, occupancy sampling might be carried out using sign surveys along trails combined with camera traps at some points along trails as well.

Finally, we note that occupancy models can also be used to address questions about species richness and associated dynamics. Studies of tigers, for example, might focus on the ungulate prey community, rather than on particular prey species. Data for such inferences are detections and nondetections of each species in the local species pool at each sampling occasion. One approach to inference about richness can be implemented at a single location. The species are treated as "sites" in standard occupancy modeling, and the ψ_i parameters reflect probability of occupancy at the location by species i (MacKenzie et al. 2006). The other approach uses multispecies detection-nondetection from a set of sites. Species richness at a site is then estimated as the sum of unconditional occupancy probabilities for all species in the community or, equivalently, as the sum of species actually detected and conditional (on nondetection) occupancy probabilities for species not detected (Dorazio and Royle 2005; MacKenzie et al. 2006).

4.6 Discussion

We believe that occupancy modeling can be useful for addressing a variety of questions relevant to tiger and associated prey communities. However, we remind the reader that occupancy (presence in, or use of, a site by a focal species) and abundance (number of animals of the focal species present in, or using, a site) are related concepts but not equivalent. If we view the abundance distribution across sites as the number of sites with 0 individuals, 1 individual, 2 individuals, etc., then we would think of occupancy as separating two parts of this distribution, places with 0 individuals and places with >0 individuals. So the concepts are clearly related. However, it is also clear that the abundance distribution contains more information than the occupancy distribution. This distinction is thus relevant to our reminders throughout this volume that the investigator or manager must select the methods that best suit his or her objectives (i.e., scientific questions or management goals).

Occupancy modeling represents one approach to the conduct of species distribution modeling (SDM), which has become a popular endeavor among ecologists. An important goal of SDM is to determine environmental and habitat determinants of species distributions. Although we agree that such questions are ecologically interesting and potentially useful for conservation, we disagree with the approaches commonly used to develop such models. The vast majority of SDM efforts use so-called presence-only data, records of where animals have been detected, without information on locations that have been surveyed but no detections recorded (e.g., Elith et al. 2006). SDM analyses using presence-only data (e.g., MAXENT, Phillips et al. 2006, and related software) require extremely restrictive assumptions that are seldom met by the data used with them (see critiques by Royle et al. 2012, Yackulic et al. 2013). We thus strongly recommend use of detection-nondetection data with the single-season occupancy models of this chapter (Sect. 4.3) as the path to strong inference for SDM, both in general and in particular for tiger and prey surveys (MacKenzie et al. 2006; Bailey et al. 2014; Guillera-Arroita 2016).

In terms of estimation methodology, we have focused on maximum likelihood as a primary inference method and have emphasized software that uses this approach (e.g., PRESENCE; Hines 2006). We have also mentioned the possible use of MCMC methods of inference (e.g., Royle and Dorazio 2008; Kery and Schaub 2012; Kery and Royle 2016). We note here that Bayesian approaches to inference such as MCMC have the advantage of permitting incorporation of prior information into inferences. The ability to use information from past surveys may be useful in some cases.

Occupancy models are expected to be especially useful for inferences about species distribution modeling over medium to large spatial scales. Karanth et al. (2009, 2010) used occupancy approaches with observation data from interviews with local naturalists to draw inferences about selected large mammals (including tigers and multiple prey species) across all India. Protected areas were found to be important for a number of species. Tigers were estimated to occur across approximately a quarter of India (Karanth et al. 2009). Long-term extinction probabilities were investigated by relating recent occupancy estimates to historical records of confirmed species presence, and protection was found to be negatively related to local extinction probabilities for tigers and several other species, as predicted (Karanth et al. 2010). The example of Sect. 4.3.4 relied on data from a large-scale field survey of the Malenad Tiger Landscape (MTL) and provided detailed inferences about distribution of tiger (and multiple prey species) across a large landscape (Karanth et al. 2011).

We can think of many additional uses of standard occupancy models that could be useful in tiger conservation. In addition, single-season multistate models could potentially be used to identify sites at which tiger reproduction occurs: the so-called source populations (Walston et al. 2010) that are critical to species recovery efforts. Two-species distribution models of tiger and leopard, or tiger and select prey species, could prove interesting and useful for drawing inferences about potential species interactions. Single-season occupancy modeling can also be usefully combined with more intensive sampling (e.g., camera trap density estimation) in an approach that is both integrated and adaptive (Conroy et al. 2008).

Although there have been some published uses of single-season occupancy modeling for tigers and associated prey species, analyses using multiseason models are currently underway (K.U. Karanth pers. comm.), but not yet published. For example, the second large survey of the MTL has been recently completed permitting multiseason modeling of tigers and prey. Such modeling will focus on the influence of covariates (e.g., protection status) on the probabilities of local tiger extinction and colonization. Multiseason modeling for tigers and associated competitor species (leopards) or prey species (sambar, chital, etc.) can be used to draw inferences about potential effects of the presence of one species on the local extinction and colonization of the other. Multistate modeling could be used to draw inferences about the dynamics of site state as defined, for example, by presence or absence of reproduction at sites.

Connectivity of subpopulations in a heterogeneous landscape has been identified as a potentially important feature of current tiger metapopulations (Chap. 13). One approach to inferences about tiger movement among different sites is based on intensive sampling of marked individuals using multistate capture-recapture modeling (Chap. 13). However, inferences at larger scales can be based on the autologistic models described in this chapter. In particular, these can be used to assess the importance of neighborhood occupancy to probabilities of local extinction and colonization of a focal site (Chap. 13).

This discussion has just scratched the surface of the kinds of questions that can be addressed about tiger and prey distribution and associated dynamics using occupancy modeling. However, we also note that occupancy can be used in novel ways to address questions that are relevant to conservation, yet go beyond the state variable of species occupancy to focus on other state variables associated with the species. Barber-Meyer (2010) described the use of occupancy modeling with species data from wildlife trade market surveys, in order to estimate the prevalence of a species in illegal trade. Goswami et al. (2015) focused on human-wildlife conflict, noting that incidents are not always reported. They used data from local surveyors to estimate prevalence and correlates of such conflict over time and space with the help of dynamic occupancy models. In summary, we see much scope for application of rigorous occupancy modeling concepts outlined in this chapter for assessing not only ecological questions but also conservation threats to tigers in the future.

References

Bailey LL, Mackenzie DI, Nichols JD (2014) Advances and applications of occupancy models. Methods Ecol Evol 5:1269–1279

Barber-Meyer SM (2010) Dealing with the clandestine nature of wildlife-trade market surveys. Conserv Biol 24:918–923

Barbraud C, Nichols JD, Hines JE, Hafner H (2003) Estimating rates of local extinction and colonization in colonial species and an extension to the metapopulation and community levels. Oikos 101:113–126
Bled F, Royle JA, Cam E (2011) Hierarchical modeling of an invasive spread: the Eurasian Collared-Dove Streptopelia decaocto in the United States. Ecol Appl 21:290–302
Bled F, Nichols JD, Altwegg R (2013) Dynamic occupancy models for analyzing species' range dynamics across large geographic scales. Ecol Evol 3:4896–4909
Brown JH, Kodric-Brown A (1977) Turnover rates in insular biogeography: effect of immigration on extinction. Ecology 58:445–449
Chambert T, Miller DAW, Nichols JD (2015a) Modeling false positive detections in species occurrence data under different study designs. Ecology 96:332–339
Chambert T, Kendall WL, Hines JE, Nichols JD, Pedrini P, Waddle JH, Tavecchia G, Walls SC, Tenan S (2015b) Testing hypotheses on distribution shifts and changes in phenology of imperfectly detectable species. Methods Ecol Evol 6:638–647
Conroy MJ, Runge JP, Barker RJ, Schofield MR, Fonnesbeck CJ (2008) Efficient estimation of abundance for patchily distributed populations via two-phase, adaptive sampling. Ecology 89:3362–3370
Dempster AP, Laird NM, Rubin DB (1977) Maximum likelihood from incomplete data via the EM algorithm. J R Stat Soc Ser B 39:1–38
Dorazio RM, Royle JA (2005) Estimating size and composition of biological communities by modeling the occurrence of species. J Am Stat Assoc 100:389–398
Eaton MJ, Hughes PT, Hines JE, Nichols JD (2014) Testing metapopulation concepts: effects of patch characteristics and neighborhood occupancy on the dynamics of an endangered lagomorph. Oikos 123:662–676
Elith J, Graham CH, Anderson RP, Dudlik M, Ferrier S, Guisan A, Hijmans RJ, Huettmann F, Leathwick JR, Lehman A, Li J, Lohmann LG, Loiselle BA, Manion G, Moritz C, Nakamura M, Nakazawa Y, Overton JM, Peterson AT, Phillips SJ, Richardson KS, Scachetti-Pereira R, Schapire RE, Sobero'n J, Williams S, Wisz MS, Zimmermann NE (2006) Novel methods improve prediction of species' distributions from occurrence data. Ecography 29:129–151
Fiske I, Chandler RB (2011) Unmarked: an R package for fitting hierarchical models of wildlife occurrence and abundance. J Stat Softw 43:1–23
Goswami VR, Sridhara S, Medhi K, Williams AC, Chellam R, Nichols JD, Oli MK (2014) Community-managed forests and wildlife-friendly agriculture play a subsidiary but not substitutive role to protected areas for the endangered Asian elephant. Biol Conserv 177:74–81
Goswami VR, Medhi K, Nichols JD, Oli MK (2015) Mechanistic understanding of human–wildlife conflict through a novel application of dynamic occupancy models. Conserv Biol 29:1100–1110
Guillera-Arroita G (2016) Modelling of species distributions, range dynamics and communities under imperfect detection: advances, challenges and opportunities. Ecography. doi:10.1111/ecog.02445
Guillera-Arroita G, Morgan BJT, Ridout MS, Linkie M (2011) Species occupancy modelling for detection data collected along a transect. J Agric Biol Environ Stat 3:301–317
Hanski I (1998) Metapopulation dynamics. Nature 396:41–49
Hines JE (2006) PRESENCE2 – Software to estimate patch occupancy and related parameters. USGS-PWRC. http://www.mbr-pwrc.gov/software/presence.html
Hines JE, Nichols JD, Royle JA, Mackenzie DI, Gopalaswamy AM, Kumar NS, Karanth KU (2010) Tigers on trails: occupancy modeling for cluster sampling. Ecol Appl 20:1456–1466
Hines JE, Nichols JD, Collazo JA (2014) Multiseason occupancy models for correlated replicate surveys. Methods Ecol Evol 5:583–591
Karanth KU, Nichols JD (2002) Monitoring tigers and their prey: a manual for researchers, managers and conservationists in tropical Asia. Centre for Wildlife Studies, Bangalore
Karanth KK, Nichols JD, Hines JE, Karanth KU, Christensen NL (2009) Patterns and determinants of mammal species occurrence in India. J Appl Ecol 46:1189–1200
Karanth KK, Nichols JD, Karanth KU, Hines JE, Christensen NL Jr (2010) The shrinking ark: patterns of large mammal extinctions in India. Proc R Soc Biol Sci Ser B 277:1971–1979

Karanth KU, Gopalaswamy AM, Kumar NS, Vaidyanathan S, Nichols JD, MacKenzie DI (2011) Monitoring carnivore populations at the landscape scale: occupancy modelling of tigers from sign surveys. J Appl Ecol 48:1048–1056

Kendall WL, White GC (2009) A cautionary note on substituting spatial subunits for repeated temporal sampling in studies of site occupancy. J Appl Ecol 46:1182–1188

Kendall WL, Hines JE, Nichols JD, Campbell Grant EH (2013) Relaxing the closure assumption in occupancy models: staggered arrival and departure times. Ecology 94:610–617

Kéry M, Royle JA (2016) Applied hierarchical modeling in ecology. Academic, Waltham

Kéry M, Schaub M (2012) Bayesian population analysis using WinBUGS – a hierarchical perspective. Academic, Waltham

Lachish S, Gopalaswamy AM, Knowles SCL, Sheldon BC (2012) Site-occupancy modelling as a novel framework for assessing test sensitivity and estimating wildlife disease prevalence from imperfect diagnostic tests. Methods Ecol Evol 3:339–348

Link WA, Barker RJ (2010) Bayesian inference with ecological applications. Academic, London

MacKenzie DI, Nichols JD, Lachman GB, Droege S, Royle JA, Langtimm CA (2002) Estimating site occupancy rates when detection probabilities are less than one. Ecology 83:2248–2255

MacKenzie DI, Nichols JD, Hines JE, Knutson MG, Franklin AB (2003) Estimating site occupancy, colonization, and local extinction when a species is detected imperfectly. Ecology 84:2200–2207

MacKenzie DI, Bailey LL, Nichols JD (2004) Investigating species co-occurrence patterns when species are detected imperfectly. J Anim Ecol 73:546–555

MacKenzie DI, Nichols JD, Royle JA, Pollock KH, Bailey LL, Hines JE (2006) Occupancy estimation and modeling: inferring patterns and dynamics of species occurrence. Elsevier, San Diego

MacKenzie DI, Nichols JD, Seamans ME, Gutierrez RJ (2009) Modeling species occurrence dynamics with multiple states and imperfect detection. Ecology 90:823–835

MacKenzie DI, Bailey LL, Hines JE, Nichols JD (2011) An integrated model of habitat and species occurrence dynamics. Methods Ecol Evol 2:612–622

Martin J, Chamaillé-Jammes S, Nichols JD, Fritz H, Hines JE, Fonnesbeck CJ, MacKenzie DI, Bailey LL (2010) Simultaneous modeling of habitat suitability, occupancy, and relative abundance: African elephants in Zimbabwe. Ecol Appl 20:1173–1182

McClintock BT, Bailey LL, Pollock KH, Simons TR (2010a) Experimental investigation of observation error in Anuran call surveys. J Wildl Manag 74:1882–1893

McClintock BT, Bailey LL, Pollock KH, Simons TR (2010b) Unmodeled observation error induces bias when inferring patterns and dynamics of species occurrence via aural detections. Ecology 91:2446–2454

McClintock BT, Nichols JD, Bailey LL, MacKenzie DI, Kendall WL, Franklin AB (2010c) Seeking a second opinion: uncertainty in wildlife disease ecology. Ecol Lett 13:659–674

Miller DAW, Nichols JD, McClintock BT, Grant EH, Bailey LL, Weir LA (2011) Improving occupancy estimation when two types of observational error occur: non-detection and species misidentification. Ecology 92:1422–1428

Miller DAW, Weir LA, McClintock BT, Campbell Grant EH, Bailey LL, Simons TR (2012a) Experimental investigation of false positive errors in auditory species occurrence surveys. Ecol Appl 22:1665–1674

Miller DAW, Talley BL, Lips KR, Grant EHC (2012b) Estimating patterns and drivers of infection prevalence and intensity when detection is imperfect and sampling error occurs. Methods Ecol Evol 3:850–859

Miller DAW, Brehme CS, Hines JE, Nichols JD, Fisher RN (2012c) Joint estimation of habitat dynamics and species interactions; disturbance reduces co-occurrence of non-native predators with an endangered toad. J Anim Ecol 81:1288–1297

Miller DAW, Nichols JD, Gude JA, Rich LN, Podruzny KM, Hines JE, Mitchell MS (2013) Determining occurrence dynamics when false positives occur: estimating the range dynamics of wolves from public survey data. PLoS One 8:e65808

Nichols JD, Karanth KU (2002) Statistical concepts: assessing spatial distribution. In: Karanth KU, Nichols JD (eds) Monitoring tigers and their prey. Centre for Wildlife Studies, Bangalore, pp 29–38

Nichols JD, Hines JE, MacKenzie DI, Seamans ME, Gutierrez RJ (2007) Occupancy estimation and modeling with multiple states and state uncertainty. Ecology 88:1395–1400

Nichols JD, Bailey LL, O'Connell AF Jr, Talancy NW, Campbell Grant EH, Gilbert AT, Annand EM, Husband TP, Hines JE (2008) Multi-scale occupancy estimation and modelling using multiple detection methods. J Appl Ecol 45:1321–1329

Phillips SJ, Anderson RP, Schapire RE (2006) Maximum entropy modeling of species geographic distributions. Ecol Model 190:231–259

Pollock KH (1982) A capture–recapture design robust to unequal probability of capture. J Wildl Manag 46:752–757

Rota CT, Fletcher Jr RJ, Dorazio RM, Betts MG (2009) Occupancy estimation and the closure assumption. J Appl Ecol 46:1173–1181

Royle JA (2004) Modeling abundance index data from anuran calling surveys. Conserv Biol 18:1378–1385

Royle JA, Dorazio RM (2008) Hierarchical modeling and inference in ecology. Academic, San Diego

Royle JA, Link WA (2005) A general class of multinomial mixture models for anuran calling survey data. Ecology 86:2505–2512

Royle J, Link WA (2006) Generalized site occupancy models allowing for false positive and false negative errors. Ecology 87:835–841

Royle JA, Nichols JD (2003) Estimating abundance from repeated presence-absence data or point counts. Ecology 84:777–790

Royle JA, Chandler RB, Yackulic CB, Nichols JD (2012) Likelihood analysis of species occurrence probability from presence-only data for modelling species distributions. Methods Ecol Evol 3:545–554

Walston J, Robinson JG, Bennett EL, Breitenmoser U, da Fonseca GAB et al (2010) Bringing the tiger back from the brink – the six percent solution. PloS Biol 8:e1000485

White GC, Burnham KP (1999) Program MARK: survival estimation from populations of marked animals. Bird Study 46(Suppl):120–138

Williams BK, Nichols JD, Conroy MJ (2002) Analysis and management of animal populations. Academic, San Diego

Yackulic CB, Reid J, Davis R, Hines JE, Nichols JD, Forsman E (2012) Neighborhood and habitat effects on vital rates: expansion of the Barred Owl in the Oregon Coast Ranges. Ecology 93:1953–1966

Yackulic CB, Chandler R, Zipkin EF, Royle JA, Nichols JD, Campbell Grant EH, Veran S (2013) Presence-only modelling using MAXENT: when can we trust the inferences? Methods Ecol Evol 4:236–243

Yackulic CB, Reid J, Nichols JD, Hines JE, Davis R, Forsman E (2014) The roles of competition and habitat in the dynamics of populations and species distributions. Ecology 95:265–279

Yackulic CB, Nichols JD, Reid J, Der R (2015) To predict the niche, model colonization and extinction. Ecology 96:16–23

Field Practices: Assessing Tiger Habitat Occupancy Dynamics

5

N. Samba Kumar, Mahi Puri, Dale G. Miquelle, Anak Pattanavibool, Abishek Harihar, Sunarto Sunarto, Srinivas Vaidyanathan, and K. Ullas Karanth

5.1 Introduction

Conservation and management planning of tigers and prey species requires basic information on the spatial distribution at regional and landscape levels, at an appropriate scale (Karanth and Nichols 2000). One of the most useful activities that can be taken up at regional or countrywide levels is the assessment and monitoring of the spatial distribution of tiger (and prey) populations using field surveys (Chap. 1). The statistical concepts underlying such surveys of spatial distribution are covered in Chap. 4, under the general estimation and sampling framework described in Chap. 3. This chapter provides details of best field practices that are essential for reliable assessment of tiger distribution at large spatial scales.

As we saw in Chap. 1, given the critical status of tigers and the substantial investments being made in tiger conservation, wildlife managers and conservation agencies primarily need clear and reliable answers to the following basic questions regarding distribution of tigers:

N.S. Kumar (✉)
Wildlife Conservation Society, India Program, Bengaluru, India
e-mail: nrao.skumar@gmail.com

M. Puri
Wildlife Conservation Society, India Program, Bengaluru, India

Centre for Wildlife Studies, Bengaluru, India

Department of Wildlife Ecology and Conservation, University of Florida, Gainesville, FL, USA
e-mail: mahi.puri@gmail.com

D.G. Miquelle
Wildlife Conservation Society (WCS), New York, NY, USA

Department of Ecology, Far Eastern Federal University, Vladivostok, Russia
e-mail: dmiquelle@wcs.org

K.U. Karanth, J.D. Nichols (eds.), *Methods For Monitoring Tiger And Prey Populations*, DOI 10.1007/978-981-10-5436-5_5

1. What is the geographical range occupied by individual tiger populations in the country or a wider region?
2. What are the ecological and management factors that influence patterns of tiger habitat occupancy?
3. Where are individual tiger populations increasing their ranges and where are the ranges fragmenting and shrinking?
4. Within a landscape of interest, what is the proportion of area occupied by tiger populations (reproductive and transient)?
5. Which are the habitat patches that could potentially enhance movement permeability and connect populations or subpopulations?

Tigers are elusive, secretive, and difficult to see in dense vegetation; hence, encounters are infrequent when they occur at low densities. These considerations typically preclude the use of sighting-based survey methods to assess distribution. However, tigers leave behind a variety of signs such as tracks, scats, and scrape marks which are readily seen in the field. Information from sign surveys can provide useful insights for management and science, even under field conditions characterized by scarcity of resources and lack of trained manpower that prevail in most parts of the tiger range. Consequently, sign-based occupancy surveys are

A. Pattanavibool
Wildlife Conservation Society (WCS) – Thailand Program, Bangkok, Thailand

Department of Conservation, Kasetsart University, Bangkok, Thailand
e-mail: anakp@wcs.org

A. Harihar
Tiger Program, Panthera, New York, NY, USA

Nature Conservation Foundation, Mysuru, India
e-mail: aharihar@panthera.org; aharihar@ncf-india.org; https://www.panthera.org/; http://ncf-india.org

S. Sunarto
World Wide Fund for Nature (WWF) – Indonesia, Jakarta, Indonesia

Forum HarimauKita, Bogor, Indonesia
e-mail: macandahan@gmail.com; http://www.wwf.id/en; https://www.harimaukita.or.id/

S. Vaidyanathan
Foundation for Ecological Research, Advocacy and Learning (FERAL), Auroville, Tamil Nadu, India
e-mail: srinivasv@feralindia.org

K.U. Karanth
Wildlife Conservation Society (WCS), New York, NY, USA

Centre for Wildlife Studies, Bengaluru, India

Wildlife Conservation Society, India Program, Bengaluru, India

National Centre for Biological Sciences-TIFR, Bengaluru, India
e-mail: ukaranth@wcs.org; https://www.wcs.org/; http://cwsindia.org/; http://wcsindia.org/home/; https://www.ncbs.res.in/

often preferred for assessing distribution and habitat use. Additionally, occupancy methods and sign surveys can be used to identify potential habitat corridors, landscape connectivity, dispersal routes, and threats faced by tigers (see Chap. 13). Furthermore, sign survey-based occupancy modeling has the potential to link occupancy data with more intensively measured abundance estimates at local scales (Royle and Nichols 2003; Conroy et al. 2008, Dey et al. 2017, see Chap. 12) to draw inference about abundance at large spatial scales.

When such surveys are conducted periodically (e.g., each year) as part of a long term monitoring program, then resulting data can be used to assess the influence of landscape characteristics on metapopulation vital rates (local extinction and colonization rates) and changes in occupancy. These surveys also help us identify and deal with prevailing or emerging threats to tigers, prey, and habitats. Over time, results from periodic monitoring can be used to evaluate the success of landscape conservation and management efforts. Such monitoring surveys, therefore, should be a part of any tiger and prey recovery program over large geographic regions.

In this chapter, we consider field practices and protocols that can be employed to implement the occupancy modeling concepts discussed in Chap. 4. We note, however, that these field protocols would require appropriate modifications to suit local conditions.

5.2 Survey Design Protocols

The size of the occupancy survey grid cells (habitat patches, sites) should be based on the questions of interest and prior knowledge of tiger ecology, which may vary from one landscape to another. For example, grid cell sizes of $\sim$200 km^2 for India and 256 km^2 or 300 km^2 for Southeast Asian sites have been used. The biological basis for this is that the largest expected range for any individual animal (a male transient tiger) is <250–300 km^2 in India and Southeast Asia. The expected home range sizes for tigers in all other age-sex classes will be smaller. Also, tigers can move anywhere between 0–40 km in 24 h. Therefore, teams searching and moving linearly along good trails are more likely to increase detections of signs (scats and tracks for tigers). This approach to survey design aims at assessing "habitat occupancy" rather than "intensity of habitat use" (see Chap. 4 and MacKenzie and Royle 2005). Grid cell size for studies of habitat use intensity might be substantially smaller than those based on home range sizes. In practice, cell sizes can be modified marginally so that cell boundaries can be made to coincide with geographic coordinates such as latitude and longitude or UTM markings on the maps being used by the field teams. This makes it easier for field teams to follow the survey design and record data with minimum error.

Once the appropriate grid cell size has been selected, the leaders of the survey should delineate the survey area on a base map showing altitudinal contours, vegetation types, human settlements, land use, and other relevant details. This base map should then be used to create a map of grid cells overlaid on the entire area of interest. Each grid cell (say of size 200 km^2) is then further divided into some number (e.g., 16) of smaller sub-cells of equal size (Fig. 5.1). To meet the randomized sampling scheme design, one of the 16 sub-cells should be randomly picked in advance, and the survey teams must ensure that the survey route passes through this randomly chosen sub-cell. The overall survey route should be planned based on ecological considerations (coverage of diverse potential tiger habitats found among the sub-cells), local expert knowledge (where tigers are likely to be), and sampling effort considerations (survey route length), besides the mandatory

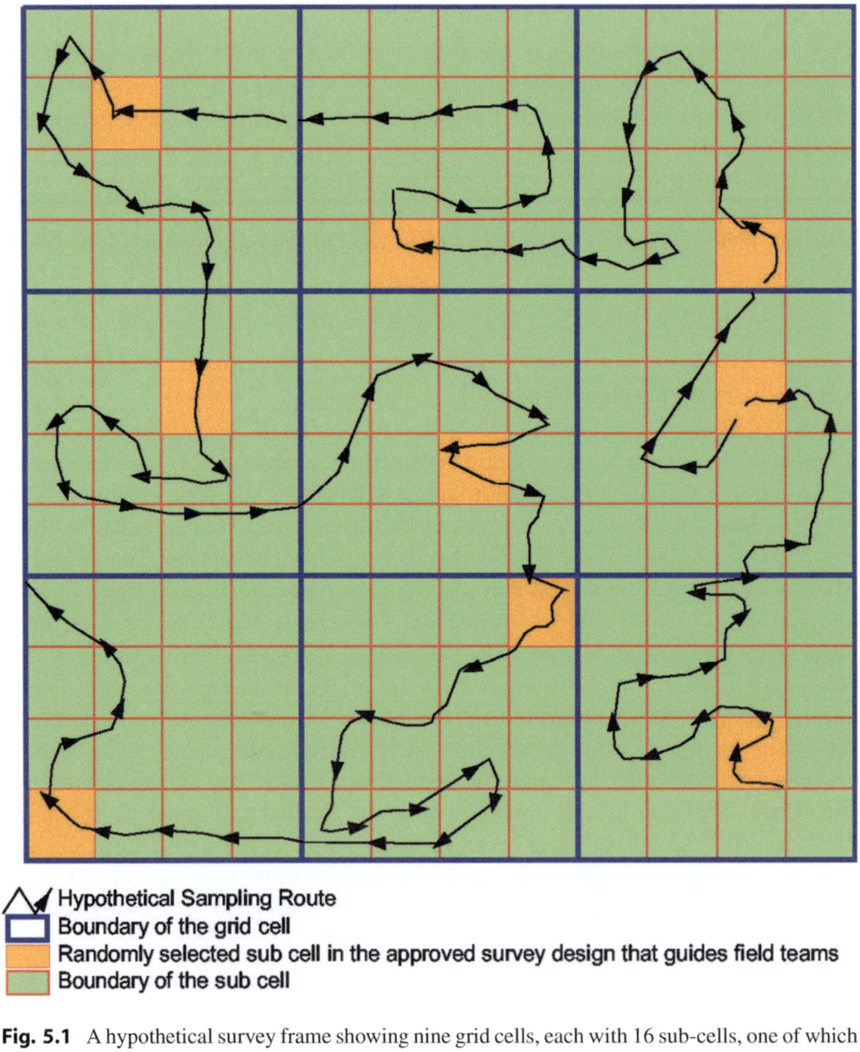

Fig. 5.1 A hypothetical survey frame showing nine grid cells, each with 16 sub-cells, one of which is selected randomly, and which must be visited by the field team

need of covering the randomly chosen sub-cell. The sampling route should also not be clumped around a few sub-cells but should aim to ensure adequate spatial coverage of the overall grid cell. Figure 5.1 gives an example of an occupancy survey with nine grid cells (all with 100% tiger habitat), each with its own randomly selected sub-cell and a hypothesized sequence of field survey effort, with 3 days (in a normal terrain; longer in more difficult areas) being spent in each grid cell to search for the tiger sign and with the sequential order of grid cells to be surveyed being based on logistics and convenience. Alternatively, in Thailand (and other Southeast Asian countries) with limited road availability and accessibility, grid cells are divided into only four smaller sub-cells, allowing for greater flexibility in designing survey routes (see Box 5.1 for summary of field protocols followed in Thailand).

Box 5.1: Summary Features of Tiger Occupancy Field Survey Protocols Followed in Thailand's Western Forest Complex

- The study area of 18,000 km^2 was divided into 104 grid cells of 16 × 16 km (256 km^2) following the grid cells on a topographic map scale of 1:50,000 in Thailand.
- These grid cells were then subdivided into four sub-grids of 8 × 8 km (64 km^2), which formed the basic template for field survey planning.
- To meet the randomization of survey locations, each 64 km^2 grid cell was divided into 16 sub-grid cell of 4 km^2 and randomly selected one sub-grid which the surveyors passed through during the field survey.
- Within each 64 km^2 grid cell, linear survey routes were delineated, which composed of 1-km spatial replicates. The number of replicates was proportional to the amount of forest habitat found within the grid cell. A maximum route length of 15 km per 64 km^2 was used if the entire grid cell was forested with no villages or agricultural land inside. Grid cells with <10% forest cover were not surveyed. During the survey, surveyors recorded detection/non-detection data of the target species at every 100-m intervals.
- The survey teams walked through habitat that was most likely used by tigers such as animal trails, mineral licks, forest roads, and river banks. Data recorded included direct sightings, scats, pugmarks/tracks, carcasses, scent marks, and vocalizations. Only fresh scats and signs (<7 days) were recorded to avoid biases that may arise from variable decay rate. Additionally, in each 100-m subsegment, the survey team recorded substrate condition, habitat type, and human activities such as evidence of domestic animals.
- Surveys were conducted in the dry season (October –December of 2010 and January –May and October–December of 2011).

If resources and logistic considerations permit, we recommend a 100% spatial sampling of all grid cells (sites) within the predefined sampling frame, to increase sample sizes for reliable occupancy modeling and to permit more direct inference about the entire study area (Chap. 4). Such an approach enables mapping tiger occupancy even if suitable covariate data that permit extrapolation of occupancy estimates to un-surveyed areas are not available. Furthermore, this approach offers some advantages for monitoring changes in occupancy across the years more reliably besides enabling better assessment of landscape-level threats to the tiger habitat as well as evaluation of potential connectivity/corridors between populations. However, when 100% sampling is not feasible, a proper spatial sampling design can be used to select cells for the survey (MacKenzie et al. 2006, see Chap. 4). The exact type of design will be governed by the objectives of the occupancy modeling. For example, a focus on occupancy of a large area may lead to systematic or random sampling of cells, whereas a focus on a question about habitat use could be based on random sampling of the focal habitats.

Once the grid array for the region of interest is finalized, survey teams should calculate the extent of potential tiger habitat available in each grid cell and eliminate those that are unlikely to hold tigers. Based on our experience with studies on tigers, all cells containing more than 10% tiger habitat should be included within the survey sampling frame, if cell sizes are 200–300 km^2 in size. The logic here is that tigers cannot "live" in patches smaller than these (20–30 km^2), although they may occasionally pass through or take refuge in them for a short period. However, this decision, which defines the sampling frame, should be based on local knowledge of tiger ecology as well as the cell sizes used in the specific survey.

Once the sampling frame has been finalized, it is necessary to decide on the number of spatial replicates per cell or the total distance to be sampled in each cell. The sampling effort required will depend on the desired level of precision needed in the occupancy estimates and can be assessed through simulation programs such as GENPRES. From experience with sign-based occupancy surveys of tigers in India (Karanth et al. 2011) and in Thailand (Pattanavibool, Unpublished data), we have found that a "spatial replicate" (Kendall and White 2009; Karanth et al. 2011; Chap. 4) can be a trail segment of about 1–2-km length. The issue of potential lack of independence among sign detection events on such replicates has been addressed by the development of a new occupancy model (Hines et al. 2010).

We suggest that the length of the spatial replicate for a tiger survey be set constant at 1 km, with the effort (total distance to be walked within each cell) proportional to the extent of potential tiger habitat available in the cell. In the above-cited example (Karanth et al. 2011), 40 km of walk/search effort was invested in a cell with 100% tiger habitat, and the effort for cells with smaller amounts of tiger habitat was scaled in proportion to the extent of habitat available in each cell. Thus, the minimum number of 1-km long replicates per grid cell was 4 for a 200-km^2 cell with 10% habitat cover and 40 for a cell with 100% habitat cover. The field survey teams should record detections of the tiger sign (or non-detections) for every 100-meter segment along each 1-km replicate, although during the analysis, the data may get aggregated as detection histories for each 1-km long replicate (see Karanth et al.

2011). This approach will provide additional flexibility for modeling covariates and to consider various analytical options based on site-specific considerations. Implementation of such an approach also allows analysis at multiple spatial scales to simultaneously assess tiger occupancy and/or habitat use (Sunarto et al. 2012).

5.3 Resource and Logistic Considerations for Conducting Sign-Based Occupancy Surveys

Similar to any other field survey, an adequate number of personnel should be available to survey the area being assessed (i.e., a "landscape"). Remember, however, that the aim is not to count every individual tiger or even every tiger sign present in the surveyed area. Rather, the goal is to estimate the proportion of habitat patches or sites occupied by tigers. Therefore, the surveys need not be conducted in all patches or sites simultaneously, as long as overall closure assumptions are met as discussed in Chap. 4. This logistical flexibility helps investigators deploy survey resources judiciously.

Skilled survey personnel who can find and identify tiger signs reliably and accurately and who can also record, map, and geo-reference the data collected are essential to the success of a survey. Typically, sign-based occupancy survey teams may have three to four observers, with at least one "expert tiger tracker," and others who possess the necessary skills to use maps and GPS and to record data.

The surveys can be conducted over shorter periods by employing a large number of government staff, local naturalists, or trained civil society volunteers, if they are available. If sufficient people are not available, smaller teams can sequentially cover the cells over a longer period.

As part of the planning, it is useful to determine the actual number of days available for fieldwork after carefully considering the availability of competent survey teams and potential logistical problems that field teams could face while carrying out surveys (safety of personnel, rivers to be crossed, topography and terrain problems, vehicles, accessibility, snow, rainfall, holidays, etc.). It is very important to be realistic and practical about the areas to be covered, distances to be walked, and more importantly how and by whom the survey will be conducted and supervised.

It is also important to choose the best season and months of the year for the survey, based on knowledge of field conditions: for example, soon after first showers or snowfall, when there is less leaf litter, when survey personnel are free of other tasks, or when volunteers are easier to find. The whole point is to carry out the survey when personnel are available, logistics are tractable, and animal signs easier to find. After you have chosen the best time for the survey, try to conduct the survey in roughly the same period periodically (e.g., every 3–4 years) thereafter. Such standardization is helpful for monitoring effort and to minimize seasonal bias. The best time for field surveys is likely to vary from region to region, even within a country.

5.4 Identification and Recording of Tiger Signs

Tigers are secretive, wide-ranging animals, distributed over an extensive area. They are difficult to observe in the field, and therefore signs such as tracks, scats, scent marks, and kills are important means for assessing the status of tiger populations over large regions. Tracks of tigers are readily detected wherever substrate conditions are appropriate. In the tropics, such conditions are met on dusty roads, sandy riverbeds, riparian deltas, and alluvial soils. Unfortunately, in most parts of the tropics, tiger tracks are difficult to find because of unsuitable soil, rainfall, and leaf-fall. Where tiger tracks are not easily detected, tiger scats can be recorded, although usually fewer scats than tracks are encountered. Additionally, tiger scats can be aged and categorized as fresh or old with relative ease, compared to other tiger signs. It is important to note, however, that in order to reduce ambiguity about the meaning of "occupancy" (see Chap. 4), only data on "fresh" (<7 days) scats should be recorded, excluding older scats that may have remained intact for many months.

Occupancy surveys must also record the signs of principal prey species (including livestock). Sightings, tracks, dung, and calls of prey species are the type of data gathered from field surveys.

Because the viability of tiger populations is chiefly a function of prey abundance (Karanth and Stith 1999; Miquelle et al. 1999), questions about tiger status can be answered better if tiger surveys are carried out simultaneously with sign surveys of the principal prey species. Such data on distribution and encounter rates of prey signs can be further used as covariate information along with other habitat and environmental factors, to model tiger occupancy.

Personnel participating in the field survey must be able to correctly recognize the signs of tigers and prey species that they encounter in the field. Misidentification of signs can lead to serious flaws at the stage of data analysis and interpretation of results. It should be emphasized here that if certain signs cannot be attributed to a species with certainty, it is better to exclude them from being recorded. It may also be useful to have a field orientation workshop before the surveys, for general skill development and as a team building exercise. For some sign such as scats, uncertainty in species assignment (e.g., leopard versus tiger) can be incorporated into the occupancy modeling (e.g., Miller et al. 2011, 2013; Chambert et al. 2015) if some scats are collected for subsequent laboratory analysis to ascertain truth.

Usually, experienced survey personnel can distinguish similar signs such as tiger from leopard. Tracks that look like leopard tracks, but accompany a definite tiger track, often indicate a tigress moving with her cubs. Several publications describe how to recognize and record tiger and prey signs in the field (van Strien 1983; Rabinowitz 1997; WWF-Nepal 1998; McDougal 1999). Illustrations of common large carnivores and prey species of tigers that occur in tropical Asia and of the tracks they leave are provided in Karanth and Nichols (2002; pp 167-175). These materials can be used to familiarize survey personnel with the survey objects. However, we reiterate that manuals are not a substitute for actually looking at animals or signs and gaining practical experience in accurate recognition.

5.5 Field Survey (Data Collection) Protocols

It is extremely important to prepare standardized data forms beforehand (see Table 5.1 for example). Clear guidelines must be given to field teams on how to fill in each column for every record. This will minimize error and unstandardized data collection and recording.

Field survey teams should make preplanned visits to cells and search for the tiger sign following a carefully worked-out survey design (see above and Chap. 4; also see Hines et al. 2010 and Karanth et al. 2011 for example). As described above, the

Table 5.1 Occupancy survey: field data form

Form No.:_______ Date:_______ Grid:_______ Survey Team:_______________

Start time:_______ End time:_________ GPS No.:_________ GPS File Name:_____ Replicate:____

Start LAT/LONG: ______________________ End LAT/LONG:______________________

Area:______________________ Survey Route:______________________________

Segment number	1	2	3	4	5	6	7	8	9	10
Segment type										
Substrate condition										
Habitat type										
TGR										
LPD										
DHL										
BER										
ELP										
CHT										
SBR										
GAR										
MJK										
PIG										
LVS										
UID										
Remarks										

Table 5.1 (continued)

SUPPLEMENTARY INFORMATION RECORDED

Species	Segment No.	No. Seen (Direct sighting)	Photo ID	Remarks

Instructions for filling the form:

Record ONLY the first encounter of TRACK and SCAT/PELLET for each species in every 100-m segment. Segment Type, Substrate Condition and Habitat Type should be entered for every 100-m segment.

Enter a "tick mark" for specific human impact categories ONCE every kilometre. Describe details of human impact under "Remarks" column of the main datasheet in the segment where the human impact sign was seen.

Details of Direct Sighting (number seen for each and every direct sighting), photos taken (photo file name) with details in the "Remarks" column should be entered in the Supplementary Information Table.

Acronyms used in the form:

Segment type: ROD=Road, **TRL**=Trail

Substrate condition: SOF=Soft Soil, **HAR**=Hard Soil, **LLT**=Leaf Litter, **GCR**=Grass Cover

Habitat type: FOR=Forest, **SCB** =Scrub, **GRS**=Grassland, **OTH**=Other

Species: TGR=Tiger, **LPD**=Leopard, **DHL**=Dhole, **BER**=Bear, **ELP**=Elephant, **CHT**=Chital, **SBR**=Sambar, **GAR**=Gaur, **MJK**=Muntjac, **PIG**=Wild Pig, **LVS**=Livestock, **UID**=Unidentified

Evidence: S=Scat, **T**=Track, **P**=Pellet, **D**=Direct Sighting, **C**=Call

Remarks: OBE=Organized Biomass Extraction, **LBE**=Local Biomass Extraction, **FIR**=Forest Fire, **MFP**=Minor Forest Products (non-timber), **PCH**=Poaching, **MIN**=Mining/Clearance of land, **MOD**=Habitat Modification

field survey teams should be directed to pass through the randomly located sub-cell in each cell. The field teams should search potential tiger habitat within each larger cell, guided by local knowledge. The survey of a cell may take 1–7 days depending on the topography, terrain, vegetation density, presence of natural barriers, and logistical constraints such as availability of manpower. Each cell should be surveyed in a predetermined, convenient sequence so that all the cells are covered within the entire occupancy survey period.

When searching for tiger signs, the field team(s) should thoroughly search at most likely places to locate tiger signs by walking along trails, mineral licks, logging roads, rivers banks, and stream beds – whatever works best in the field situation. Linear daily movement of survey teams should aim to mimic the movement patterns of tigers. Each trail segment should be surveyed only once, and "return journeys" made along it should be excluded for data collection purposes.

We note that, sometimes, after completing the predetermined sampling effort within a cell, substantial distances may have to be walked within that same cell to the next cell to be surveyed. Similarly, long distances may have to be walked to reach overnight camping points. We recommend that all such walking effort expended within the "defined tiger habitat inside each grid cell" should not be considered "wasted effort" or "down time," as far as possible. With careful planning, even these distances walked can be combined to meet the required survey effort for each cell, under the specified survey design. Thus, all tiger sign data should be collected and recorded meticulously, including the survey effort, within all tiger habitat traversed. The only time the tiger sign should not be recorded is when the team ends up completely out of the tiger habitat, for example, while moving to the camp or to the next cell. Only such "real wasted effort" should be excluded from the survey sampling effort. It should be noted here that fatigue can sometimes cause teams to miss signs. Therefore, in order to not compromise on the data quality, it is important to carefully plan the survey effort/distance to be walked per day depending on teams' abilities, terrain, etc.

5.6 Protocol for Collecting Covariate Data

Different factors can influence the occupancy of tigers in a given site depending on habitat conditions and refuge, availability of sufficient prey, and anthropogenic disturbances. Indeed, assessment of the importance of such factors may be an important objective of the survey and modeling effort. Concurrently, certain factors may also affect the detectability of tiger and prey signs such as substrate and weather conditions. Tremendous value can be therefore added to a survey if appropriate covariate data are collected during a sign-based occupancy survey. Thus, in addition to data on tiger signs, data on predefined covariates should be recorded for every 100-m or 1-km segment (see the datasheet in Table 5.1). As an example, we provide in Box 5.2 covariate data that were gathered during tiger occupancy surveys in India (Karanth et al. 2011; also see Sunarto et al. 2012, Harihar and Pandav 2012, Chanchani et al. 2016 for additional examples). It is important to note that covariate data need to be collected for every search path segment in every cell irrespective of whether tiger signs were encountered in the segments. Moreover, it is imperative to train surveyors and standardize the observation process and recording of the covariate so as to minimize observer bias. Apart from collecting ground-based covariates, data on certain covariates such as terrain, forest type, etc., can be obtained from remotely acquired satellite imageries. It may also be useful, where possible, to consider covariates that are continuous or ordinal rather than categorical variables to have more flexibility and power in data analysis.

Box 5.2: Example of Covariate Data That Can Be Collected During a Sign-Based Occupancy Survey (from Karanth et al. 2011)

Segment Type ROD = Road, TRL = Trail

Segment-type covariate is to indicate the width of the 100-m sample segment being surveyed. Road means a forest road which is wide enough for a four-wheel vehicle to pass. Road may have either two visible or no tracks in it. Trail means a narrow forest or animal trail used either by humans or other animals.

Substrate Condition SOF = Soft Soil, HAR = Hard Soil, LLT = Leaf Litter, GCR = Grass Cover

Substrate condition covariate is to indicate the prominent substrate condition for every 100-m sample segment. For example, if you are walking on a road, the substrate condition for the tracks on the road should be recorded and not for the center of the road which is usually covered with grass. The dominant substrate condition of the sample segment should also be recorded. For example, if the substrate condition for the tracks is soft for more than 50% of the 100-m segment and the rest is hard, then it should be recorded as soft (SOF). Similarly if the tracks on the road or trail are covered with leaf litter for more than 50% of the 100-m segment, then they should be recorded as LLT.

Habitat type MDF = Mixed Deciduous Forest, EVG = Evergreen Forest, GRS = Grassland, OTH = Other

Habitat-type covariate is to indicate the prominent habitat found in the 100-m segment surveyed. MDF indicates mixed deciduous forest (both moist and dry deciduous forest types are included in this category), EVG indicates evergreen forest, GRS indicates grassland, and OTH indicates any other type of habitat that might be found (e.g., plantations). The prominent habitat type for the 100-m segment walked should be recorded. For example, if more than 50% of the 100-m segment was deciduous and less than 50% was evergreen, then the habitat type should be recorded as MDF for this 100-m segment. The habitat-type categories have to be decided *a priori* based on information available on the forest vegetation types prevalent in the overall study area.

5.7 Collection of Ancillary Data

As mentioned earlier, misidentification of signs and generating false positives can result in severe misinterpretations of the results. If possible, it is always useful to collect fecal/scat samples to extract fecal DNA. Fecal DNA-based genetic analysis facilitates the unambiguous identification of species, thus eliminating misclassification errors. It is also useful for identification of individual tigers, which permits application of advanced modeling to derive tiger abundance estimates and other population dynamic parameters under appropriate survey designs. Furthermore, it can be used to investigate genetic relatedness between populations over large regions (see Chap. 13 for details). However, to ensure the scat samples are useful, it is necessary to follow carefully standardized protocols (e.g., to avoid cross contamination of DNA, which may render the analyses futile).

DNA is best obtained from fresh <24-h scats. However, we have had some success in getting DNA from even older scats. Therefore, it is advisable to collect scats that are judged to be <7 days old, in reasonably intact, not totally washed, or dried-out conditions. However older scats including the deteriorated ones can be collected for other non-DNA, dietary studies. Furthermore, new or future analytical techniques such as e-DNA may allow DNA extraction from old scats or other low quality samples. If logistics permit, scats can be frozen, and DNA can be extracted later as well. For further details on collection and preservation protocols, please see Chap. 11.

5.8 Additional Methods for Assessing Occupancy

Before we describe alternate methods to assessing occupancy of tigers, please note that these are not substitutes for conducting field surveys at finer scales as described earlier in this chapter and elsewhere in this manual. The objective of this section is to provide an overview of other methods that have used the occupancy framework to address similar questions.

5.8.1 Questionnaire Surveys

If very large regions are to be surveyed in a short period without investing tremendous amounts of effort in the field, occupancy surveys based on "expert information" can be considered. In such a case, each reliable informant or expert consulted becomes a "replicate sample." See Karanth et al. (2009) for examples of occupancy modeling of large mammals, including tigers, based on expert information surveys. In order to be most useful for occupancy modeling, such interviews should focus on recent time spent by the informant in the field and any actual detections of tigers or their signs. Note that this kind of specific information is very different from simply asking informants about whether they believe tigers to be present in an area or not. Information from reliable local hunters, naturalists, researchers, and wildlife personnel can be useful, but soliciting information from unreliable informants – via random sampling or the like – might not provide valuable data for analysis. We note that quite often, such knowledgeable informants may not be literate or educated and may be even suspicious about the surveyors' motives, and hence special communication skills may be required to elicit reliable information from them (e.g., making them first identify a species from photographs). Questionnaire surveys should be conducted only by trained personnel capable of assessing the quality of information. If the informants are ignorant or untruthful, the data should be discarded. Questionnaire surveys with reliable informants or experts may also yield information on certain aspects of populations that are difficult to obtain through sign-based occupancy surveys. For example, one may be able to assess where the population is breeding over large regions by gathering information on sightings of mothers with cubs by reliable informants.

Each individual informant must be treated as a distinct survey (replicate), and his or her information must be recorded on a fresh questionnaire survey form. As in the case of field surveys of signs, these forms must be systematically numbered and cross-linked to geo-referenced maps. In such studies, one must carefully delineate/identify the areas, represented by a single or a group of sampling units (such as grids), that the interviewee is knowledgeable about (area of knowledge). Several criteria can be used to determine the interviewees' familiarity with the area (e.g., interviewee must have either lived in or visited or "surveyed" a sampling unit at least twice per month for a year; Zeller et al. 2011); however, these will have to be defined keeping in mind the specific objectives of the study. Also it is critical to ensure that the detection of species is based on first-hand information and not hearsay (as this could violate the assumption of independence).

5.8.2 Media Reports

Certain species such as tigers are able to generate substantial media attention, especially if they are detected outside, or on the fringes of forested habitats. Athreya et al. (2015) used media reports collected over a 1-year period to determine the occupancy of leopards across a 191,791-km^2 landscape. Since such an occupancy analysis is based on secondary data, certain precautions must be taken. First and foremost, it is of utmost importance to check the reliability and authenticity of the news report as there are chances of obtaining a false presence. To ensure this, one should rely on trusted news sources, as well as checking if there are multiple reports about the same incident. Additionally, there may be a bias in the process of selecting media reports – certain areas may have better reportage in comparison to isolated or rural areas. It is therefore important that this detection bias is considered during the modeling process (e.g., via use of distance to a human population center as a covariate for detection probability). Care must also be taken while aggregating data – multiple news reports about the same incident/sighting should be combined and treated as unique records. It is vital to consider the spatial scale of occupancy, i.e., size of the grid cells or administrative units should be coarse enough to reliably attribute news reports to a location. For example, in the above study, while incidents/sightings pertained to specific villages, assessments were made at the subdistrict level. This was because at the subdistrict level, the spatial locations could be unambiguously assigned to the correct administrative unit (as there are several villages in India with the same names). While each month was considered as an independent replicate, one should be careful to distinguish between the date of news report and the date of incident/sighting when attributing the detections.

5.8.3 Historical and Museum Records

For the purpose of monitoring, and to assess range contraction over long periods of time (decades or even centuries), it may be useful to use either historical literature (such as hunting records or natural history notes) or museum records to provide information on species distribution. However, such records provide information on the distribution of detections, but not on what areas were "sampled" but provided no detections. Such presence-only data cannot be used with standard occupancy modeling and require extremely restrictive assumptions for any sort of distribution modeling (see Royle et al. 2012; Yackulic et al. 2013). Such distribution surveys can also be useful for estimating extinction rates and predicting species extinctions over longer time periods. For example, Karanth et al. (2010) conditioned on locations of known historical occurrences and then conducted present-day sampling to estimate the fraction of these locations that were still occupied by 25 species of Indian mammals. The complements of these persistence proportions were extinction rates, which were modeled as functions of relevant covariates. However, extra care must be given as old records do not typically have accurate location information and only have an associate position with nearest village or town.

5.9 Organizing Field Data for Mapping and Analysis

The data from field surveys or questionnaires will be in the form of several maps and "tiger detection" data on field forms linked to these maps. The investigator should examine these forms to correct errors, remove ambiguities, discard questionable data, and fill in missing information by interviewing the field survey team members. This must be done immediately after the surveys. It is often a good idea for the investigators to physically check a certain proportion of the data through random field visits, as this will enhance data quality. Thereafter, the investigator must ensure data forms and maps are intelligible to persons entering the data into computers, preparing spatial distribution maps, or performing other analyses. As data volumes can be very large, we recommend using a database (e.g., MS Access, PostgreSQL) to store and retrieve data for analysis. Additionally, database software allows creation of forms to allow data entry which eliminates entry mistakes that can prove costly while carrying out analysis.

References

Athreya V, Srivathsa A, Puri M, Karanth KK, Kumar NS, Karanth KU (2015) Spotted in the news: using media reports to examine leopard distribution, depredation, and management practices outside protected areas in southern India. PLoS One 10(11):e0142647

Chambert T, Miller DAW, Nichols JD (2015) Modeling false positive detections in species occurrence data under different study designs. Ecology 96:332–339

Chanchani P, Noon BR, Bailey LL, Warrier RA (2016) Conserving tigers in working landscapes. Conserv Biol 30(3):649–660

Conroy MJ, Runge JP, Barker RJ, Schofield MR, Fonnesbeck CJ (2008) Efficient estimation of abundance for patchily distributed populations via two-phase, adaptive sampling. Ecology 89(12):3362–3370

Dey S, Delampady D, Parameshwaran R, Kumar NS, Srivathsa A, Karanth KU (2017) Bayesian methods for estimating animal abundance at large spatial scales using data from multiple sources. J Agric Biol Environ Stat. doi:10.1007/s13253-017-0276-7

Harihar A, Pandav B (2012) Influence of connectivity, wild prey and disturbance on occupancy of tigers in the human-dominated western Terai Arc landscape. PLoS One 7(7):e40105

Hines JE, Nichols JD, Royle JA, MacKenzie DI, Gopalaswamy AM, Kumar NS, Karanth KU (2010) Tigers on trails: occupancy modelling for cluster sampling. Ecol Appl 20:1456–1466

Karanth KU, Nichols JD (2000) Ecological status and conservation of tigers in India. Final technical report to the US Fish and Wildlife Service (Division of International Conservation), Washington, DC, and Wildlife Conservation Society, New York. Centre for Wildlife Studies, Bangalore, India

Karanth KU, Nichols JD (2002) Monitoring tigers and their prey: a manual for researchers, managers and conservationists in tropical Asia. Center for Wildlife Studies, Bangalore

Karanth KU, Stith BM (1999) Prey depletion as a critical determinant of tiger population viability. In: Seidensticker J, Christie S, Jackson P (eds) Riding the tiger: Tiger conservation in human-dominated landscapes. Cambridge University Press, Cambridge, pp 100–113

Karanth KK, Nichols JD, Hines JE, Karanth KU, Christensen NL (2009) Patterns and determinants of mammal species occurrence in India. J Appl Ecol 46:1189–1200

Karanth KK, Nichols JD, Karanth KU, Hines JE, Christensen NL (2010) The shrinking Ark: patterns of mammal extinctions in India. Proc Trans R Soc B 227:1971–1979

Karanth KU, Gopalaswamy AM, Kumar NS, Vaidyanathan S, Nichols JD, MacKenzie DI (2011) Monitoring carnivore populations at the landscape scale: occupancy modelling of tigers from sign surveys. J Appl Ecol 48:1048–1056

Kendall WL, White GC (2009) A cautionary note on substituting spatial subunits for repeated temporal sampling in studies of site occupancy. J Appl Ecol 46:1182–1188

Mackenzie DI, Royle JA (2005) Designing occupancy studies: general advice and allocating survey effort. J Appl Ecol 42:1105–1114

MacKenzie DI, Nichols JD, Royle JA, Pollock KH, Bailey LL, Hines JE (2006) Occupancy estimation and modelling: inferring patterns and dynamics of species occurrence. Academic, Burlington

McDougal C (1999) You can tell some tigers by their tracks with confidence. In: Seidensticker J, Christie S, Jackson P (eds) Riding the tiger: Tiger conservation in human-dominated landscapes. Cambridge University Press, Cambridge, pp 190–191

Miller DA, Nichols JD, McClintock BT, Grant EHC, Bailey LL, Weir L (2011) Improving occupancy estimation when two types of observational error occur: nondetection and species misidentification. Ecology 92:1422–1428

Miller DAW, Nichols JD, Gude JA, Rich LN, Podruzny KM, Hines JE, Mitchell MS (2013) Determining occurrence dynamics when false positives occur: estimating the range dynamics of wolves from public survey data. PLoS One 8(6):e65808

Miquelle DG, Smirnov EN, Merrill TW, Myslenkov AE, Quigley HB, Hornocker MG, Schleyer B (1999) Hierarchical spatial analysis of Amur Tiger relationships to habitat and prey. In:

Seidensticker J, Christie S, Jackson P (eds) Riding the Tiger: Tiger conservation in human dominated landscapes. Cambridge University Press, Cambridge, pp 71–99
Rabinowitz A (1997) Wildlife field research and conservation training manual. Wildlife Conservation Society, New York
Royle JA, Nichols JD (2003) Estimating abundance from repeated presence-absence data or point counts. Ecology 84:777–790
Royle JA, Chandler RB, Yackulic C, Nichols JD (2012) Likelihood analysis of species occurrence probability from presence-only data for modeling species distributions. Methods Ecol Evol 2012:545–554
Sunarto S, Kelly MJ, Parakkasi K, Klenzendorf S, Septayuda E, Kurniawan H (2012) Tigers need cover: multi-scale occupancy study of the big cat in Sumatran forest and plantation landscapes. PLoS One 7(1):e30859
van Strien NJ (1983) A guide to the tracks of mammals of W. Indonesia (Unpublished)
WWF-Nepal (1998) Tiger manual: indirect field study techniques for the kingdom of Nepal. World Wildlife Fund-Nepal, Kathmandu
Yackulic CB, Chandler R, Zipkin E, Royle A, Nichols JD, Grant EHC, Veran S (2013) Presence-only modeling using MAXENT: when can we trust the inferences? Methods Ecol Evol 4:236–243
Zeller KA, Nijhawan S, Salom-Pérez R, Potosme SH, Hines JE (2011) Integrating occupancy modeling and interview data for corridor identification: a case study for jaguars in Nicaragua. Biol Conserv 144(2):892–901

Concepts: Estimating Abundance of Prey Species Using Line Transect Sampling

6

Samantha Strindberg, N. Samba Kumar, Len Thomas, and Varun R. Goswami

6.1 Introduction

The principal prey of tigers are typically large, diurnal ungulates and can be visually detected and counted by observers while walking, riding domesticated elephants, or in very open habitats from aircraft, moving along a straight line. At the same time, the observers can also record distances to observed animals from this line. Therefore, line transect sampling is a method frequently used to estimate their density and abundance, as with many other wildlife species. It is a special case of a wider class of methods called distance sampling: in line transect surveys, data are collected along transect lines (as opposed to point transects, where data are collected from points). The basic distance sampling methods are described comprehensively in the book by Buckland et al. (2001), which is an update to a previous book by Buckland et al. (1993) and is predated by earlier work (e.g., Burnham et al. 1980). More advanced elements are described by Buckland et al. (2004) and recent

S. Strindberg (✉)
Wildlife Conservation Society (WCS), New York, NY, USA
e-mail: sstrindberg@wcs.org; http://www.wcs.org

N.S. Kumar
Wildlife Conservation Society, India Program, Bengaluru, India
e-mail: nrao.skumar@gmail.com

L. Thomas
Centre for Research into Ecological and Environmental Modelling, University of St Andrews, Fife, UK
e-mail: len.thomas@st-andrews.ac.uk; https://creem2.st-andrews.ac.uk/

V.R. Goswami
Wildlife Conservation Society, India Program, Bengaluru, India

Centre for Wildlife Studies, Bengaluru, India
e-mail: varunr.goswami@gmail.com; http://wcsindia.org/home/; http://cwsindia.org/

K.U. Karanth, J.D. Nichols (eds.), *Methods For Monitoring Tiger And Prey Populations*, DOI 10.1007/978-981-10-5436-5_6

developments can be found in Buckland et al. (2015) and Kéry and Royle (2015). These references should be consulted to obtain an in-depth understanding of theory involved.

Thomas and Karanth (2002) provided a concise summary of the line transect sampling method as applicable to tiger prey species earlier. The current chapter explains fundamental concepts of line transect sampling together with the summary of recent advances that are relevant for sampling tiger prey populations in forested habitats. Chapter 7 that follows describes field practices.

In the line transect method, detections are made of individual animals (or groups of animals) of the target species; this requires habitat in which individual animals can be seen before they respond and move. Here we provide a basic introduction to line transect sampling focusing on visual detection of tiger prey species to estimate their density, abundance, and factors influencing their spatial distribution. Throughout, we assume that the target during any survey is one or several tiger prey species, such as gaur, Asiatic water buffalo, sambar, chital, muntjac, and wild pig that typically live in forested habitats (Chap. 2).

Line transect sampling can be seen as a generalization of strip transect sampling. During strip transect sampling, observers traverse a set of randomly located lines, detecting and recording all animals of the target species within a set distance w on either side of the line. To ensure that no animals within the strip of width $2w$ are missed, the strip must be reasonably narrow. This is fine for animals that occur at high density and are easily detectable. Strip transect sampling, however, is inefficient and leads to biased estimates in most surveys of tiger prey species. In a line transect, a much wider strip can be searched, potentially as far as the observer can see from the line, because perfect detectability is assumed only on the line itself and the assumption that all animals in the strip are counted is relaxed (Fig. 6.1).

Additional data, namely, the perpendicular distance between each observation and the transect line, are used to estimate the detection probability. In the case of tiger prey surveys, observers typically record the observer-to-animal distance (the "radial distance") and the compass bearings to the animal and line, which are used to calculate perpendicular distance. It is standard practice in surveys of prey species to record only visual detections, because aural detections tend to be inaccurate in terms of distance, species, and group size estimates (Alldredge et al. 2007a).

To gain an understanding of how the line transect method fits into the general statistical framework described in Chap. 3, let us revisit the canonical estimator:

$$\widehat{N} = \frac{C}{\widehat{p}\alpha} \tag{6.1}$$

where $\widehat{N}$ is estimated abundance, C is the number of animals counted, $\widehat{p}$ is the estimated proportion of animals counted, and α is the proportion of the study area surveyed; $\widehat{p}$ relates to detectability and α relates to spatial sampling.

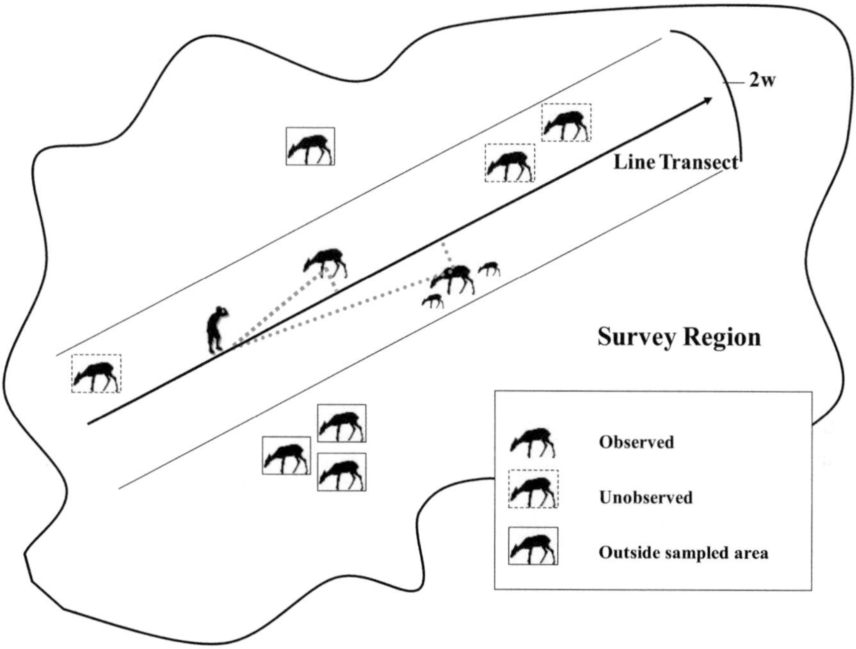

Fig. 6.1 During a line transect survey of tiger prey species, observers move along a set of randomly located line transects. The unit of observation is the prey group. Group size and radial distance to the center of the groups is recorded. Some of the animal groups are seen by the observers, while others are not; some fall outside the sampled area – the subregion, a distance w from the line. Methods are used to correct for these animals missed by the observer

In strip transect sampling, we assume all animals within the surveyed strips are detected and hence $\widehat{p}=1$, whereas in line transect sampling, we use the perpendicular distances to detected animals to estimate p. Regarding spatial sampling, if L is the total length of all the transects, then we can conceptualize the area surveyed as a rectangle of width $2w$ and length L. Thus the area surveyed is $2wL$ and the proportion of the study area, A, surveyed is $\alpha = 2wL/A$. Consequently, Eq. (6.1) can be written as

$$\widehat{N} = \frac{AC}{\widehat{p}2wL} \tag{6.2}$$

Density and abundance are related as follows $N = D \times A$, so the equation to estimate density is

$$\widehat{D} = \frac{C}{\widehat{p}2wL} \tag{6.3}$$

Given that tiger prey species often occur in groups (referred to as "clusters" in the distance sampling literature), it is the group that constitutes the unit of detection during line transect surveys. Thus, the perpendicular distance to the center of the group is recorded, as well as the group size. Once group density in the study area has been estimated using Eq. (6.3), an estimate of individual animal density can be obtained as follows:

$$\widehat{D} = \frac{C\widehat{E}(s)}{\widehat{p}2wL} \tag{6.4}$$

where C is now the number of detected groups and $\widehat{E}(s)$ denotes the estimated population mean group size.

In the remainder of the chapter, we will focus on (1) the theory behind the line transect sampling method, (2) the critical assumptions underlying density or abundance estimation with line transect sampling and the potential biases that arise if these are violated, (3) survey design options with an emphasis on reducing variability and other alternatives for improving precision, and (4) data analysis with an overview of the software tools available to design line transect surveys and analyze the resulting data. In this last section, we focus on the heart of distance sampling (of which line transect sampling is a special case), namely, modeling the detection function, and revisit some of the more advanced topics introduced in earlier sections, such as multiple covariate distance sampling and spatial modeling using distance sampling data and other covariates.

6.2 Line Transect Sampling Theory

In this section we outline how density estimates are obtained by accounting for the proportion of animals counted and the proportion of the study area sampled. We also describe how associated variances and confidence intervals for density and abundance are estimated.

6.2.1 The Distance Sampling Survey

Prior to conducting the line transect survey, k transect lines of length $l_1, \ldots, l_k$ (with total length $L = \sum_{j=1}^{k} l_j$) are randomly located within the study area (survey design is covered in more detail in Sect. 6.4). During the survey, observers traverse these lines and record clusters of all tiger prey species seen. A "cluster" consists of one or more individual animals where group membership is clearly defined in the field protocol ahead of the survey (see Chap. 7). Observers record data necessary to obtain the perpendicular distance to the center of the group and the group size. Sometimes only observations out to a distance w from the line are recorded, but more frequently all observations are recorded and w is set during analysis (see Sect. 6.5).

Line transect surveys for tiger prey species can be conducted from an airplane (in very open habitats), from elephant back, or on foot. Generally, for these types of survey, different kinds of measurements are taken to calculate the perpendicular distance x from the center of the animal group to the transect line during analysis: frequently for an aerial survey, a clinometer reading to obtain the angle of declination ϕ to the center of the group as it passes abeam is taken (where 0° is at the horizontal plane and 90° is directly below the aircraft) and the height of the airplane h is recorded; for a ground-based survey, it is more usual to obtain a radial (sighting) distance r and sighting angle θ. The sighting distance is the distance from the observer on the transect line to the center of the group of animals, and the sighting angle is the angle between the transect line and an imaginary line drawn between the observer and the center of the group (Fig. 6.2). For the aerial survey $x = h/\tan\phi$ and for the terrestrial survey $x = r\sin\theta$.

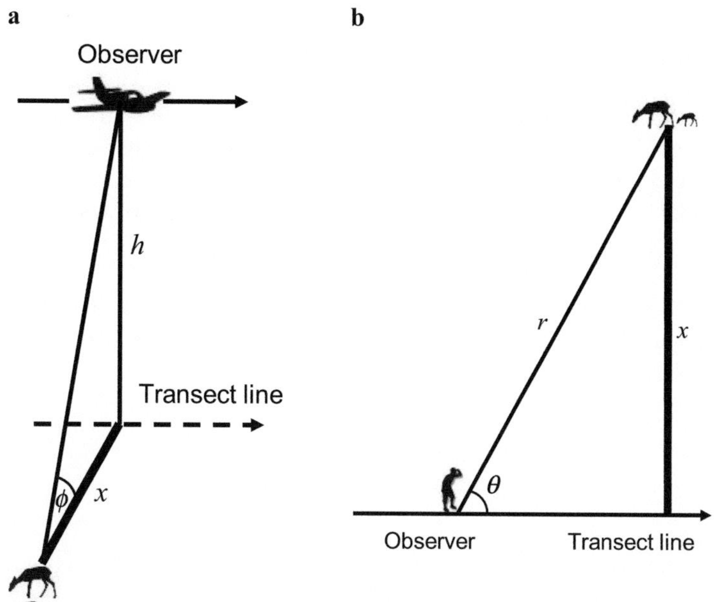

Fig. 6.2 Perpendicular distance x to the center of the animal group is calculated (**a**) from the height of the airplane h and angle of declination φ by applying the formula $x = h/\tan\phi$, for an aerial survey (note that the alternate interior angle that has the same degree measurement as the angle of declination, which is the angle between the horizontal plane and the sighting, is labeled on the figure to make the formula more intuitive), and (**b**) from the radial (sighting) distance r and sighting angle θ by applying the formula $x = r\sin\theta$, for a terrestrial survey conducted from elephant back or on foot

6.2.2 Estimating Detectability

When the transect lines are located randomly, there will be, on average, an equal number of animal groups close to the transect lines as there are far away. However, the number of *detected* groups drops off with increasing perpendicular distance (Fig. 6.3a), because animals farther from the line are more likely to be obscured by vegetation. This decrease in detected groups with distance from the line therefore tells us about how detectability changes with distance. During a standard distance sampling analysis, we exploit this fact to estimate p the average proportion of animal groups in the surveyed strips that is counted. We define a detection function $g(x)$, which gives the probability of detecting a group when it is at a distance x from the line, where x is between 0 and w. $g(x)$ will tend to decrease with increasing x. We can conceptualize this by considering the detection curve that has been fitted by eye to the data in Fig. 6.3a. To estimate p, the area under the curve corresponding to the average number of detected animal groups at each distance is divided by the area of the rectangle corresponding to the actual average number of groups at each distance (i.e., the number we would detect, on average, if we could detect them all). Note that this requires us to assume all animals at zero distance are detected (i.e., that $g(0) = 1$), in order to determine the height of the rectangle.

Figure 6.3b shows the same curve but this time rescaled so that the y-axis is the probability of detection, $g(x)$, rather than average number of detections. We introduce another term commonly found in the line transect literature: the "effective strip width" (ESW) (or, more correctly, the "effective strip half width") denoted by μ, as the area under the detection function: $\mu = \int_{x=0}^{w} g(x)dx$. Since the proportion of animal groups counted, p, is equal to the area under the curve divided by the area of the rectangle, and in this case the height of the rectangle is $g(0)$, which is 1, this implies that

$$\widehat{p} = \frac{\int_{x=0}^{w} \widehat{g}(x)dx}{1.0 \times w} = \frac{\widehat{\mu}}{w}$$

We can think about the ESW as the distance at which as many animals are expected to be seen beyond μ as are missed within μ (note the two hatched areas in Fig. 6.3b are equal in size).

One last piece of terminology is helpful to understand how the detection function $g(x)$ is actually fitted to the observed distances: the probability density function (pdf) of the observed distances, denoted $f(x)$. This can be loosely thought of as the probability that a detected animal is detected at distance x. It is obtained by rescaling the detection function: $f(x) = \frac{g(x)}{\mu}$.

The area under the pdf must be 1 (i.e., all detected animals are detected somewhere between distance 0 and w), and if we define $f(x)$ as above, we obtain $\int_{x=0}^{w} f(x)dx = \frac{1}{\mu}\int_{x=0}^{w} g(x)dx = \frac{\mu}{\mu} = 1$.

We now can write down the statistical likelihood of the data, given some model for the detection function:

$$\mathcal{L} = \prod_{i=1}^{C} f(x_i)$$

where x_i is the observed distance of the ith detected animal, there are C animals detected in total, and $f(x)$ denotes the model for the detection function. This in turn is the key to estimating the parameters of the detection function model using well-known methods such as maximum likelihood or Bayesian inference. The software packages used for model fitting are more fully described in Sect. 6.5.

Once we have an estimate of $f(x)$, it is easy to derive estimates of μ or p: since we assume $g(0) = 1$, it follows that the pdf evaluated at $x = 0$ is given by $f(0) = \frac{1}{\mu}$ (see Fig. 6.3c) and hence $\mu = \frac{1}{f(0)}$; hence $p = \frac{1}{f(0)w}$.

6.2.3 Estimating Abundance and Density

Once we have an estimate of the detection probability, p, animal abundance can be estimated using Eq. 6.2. and density using Eq. 6.3. If animals are detected in groups, we can use Eq. 6.4 – but we must first obtain an estimate of the population mean group size, $E(s)$. If large and small groups are equally visible at any distance from the transect line, $E(s)$ can simply be estimated by taking the mean of the observed group sizes. Frequently, this is not the case, as large groups tend to be more visible especially as distance from the line increases. This phenomenon leads to size bias, because large groups are overrepresented in the sample. There are a number of approaches for dealing with this type of size-biased sampling, where the detection probability is a function of both distance from the observer and group size, and they are briefly covered in Sect. 6.3.5.

6.2.4 Estimating Variance and Confidence Intervals

Quantifying the uncertainty in an estimate is just as important as calculating the estimate itself. For example, you would place a very different interpretation on the statement that there are 2.1 muntjac per square kilometer of forest if you were also told that the 95% confidence interval on this estimate was 1.7–2.6 versus if this confidence interval was 0.4–10.7! Here we outline how to quantify the uncertainty on a density estimate in terms of variance, coefficient of variation (CV), and confidence interval.

The variance of the density estimate, $\text{var}\left(\widehat{D}\right)$, has three components: variance in number counted (C), variance in the estimation of detectability (p), and variance in the estimation of average group size $E(s)$. Typically, these three components

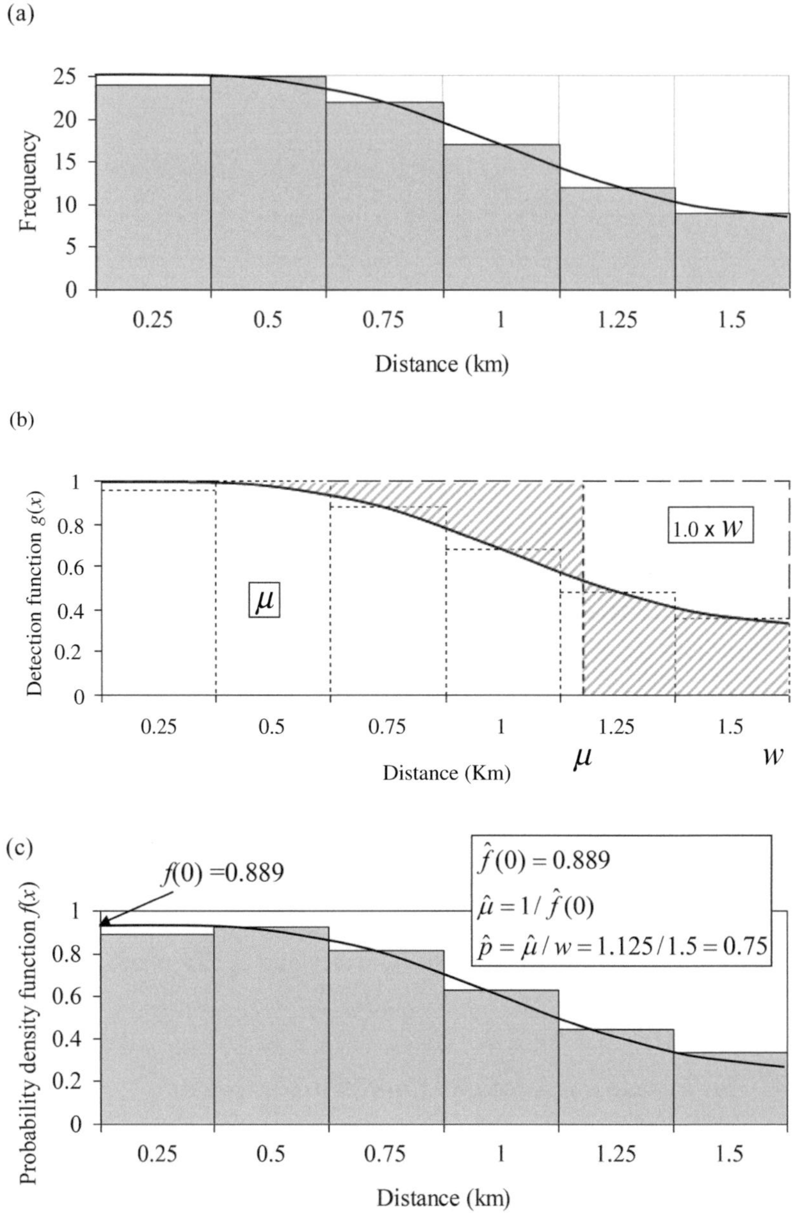

Fig. 6.3 (**a**) An example of plotting the distances recorded to animal groups during a distance sampling in six distance intervals. The dashed lines show the expected number of objects in each interval. The curve has been fitted to these data by eye, and the area under the curve represents the number of observations that were made, while the area above the curve represents the missed observations. Thus the proportion of animal groups counted is equal to the area under the curve (27.25) divided by the total area of the rectangle ($1.5 \times 25 = 37.5$), i.e., $\widehat{p} = 27.25/37.50 = 0.7676$, approximately. (**b**) A detection function is defined such that $g(0) = 1$, and the effective strip width μ is the distance at which as many animal groups are seen beyond μ as are missed within μ (*shaded regions*). The proportion of groups detected and counted $\widehat{p} = \int_{x=0}^{w} \widehat{g}(x)dx/\ (1.0 \times w) = \widehat{\mu}/w$. (**c**) μ can be estimated by fitting a probability density function $f(x)$ of perpendicular distances to the data to obtain $\widehat{\mu} = 1/\widehat{f}(0)$

are statistically independent, so var $\left(\widehat{D}\right)$ can be calculated using the delta method approximation (Powell 2007, Oehlert 1992):

$$\widehat{\text{var}}\left(\widehat{D}\right) = \widehat{D}^2 \cdot \left\{ \frac{\widehat{\text{var}}(C)}{C^2} + \frac{\widehat{\text{var}}\left(\widehat{p}\right)}{\widehat{p}^2} + \frac{\widehat{\text{var}}\left(\widehat{E}(s)\right)}{\widehat{E}(s)^2} \right\} \tag{6.5}$$

The variance of the counts among transect lines, after taking into account the potentially different line lengths, is usually used to estimate the variance in number of animal groups seen, var(C) (Buckland et al. 2001: pp. 78–79, 108–109, 154–155; Fewster et al. 2009). This is usually the largest component of the variance (70–80% of the total), and to obtain a reliable estimate of this variance at least 15–20, but preferably more than 25, replicate lines are required. A less than ideal solution that can be applied when there are insufficient transect lines to estimate var(C) empirically is to estimate this variance by assuming that it is proportional to the expected value of C. The detection function is fitted to the distance data, and thus a likelihood-based estimate of var $(\widehat{p})$ is produced. The approach for estimating $E(s)$ will determine how var $\left(\widehat{E}(s)\right)$ is estimated (Buckland et al. 2001: pp. 72–74, 120, 164).

If the components comprising Eq. 6.5 are not independent (e.g., when animals can be found in few large groups or many small groups and this tendency varies spatially), then a nonparametric bootstrap can be used to estimate var $\left(\widehat{D}\right)$. This statistical technique does not require any distributional assumptions and involves randomly resampling the data with replacement to obtain a large number of estimates of D from which var $\left(\widehat{D}\right)$ is then estimated. The resamples are taken at the level of the transect line, as these are considered to be independent, rather than at the level of the observation (Buckland et al. 2001: pp. 82–84, 117, 161–164).

When considering precision of a density estimate, it is convenient to use the coefficient of variation (CV), where $CV\left(\widehat{D}\right) = \frac{\sqrt{\text{var}\left(\widehat{D}\right)}}{\widehat{D}}$. $CV\left(\widehat{D}\right)$ gives the size of the standard error of the density estimate $\sqrt{\text{var}\left(\widehat{D}\right)}$ relative to the size of the estimate $\widehat{D}$. As a unit-less quantity, it can be used to compare different studies that may use different units or have very different estimates of density or abundance. Thus Eq. (6.5) can equivalently be written in terms of $CV\left(\widehat{D}\right)$ as follows:

$$\widehat{CV}\left(\widehat{D}\right) = \sqrt{\left[\widehat{CV}(C)\right]^2 + \left[\widehat{CV}\left(\widehat{p}\right)\right]^2 + \left[\widehat{CV}\left(\widehat{E}(s)\right)\right]^2} \tag{6.6}$$

Burnham et al. (1987: pp. 211–213) showed that log-based confidence intervals give a better measure of the precision of $\widehat{D}$ than the standard symmetrical 95% confidence intervals. Thus the approximate asymmetric 95% confidence intervals are given by $\widehat{D}/C$ and $\widehat{D} \cdot C$ where $C = \exp\left\{1.96\sqrt{\widehat{\text{var}}\left[\ln\left(\widehat{D}\right)\right]}\right\}$ with $\widehat{\text{var}}\left(\ln\left(\widehat{D}\right)\right) = \ln\left\{1 + \left[\widehat{CV}\left(\widehat{D}\right)\right]^2\right\}$. In practice the 1.96 that corresponds to a normal distribution is generally replaced by another constant (Buckland et al. 2001: pp. 77–88, 118–119). These calculations are done automatically by the standard software packages. Alternatively, a nonparametric bootstrap can be used to estimate confidence intervals without making any distributional assumptions.

6.3 Assumptions of Line Transect Distance Sampling and Biases Arising from Their Violations

In this section we look at the four critical assumptions underlying distance sampling along line transects and consider the biases that arise if these assumptions are not met. These assumptions are covered in considerably more detail in Buckland et al. (2001: pp. 29–37, 130–133). We also look at some other issues that are important to consider to successfully apply distance sampling methods.

6.3.1 An Adequate Number of Line Transects Is Located Randomly with Respect to the Distribution of the Animals

Conventional distance sampling uses both model- and design-based inference (Thompson 1992). The former is used when fitting the detection function. The latter pertains to the random placement of the transect lines. Distance sampling requires independence between the animal distribution and the location of the line transects. This can be achieved through a survey design that ensures random placement of the transect lines. A random survey design permits valid inference at two levels:

1. *Reliable estimation of the proportion of animals counted* $\widehat{p}$ *from the observed distances. In order to interpret the pattern of observed distances, we must assume that the true distribution of animals does not change with distance (i.e., the rectangle in* Fig. 6.3a*) – in other words that animal groups are uniformly distributed in the interval* [0, w]. *Although this is an assumption about the animal distribution (and so model-based inference), it is met if the lines are placed at random – so having a good survey design ensures this model-based assumption is met.*
2. *Reliable extrapolation from observations made during the survey in the sampled area to the entire study area. This requires that the surveyed lines are representative of the study area as a whole. A sufficient number of randomly located lines is likely to pass through areas with densities representative of the entire region of interest rather than some smaller set of areas with possibly atypical densities.*

Without a random design, one can resort to model-based inference, which relies on the possibility of fitting a model that reasonably approximates the processes that gave rise to the data. Thus, the simplest and most robust alternative is to use a random survey design. In Sect. 6.4 we focus on various survey design options.

If the assumption of random transect placement is violated, then the resulting density estimates have the potential to be either positively or negatively biased. For example, if the transects are placed along or in the vicinity of trails and animals either use those trails preferentially or avoid them, then density and abundance will be overestimated or underestimated, respectively, if applied to a larger region.

In addition, with an insufficient number of randomly placed lines, the potential exists to sample only areas with atypical densities by chance. In addition inadequate replication leads to poor estimates of precision. The exact number of transect lines required for adequate replication depends on the variability in tiger prey density over the region of interest; 25 spatially replicated lines are a reasonable recommendation, but sometimes 15–20 lines may suffice.

Another important consideration concerns the probability of sampling a particular location (referred to as the coverage probability) for a given type of random survey design. Ideally, every location in the survey stratum (see Sect. 6.4.1 on how to stratify your survey area to improve precision) should have the same probability of being sampled ("covered" by a transect line and its associated search area, hence the term "coverage probability"), if the standard analysis technique is to be applied. Those types of designs where the coverage probability is variable have the potential to produce biased estimates. If standard methods are applied during the data analysis phase and coverage probability is assumed to be even when it is not, then if high (or low) density areas were sampled more intensively this would lead to a positive (or negative) bias. If the differences in coverage probability are extreme, then it may be advisable to use an estimator that takes account of this. However, this type of estimator is likely to increase the variance of the estimate (Strindberg 2001; Strindberg and Buckland 2004).

6.3.2 Animals on or Very Near the Line Are Detected with Certainty

For distance sampling along line transects, the derivation of the density estimator is based on the assumption that all animal groups are detected at zero perpendicular distance from the line, i.e., $g(0) = 1$. If this assumption does not hold because animal groups whose centers are on or very near the line are missed, then estimates of density or abundance will be negatively biased, as the proportion of animal groups counted, p, will be underestimated. When only the assumption of perfect detection on or near the line fails, the negative bias is a simple function of the proportion of objects on the line that is missed. For example, if 20% of groups with center on or near the line are missed, then the estimate of animal density or abundance will be 20% lower than it should be. In particular, when dealing with animal groups, this assumes no movement, so that the entire detection function is effectively scaled by 0.8. Depressed $g(0)$ could also occur due to evasive movement, but this would not scale the entire detection function by 0.8. In reality it should be possible to meet this assumption in terrestrial surveys of tiger prey species through good field methods as outlined in Chap. 7. However, this assumption is difficult to meet in some line transect surveys (e.g., aerial surveys of tiger prey species in the Russian Far East), where advanced methods must be used to estimate $g(0)$ (see Laake and Borchers 2004, Borchers et al. 2006, 2013, Fewster and Pople 2008, Burt et al. 2015, Borchers and Langrock 2015 for details).

6.3.3 Animals Are Detected at Their Initial Location

In line transect surveys, slow nonresponsive movement of the animals relative to the speed of the observers is generally not problematic. The general recommendation is that observers should be moving 2–3 times the average speed of their target animals to avoid biases associated with animal movement (Buckland et al. 2001).

If animals systematically move, away from the observers, as they typically do, and if such responsive movement takes place before the animals are detected, then estimates of density or abundance will be negatively biased. Such evasive movement is likely for prey surveys, and it is difficult to identify when it is moderate, as it may just make the detection function look as though it has a wide shoulder. Note that a "spike" at zero distance can have several causes, including unlikely attractive movement, rounding distances or angles to zero (in the case of attractive movement, there will be a surfeit of small values, but zeroes will not be overrepresented), guarding the transect line, or cutting wide straight transects, so that (a) animals start to use them as trails and (b) one can see a long way ahead (see Chap. 7). See Fig. 6.4a for an example of how data might look with responsive animal movement.

It is difficult to obtain reliable estimates of density if there is responsive movement, emphasizing the importance of sound field procedures that minimize disturbance as observers move along line transects (see Chap. 7 for details).

Certain tiger prey species may be extremely wary, making it impossible for foot-based observers to meet this assumption. In such a case, elephant-back surveys may provide an alternative (Karanth 1988, Wegge and Storaas 2009).

6.3.4 Measurements from the Line to the Center of Each Detected Animal Group Are Exact

For a terrestrial survey, the radial distances and sighting angles should be recorded correctly, similarly, for the clinometer and height readings taken during an aerial survey. It is especially important that distances to animals near the line transect should be measured both precisely and accurately. A prerequisite for this is that the location of the line is clearly defined and known to the observers.

If distances to observations are rounded to convenient values (see Fig. 6.4c), then it is possible to deal with these "heaped" data during analysis by grouping data into distance intervals, where cut points for the intervals are chosen so that heaps fall

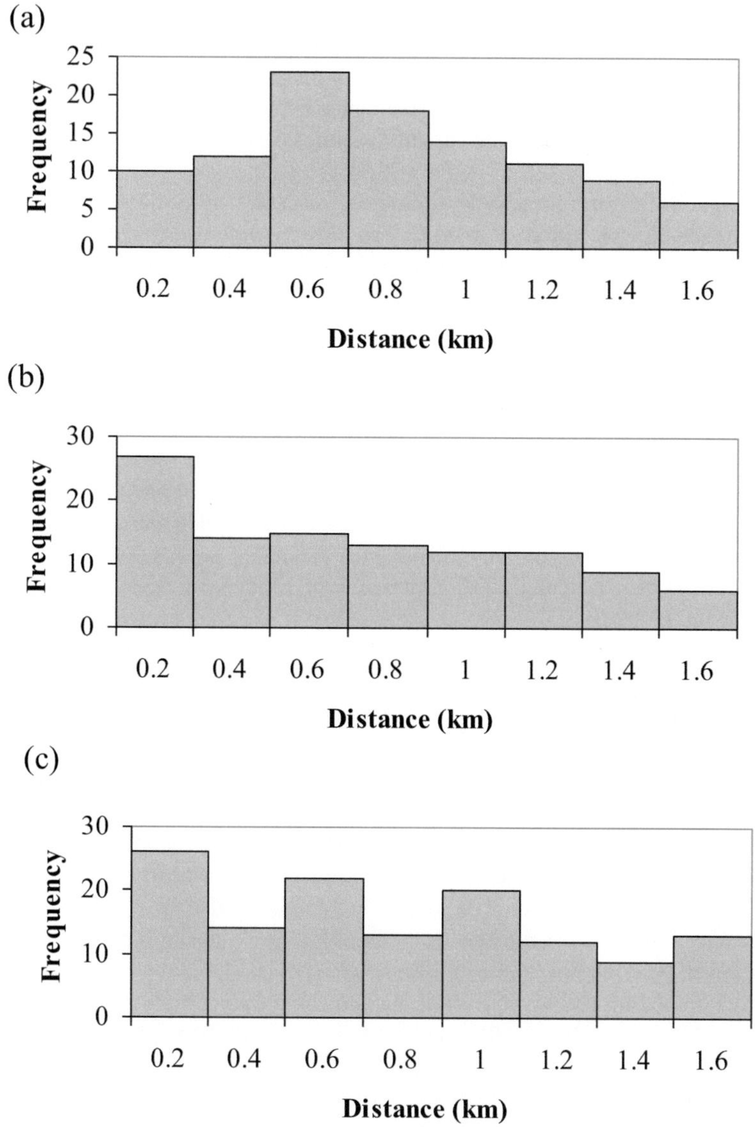

Fig. 6.4 Examples of (**a**) evasive movement where animals have moved away from the observers prior to detection; (**b**) spiked data potentially caused by rounding perpendicular distances or angles to zero, guarding the transect line, extensively clearing the transect line, or by movement toward the observers prior to detection (less likely for tiger prey surveys); and (**c**) heaped data where distances are rounded to convenient values

approximately at the midpoints of the intervals. However, systematic bias in distance measurements cannot be dealt with unless bias correction factors are estimated by means of experiments. Consistently overestimated distance measurements will lead to underestimation of densities and vice versa. This is also the reason why observers should measure the distance to the center of the animal group rather than to the closest or first individual seen; the latter option may seem more convenient but will lead to positively biased density and abundance estimates. Hence, it is important to follow good field procedures to meet this assumption (see Chap. 7 for details).

Care must also be taken to avoid a "spike" of detection data at zero distance, which occurs if sighting angles to an animal or center of a group, a considerable distance ahead, are rounded to zero (see Fig. 6.4b). Technical aids (reticule binoculars, optical and laser range-finders, compasses, clinometers, etc.) should be used to improve the accuracy and precision of measurements.

If the survey conditions make it especially difficult to record detection distances precisely, then an option is to record data by distance interval (this is particularly relevant for aerial surveys). For line transects, intervals are usually narrower near the line and increase in width with increasing distance from the line, with 5–7 distance intervals being recommended and careful measurement near the interval cut points being required. It is best to avoid measurement errors whenever possible, because the options are either a biased estimate of density and abundance or an attempt to accommodate these measurement errors with the added expense of additional data collection and increased analysis complexity with likely increased variances as well (Borchers et al. 2010).

6.3.5 Other Important Considerations

To reliably model detection, one should aim for a sample size of 60–80 animal groups per species (and per stratum) in line transect surveys. The exact sample size required depends on the nature of the data; usually fewer observations are needed if detectability is certain near the line and remains certain or nearly certain for some distance from the line, whereas a larger number is needed if detection drops off rapidly with distance from the line. Sometimes species with similar detectability can be combined when sample sizes are limited per species (Durant et al. 2011). In such instances, including species as a covariate and using the multiple covariate distance sampling (MCDS) analysis option is likely to be particularly helpful (e.g., Alldredge et al. 2007b; see Marques et al. 2007 for an introduction to MCDS). Another option is to pool data across multiple surveys to obtain more reliable estimates of detectability when sample sizes from a single survey are small, again potentially including "survey" as a covariate. In general, the MCDS analysis approach should be considered, as it is likely to provide more accurate results than simply pooling the data, which will only provide an average estimate of detectability across species or surveys with associated inaccuracy for the specific species or survey density or abundance estimate (Buckland et al. 2015).

There are other considerations that are less critical but nonetheless important. One of these is whether or not detections are independent events. When detections are dependent (e.g., animals fleeing and disturbing others that are subsequently detected), this has little effect on the point estimate of density or abundance.

However, analytical estimates of sampling variance will be negatively biased. This problem can be alleviated by using empirical estimators or resampling methods for variance estimation (e.g., using the bootstrap or jackknife that only assume independence between transect lines).

An obvious case where this assumption is violated is when animals tend to aggregate and occur in groups. If animals aggregate in loose, poorly defined groups, then it may be necessary to treat each individual animal as an observation. Distance sampling is particularly robust to the lack of independence between sightings of individuals that do in fact belong to the same group, providing an unbiased estimate of animal density or abundance as long as all individuals on or near the line are detected. Otherwise, as we described previously, we treat the group as the object of interest and measure the distance to the center of the group, as well as the group size.

It is also assumed group (cluster) size is recorded correctly when this is the unit of observation. When size-biased sampling occurs, then the detection probability of the group is a function of both distance from the observer and its size, which can lead to biases. The recommended option for dealing with this problem is size bias regression where distance or detection at a given distance is regressed against either group size or the logarithm of group size when group size is highly variable, and mean group size is then taken to be the size at the line. The slope of the regression of group size (or log group size) on distance tends to have a positive slope (as group size increases with distance) and on detection probability a negative slope. Sometimes the sign of the slope is reversed. This happens when observers underestimate the size of groups and the degree of underestimation increases with distance. Another option is to again turn to MCDS analysis and treat group size as a covariate when fitting the model for the detection function. For more details and other options for dealing with size bias, see Buckland et al. (2001: pp. 71–76, 122–130, 164–171).

The multi-analysis (MA) feature in the Distance software (Thomas et al. 2010) allows for uncertainty in covariate measurement, including group size. It also allows for sightings where the species was not identified with certainty, as long as there is a hierarchical structure of species sighting codes (see Gerodette and Forcada 2005).

If animals move in response to the observers and are thus detected several times on the same or adjacent transect line, it can cause substantial positive bias (assuming repeat counting is common during the survey). If the same animal is detected more than once while sampling the same transect *during different sampling occasions,* this is not a problem. Distance sampling theory also allows for an animal to be detected from different transects due to random movement of the animal.

Distance sampling theory performs well when detectability is certain near the line and remains nearly certain for some distance from the line. Thus, the potential for detecting animal groups should not drop off abruptly at a short distance from the line transect for ecological reasons. Although this shape criterion is not an assumption, it is a practical consideration required to provide reliable estimates of density and abundance.

6.4 Survey Design and Other Considerations to Improve Precision

For line transect surveys of tiger prey, variability in the density and abundance estimates is caused by (i) variance in observed sample size C or encounter rate C/L, (ii) variance in the estimated detection probability $\widehat{p}$ or equivalently $\widehat{f}(0)$ or $\widehat{\mu}$, and (iii) variance in the estimated expected group size $\widehat{E}(s)$. This emerges directly from the equations used to estimate density (Eqs. 6.3 and 6.4) that result in the estimate of variance given by Eq. 6.5. Variance in encounter rate is often the largest contributor to the variance of the density estimate (often 70–80% for line transects). Thus, we can reduce variability in the density and abundance estimate by improving the precision of these three components that contribute to the overall variability but particularly encounter rate.

6.4.1 Reducing Variance in Encounter Rate

6.4.1.1 Stratification

Spatial variation in animal group density between transect lines causes variance of observed sample size C or encounter rate C/L. Precision can potentially be improved by stratification. If spatial heterogeneity in population density exists, then defining strata that are internally homogeneous reduces variance. By means of stratification, we attempt to make encounter rates along transects corresponding to a particular stratum as similar as possible and encounter rates along transects corresponding to different strata as different as possible. To improve overall precision, different stratifications may be selected for different components – encounter rate, detection probability, and mean group size – of the density estimator. Stratification by habitat type (open grassland, forest, etc.) is often sensible as one might expect both density and the probability of detection to change by habitat type. This type of stratification during the design phase of the survey is only possible if the habitat types are not too fragmented and intertwined. An option for regions with patchy (fine-grained) variation in habitat types is to keep a record of when the habitat type changes along the transect line, subdividing it into smaller line segments that have an associated habitat classification. One would then have a total for the amount of effort spent in each habitat type, which would allow for post-stratification by habitat type during analysis. It is important that such a strategy is set out before data are collected and not in response to a noted variation in encounter rate during initial data analysis – the latter leads to serious underestimation of the true uncertainty. Variables such as season or time of day might also affect encounter rate, or the other components of the density estimator, and stratification by these variables should be considered.

If something is known about the relative number of animals within each stratum, then an approximate rule of thumb is to allocate effort proportional to abundance to achieve the best overall precision. It may also be necessary to do this to ensure sufficient replication per stratum (see below). However, one needs to keep in mind that pooling robustness (see Sect. 6.5.3) does not apply when sampling intensity differs by stratum and thus care needs to be taken to ensure sufficient observations are obtained, even in the low-density strata, so that a separate detection function can be fit for each stratum. Hence a safer (but less efficient) strategy is to stick to equal sampling intensity in each stratum, i.e., allocating effort proportional to sample area. See Buckland et al. (2001: pp. 246–248) and Cochran (1977: pp. 96–98) for a description of optimal effort allocation between strata. Thus precision can be improved by allocating more survey effort to those strata that have more animals, which becomes challenging for multispecies surveys. However, the distribution of effort by stratum is not as important as the total line length *L*, and clearly the more overall effort, the greater the overall improvement in precision.

Stratification may also be used for logistical reasons or because density estimates are required for certain subregions of the survey area for management purposes (e.g., inside a national park versus outside). In the former case, for example, the study area may be stratified according to ease of access, and less effort may be allocated to hard-to-access strata for reasons of cost efficiency. This may lead to some loss in precision, but logistics or the need for estimates by predefined survey units may require such stratification nonetheless. If the stratification occurs for other than logistical reasons and if nothing is known about density in each of the strata, then effort should be allocated in proportion to stratum size.

6.4.1.2 Replication

As mentioned previously, at least 15–20 replicate lines per stratum, preferably 25 or more, are needed to ensure a representative sample and to get a reliable estimate of variance in observed sample size. This consideration usually limits the number of strata that can be defined, as each needs an adequate number of transects. In general, a design that has a larger number of shorter lines is preferable to one with a smaller number of longer lines when it comes to obtaining a representative sample and also a more precise estimate of the variance of the density and abundance estimates. Transect length may be dictated by the size and shape of the survey area. Lines might cross the entire survey area (e.g., Fig. 6.6a) or may be shorter lines such that several or a single transect can be covered in a day with time to move to the next transect before setting up camp (e.g., Fig. 6.6b). Logistics might dictate repeated sampling along the same transect lines during a given survey period (e.g., if the cost of adding additional transects is prohibitively expensive compared to repeated visits to the same transect). Repeated visits to the same transects, however, do not constitute replication, but they help accrue sufficient effort to improve precision and obtain a sufficient number of observations to reliably fit the detection function.

6.4.1.3 Line Transect Orientation and Spatial Distribution

If strata are defined or if there is simply a single stratum corresponding to the entire survey region, then to further improve precision one should orient transect lines parallel to any gradients of density within each stratum. In this way, variation in encounter rate is maximized within transects and minimized between them. So, for

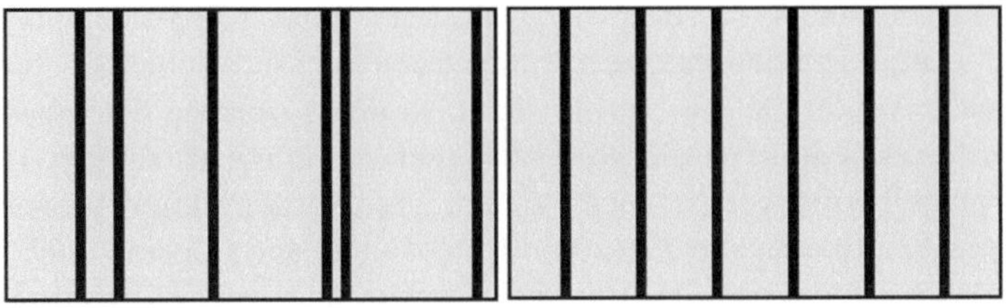

Fig. 6.5 A line transect survey comprises a series of lines. Illustrated here within a simple rectangular survey region are examples of two commonly used designs: randomly spaced parallel lines (*left*) and systematically spaced parallel lines with a random start (*right*)

example, if one suspects that density decreases with increasing distance from the edge of a habitat, a topographic feature such as a river, or a human modification to the landscape such as a road, then transects would be placed approximately perpendicular to the edge of the habitat type, river, or road.

For each survey stratum, not only the orientation of the lines, their number, and length but also their relative location to one another has implications for the precision of the density estimate. Survey designs that locate line transects systematically with a random start give a more even spatial spread of survey effort over the survey region than their nonsystematic counterparts, where each transect line is randomly located (see Fig. 6.5 the design on the left versus the right). A systematic survey design whose line transects have a more even spatial distribution across the survey area is more robust and likely to lead to less variation in the estimates of density or abundance, as it is less susceptible to variations in population behavior (Strindberg 2001; Strindberg and Buckland 2004). In other words, transect lines that have a more even spatial distribution tend to improve precision, as they ensure that a more representative sample is selected from the population. Advances in estimating the variance of encounter rate for all types of designs, including systematic designs, make it possible to exploit the greater precision these tend to afford (Fewster et al. 2009). Aside from the issue of precision, systematic designs are also more efficient and consistent (i.e., more expedient in terms of survey logistics). These are described in detail by Strindberg et al. (2004).

6.4.1.4 Randomization

Accuracy and precision of density estimates, as well as the efficiency achieved, are determined to a large extent by the survey design that dictates how the sample is collected. To obtain estimates of animal density based on valid statistical inference, the observations should be obtained by means of a probability sample. This requires that the line transects be located randomly. All survey design options discussed above presuppose a randomized sampling design. This design process is facilitated by the development of automated design algorithms that randomly superimpose line transects on the survey region of interest (Strindberg 2001; Strindberg et al. 2004).

By automating the survey design process, it is also possible to contrast designs with regard to properties such as the spatial distribution of sampling locations within the survey region, the distances covered by observers to obtain the sample data, and the probability of a particular location being included in the sample (coverage probability).

The Distance software has an automated survey design component and geographic information system (GIS) functionality that can be used for the design of distance sampling surveys (Thomas et al. 2010). In order to design a survey using this software, one has to define the survey area in a spatially explicit manner by means of an ESRI shapefile. This component permits the selection of a design from among a number of different possibilities and the exploration of the design properties given the logistical constraints for the survey in question. A number of frequently used line transect designs, both systematic and nonsystematic, with discrete or continuous transect lines, have been implemented within the automated survey design component of the Distance software. This component also provides a simulation option to check whether or not even coverage probability is achieved by a given design, which avoids potential biases when using a standard analysis. The designs shown in Figs. 6.6 and 6.7 are examples of survey plans that can be produced using this feature of Distance. In addition, the simulation feature (Marshall 2016) available from Distance version 7 onward allows one to define a study area and provide potential characteristics of the wildlife population and a possible survey design to compare the efficiency of a variety of designs under different scenarios.

6.4.2 Reducing Variance in Detection Probability Estimation

As for encounter rate, stratification (or post-stratification) can be used to improve the precision of p. For a given sample size, the best way to reduce the variance of p is to have a detection function with a wide shoulder. If detection changes by habitat type, observer, environmental conditions, etc., then estimating the detection function separately by strata defined by these variables should decrease its variance. However, as adequate sample sizes are required for reliable estimation, the number of strata should be such that enough observations occur in each stratum.

An efficient way to improving the precision of p when it is not possible to obtain sufficient sample sizes for stratified estimation of detection is through MCDS analysis. Variables (e.g., habitat type, season, observer, group size, or environmental conditions) are incorporated as covariates when fitting the model for the detection function (Marques and Buckland 2004a, b; Marques et al. 2007; Buckland et al. 2015). The methods assume that these types of covariates influence the scale of the detection function, but not its shape. Thus animal groups at the same distance from the transect line can have different probabilities of detection depending on their associated covariate values. Both the conventional distance sampling (CDS) and MCDS analysis options are implemented within the Distance software.

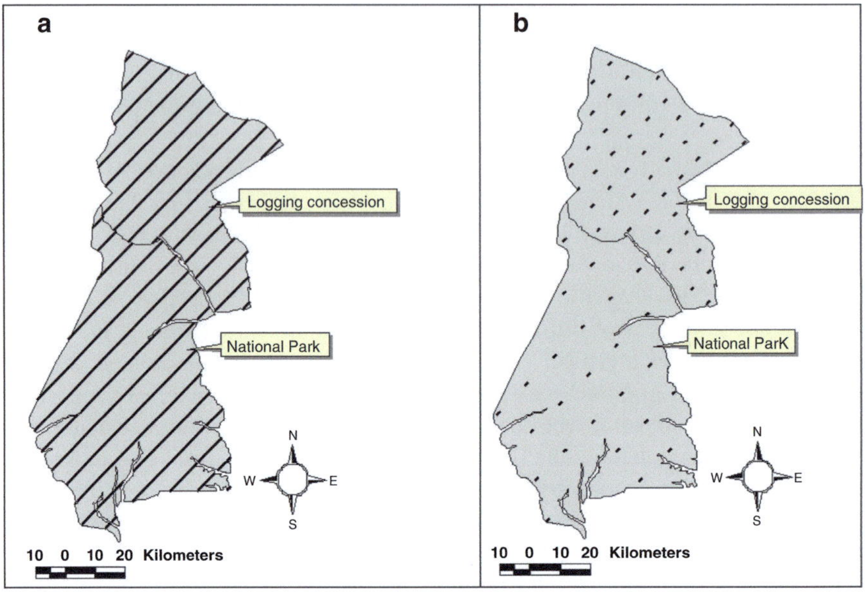

Fig. 6.6 Consider a survey region that comprises two strata, namely, a national park and a logging concession, where the latter is more heavily impacted by human activity. Transect lines are oriented in a northeast-southwest direction as this is suspected to coincide with the gradient in animal density. (**a**) If the habitat type were open and suitable for an aerial survey, then a design with systematic parallel lines located with a random start in each stratum might be appropriate. Note that the spacing between sequential line transects is 6.5 km in the national park but 6 km in the logging concession to ensure sufficient replication (15 lines in each stratum). (**b**) In the case of a more closed forest habitat type, a terrestrial survey would be a likely option. In both cases, the amount of time it takes to cover a kilometer of transect line and to move between transects is an important design consideration, as total survey time is always limited by cost and other logistic constraints. In closed habitat types, it is often difficult to cut transects and to move along them in a straight line, whereas ground can be covered much more quickly when moving between transects, during which time animal paths or existing roads or rivers can be used. Hence, a design comprising 1–2-km-long line transects systematically spaced with a random start with a larger spacing between transects could be a good design option. In this example, the 25 and 47 transect lines are 1 km long and have a systematic spacing of 12 km and 7 km in the national park and logging concession, respectively. Given the roads in the logging concession, it is easier to access this stratum and thus it is more cost-efficient to allocate more sampling effort to this stratum

6.4.3 Reducing Variance in Expected Group Size

An option for reducing the variance in estimated expected group size, $\widehat{E}(s)$, if group sizes change seasonally, is to survey when group sizes are smaller. This facilitates group size estimation and also increases encounter rates (if there is interest in obtaining seasonal estimates of density, however, then one would have to survey during the various seasons regardless of expected group sizes). Additionally, if sample sizes are adequate, then it may be possible to post-stratify by group size during analysis to improve precision.

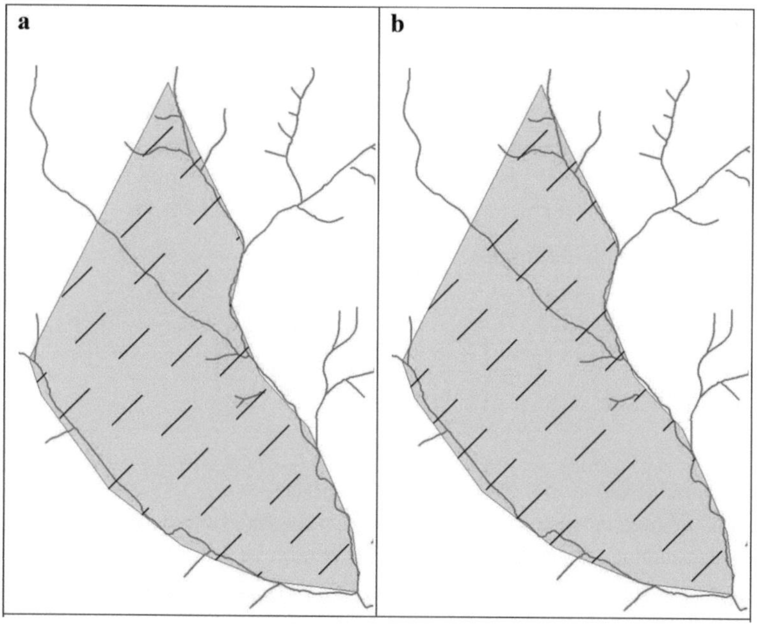

Fig. 6.7 Consider a small survey region that has large rivers along its boundaries and running through its interior. The survey region also has a ridge running in NW-SE direction through its central part. Due to the small size of the survey area, a segmented transect design was deemed most appropriate to ensure that a sufficient number of replicate transect lines could be generated within the survey region. Segments of length 1.5 km at a spacing of 1.5 km were generated according to a systematic survey design with a random start within the survey allowing for incomplete segments where the segments intersected the boundary. Two types of designs were considered: (**a**) systematic segmented trackline sampling and (**b**) systematic segmented grid sampling. The first design was selected for implementation in the field, as it is likely to provide a more representative sample. The second design tends to place a majority of transects either along a river or along the ridge where the vegetation and potentially also species density is different to what is found in the remainder of the survey region. Both designs provide approximately even coverage probability, as they allow for shorter line transects along the region boundary

6.4.4 Precision and Bias Versus Available Resources

Generally, when designing a survey, a balance has to be struck between the precision of the density estimate and the resources available for the survey in terms of time and money. This trade-off between desired precision and the cost of implementing the survey usually dictates the survey effort and design used in sampling a particular study area. A pilot survey is the best way to estimate the amount of survey effort required to achieve a desired precision. During a pilot survey, one should avoid clustering the lines or choosing the most convenient locations for their placement. The lines should be spread throughout the study area and cover different habitats or areas under different degrees of human influence to ensure that the results are representative and able to correctly inform planning of the main survey. The time and cost constraints associated with a particular type of survey in a given study area will usually dictate whether the desired precision is feasible and which survey design is most suitable for the given circumstance.

As described in the introduction, the CV is a useful unit-less metric that can be used to compare different studies. For a distance sampling survey of tiger prey, where detection on the line is certain, one can estimate the total length of transect line required for a given encounter rate $\frac{C_0}{L_0}$ and a target CV for the density estimate $CV_t\left(\widehat{D}\right)$ by applying the formula

$$L = \left(\frac{b + \left[\widehat{SD}(s)/\bar{s}\right]^2}{\left[CV_t\left(\widehat{D}\right)\right]^2} \right) \left(\frac{L_0}{C_0} \right) \tag{6.7}$$

The standard deviation of group size is $SD(s) = \sqrt{\sum_{i=1}^{C'} (s_i - \bar{s})^2 / (n-1)}$, where $\bar{s}$ is the mean group size and s_i the size of the ith group, which assumes group size is independent of detection distance. The parameter b is known as the dispersion parameter or variance inflation factor and is approximated by $\frac{\text{var}(C)}{C} + C\frac{\text{var}\left(\widehat{f}(0)\right)}{\left[\widehat{f}(0)\right]^2}$. The dispersion parameter generally takes a value in the range 1.5–3. It would take on its smallest value if the spatial distribution of the animals were random, as then one would expect the count on each line to approximately follow a Poisson distribution (i.e., $\widehat{\text{var}}(C) \cong C$). If the population is highly aggregated, then b takes on larger values. Although, there is site-specific variation, estimated values of b from previous surveys indicate that tiger prey species such as gaur, sambar, muntjac, and wild pig tend to be less patchily distributed compared to chital, for example. Thus, to avoid underestimating L for planning purposes, it is suggested that one use a value of at least 3.5 for gaur, sambar, muntjac, and wild pig and double that value for chital (assuming it is not possible to estimate b from a pilot study or use a value calculated previously from a similar study). Ideally, a pilot study would be carried out to estimate the encounter rates to be expected during the actual survey and the mean and standard deviation of group size. These values can then be plugged into the above equations to estimate the amount of effort required to achieve the desired precision. If the pilot study is more comprehensive and distances are measured, then the dispersion parameter b can be approximated by $C_0 \cdot \left\{CV\left(\widehat{D}_0\right)\right\}^2$, where C_0 is the number of animal groups counted during the pilot survey and D_0 the corresponding estimated density.

If the available resources determine the total effort in terms of line length, L, then it is possible to estimate $CV\left(\widehat{D}\right)$ using the formula

$$CV\left(\widehat{D}\right) = \sqrt{\frac{\left(b + [SD(s)/\bar{s}]^2\right) L_0}{LC_0}} \tag{6.8}$$

If $CV\left(\widehat{D}\right)$ is too large, then it may not be worthwhile conducting the survey, if a certain precision is required. Similarly, we can calculate the amount of effort, L, required to achieve our desired $CV_t\left(\widehat{D}\right)$ and possibly conclude that we do not have the resources; then it is necessary to decide whether a reduction in precision is feasible given the goals of the survey. All of these equations assume that the lines are distributed according to a random design. Additionally, if detection on the line is not certain and $g(0)$ needs to be estimated, then greater effort is required to achieve a target precision (equivalently the same amount of effort will give lower precision). For more detailed explanations and example calculations, see Buckland et al. (2001: pp. 241–244).

For animals that live at low density but are also aggregated into groups, the number of groups observed may be small, even if a great deal of sampling effort is invested in the survey. This often results in imprecise estimates that are potentially biased given the unreliability associated with fitting the detection function. For a set total survey effort L and a given encounter rate from a pilot survey of C_0/L_0, the resulting sample size can be estimated as $C = L \times C_0/L_0$, which should ideally be 60–80 animal groups. If this target sample size is not attained, then other options that can be considered include pooling data across species, surveys, or sites, as described previously (Sect. 6.3.5). Adaptive sampling is one way to increase the sample size, thus also increasing precision and reducing bias (Thompson and Seber 1996). An inherent problem with adaptive sampling is that usually the total survey effort required to complete the survey is unknown in advance, which can create logistical problems. However, an adaptive line transect sampling method that allows the amount of effort (in terms of survey time) to be fixed in advance has been developed (Pollard et al. 2002; Pollard and Buckland 2004). The details of adaptive sampling are beyond the scope of this chapter.

6.5 Data Analysis Using the Available Software Tools

Here, we consider the steps one might follow during an analysis of distance sampling data, including data entry and validation, data exploration, model fitting and selection, final analysis, and inference. Again, Buckland et al. (2001, 2015) should be consulted for more detailed information. Fortunately, there are several software tools available to support analysis and these are introduced. We focus on modeling the detection function, which lies at the heart of distance sampling. We also introduce spatial modeling based on distance sampling data.

6.5.1 Data Entry and Validation

If data are recorded on paper forms (see Chap. 7 for an example), they should be stored electronically as soon as possible (e.g., after each day of surveying) – entered into a spreadsheet, such as MS Excel, and validated to ensure that transcription errors are corrected and feasible data values are entered. If data were entered electronically in the field, e.g., using CyberTracker (cybertracker.org), SMART-ER (smartconservationsoftware.org), or another data entry application on some type of handheld device, then some reformatting may be required before the data are ready for analysis, but validation can occur as the data are collected. It is extremely important that data validation occurs as soon as possible, as it becomes harder or impossible to sort out data errors later during analysis. It is also vital that all field data are backed up, by copying the paper forms or electronic files and storing them at a different site.

6.5.2 Software for the Analysis of Distance Sampling Data

Distance is the custom windows-based computer package for the design and analysis of distance sampling surveys of wildlife populations (Thomas et al. 2010), which has evolved substantially from the program Transect and early DOS versions of the software (Laake et al. 1993). The software comes with context-sensitive online help and a comprehensive user's guide, so we only give a brief overview in this section.

Distance projects are made up of a project file with a "dst" file extension and an associated data folder with a "dat" suffix. Internal data are stored in DistData.mdb file within the data folder. The data folder is also used by default to store the GIS information (ESRI shape files) for those projects using the *Maps*, *Designs,* and *Surveys* features to create maps, explore design options for the study area, and eventually generate a survey plan. Many of the advanced features in Distance are implemented as R libraries that are called by Distance. The freely available R statistical software (www.r-project.org) needs to be installed to conduct mark-recapture distance sampling (MRDS) analysis, density surface modeling (DSM), and multi-analysis (MA) or use the simulation features available in Distance. An R folder is created automatically within the Distance data folder the first time the *mrds*, *dsm*, *mads*, or *DSsim* libraries are called. It contains the R object file (.RData) and image files generated by the R statistical software package. It is worth noting that only a subset of features available in these R libraries can be accessed via the Distance's graphical user interface and many more options are available if the libraries are called directly in the R software.

Besides creating a new project, it is possible to import data or project files from earlier versions of the software. If the survey data are stored in a package such as MS Excel, then it is relatively straightforward to import the data into Distance from such software by saving the data in a predefined order as a tab delimited text file. This is much easier and less error prone than typing the data into Distance. The data should be imported in their disaggregated form, i.e., if distance data were collected ungrouped, then they should not be grouped even if they are subsequently analyzed that way. Data should be entered taking into account stratification and sampling units (usually lines). Data can also be linked from external databases in a variety of formats. It is possible to export Distance projects to zip archive files, facilitating transfer to other computers or users, and to open projects directly from the archive file. The most current version of the software can be downloaded at no cost from the Distance website (www.ruwpa.st-and.ac.uk/distance).

To allow for analyses that use the entire power of R, an R package *distance* has been written to facilitate straightforward distance sampling analysis in R (Miller 2013). An alternative package, *unmarked*, developed by a different group (Fiske and Chandler 2011), provides a common hierarchical modeling framework not only for analyses of distance sampling data but also analyses of other types of data (e.g., occupancy) from unidentified individuals.

When advanced analyses are conducted for which there is no standard software package or library available, it is sometimes easier to use Bayesian methods. Packages such as JAGS (http://mcmc-jags.sourceforge.net/), WinBUGS (www.mrc-bsu.cam.ac.uk/bugs/winbugs/contents.shtml), or the more recent version OpenBUGS (www.openbugs.info/w/) provide users with the flexibility to conduct a multitude of analyses and can also be used in the form of the *R2WinBUGS* library (Sturtz et al. 2005; Kéry and Royle 2015).

Although there is no single approach to completing an analysis of distance sampling or associated data, it should start with a good deal of data exploration before model fitting and selection take place and the final estimates are obtained and inferences are drawn.

6.5.3 Modeling the Detection Function

6.5.3.1 Exploratory Phase

Histograms of the data should be plotted under several different groupings either using the Distance software or any other package that has graphing facilities. It can even be done by hand using pen and paper, and this may be particularly beneficial for novices to distance analyses. They can then gauge whether there are any problems with the data and estimate what the probability detection function might look like by eye (see Fig. 6.3a). Results obtained by means of this simple analysis could be compared to the results produced by Distance.

An initial analysis with many (10–20) cut points and a simple model (e.g., half-normal) should be carried out. During this phase it's best to not try to estimate density, but simply look at the histograms and quantile–quantile (qq) plots to see whether assumptions have been violated and there are problems with the data, such as heaping, a spike at zero distance, evasive movement, or outliers.

The problems caused by heaping can potentially be reduced with appropriate grouping (note that goodness-of-fit tests are sensitive to heaping and data should be appropriately grouped when performing these tests). It will be difficult to obtain reliable density estimates if evasive movement has occurred. Outliers that are caused by incorrect data entry should be corrected if possible during data validation. Then truncation can be used to eliminate the remaining outliers. Right truncation is generally recommended for robust estimation of the detection function. For line transects a rule of thumb is to truncate when $g(x) \approx 0.15$ (truncating approximately the largest 5% of distances usually works too).

One needs to decide whether to analyze data as grouped or ungrouped. If the exploratory phase of the analysis shows signs of rounding at convenient values, then the former option should be selected with cut points defined in such a manner that rounding distances lie approximately at the midpoints of intervals; this way observations will tend to fall within the correct distance interval. For aggregated populations, check for evidence of size bias and apply one of the methods described previously to take care of this problem if necessary.

6.5.3.2 Select a Model for the Detection Function

The detection function $g(x)$ gives the probability that an animal at distance x is detected from the line. To estimate density a distance sampling analysis relies on fitting a model of $g(x)$ to the observed distances (where x corresponds to the perpendicular distances $x_1, \ldots, x_C$ recorded during the line transect survey or calculated from the radial distances and angles or angles of declination and height), which allows one to estimate the proportion of animal groups within surveyed strips that are detected and counted $\widehat{p}$. Once a sufficient number of animal groups are detected to allow reliable modeling of the detection function, a precise and unbiased estimate of animal density relies on the selection of an appropriate model for $g(x)$. Such models have certain desirable properties, namely (in order of importance), model robustness, a shape criterion, and efficiency.

Model Robustness A robust model needs to be general and flexible so that it can fit a variety of shapes for the detection function. Models should also be pooling robust, which means that the data can be combined (pooled) over different factors that affect detectability (habitat, observer, weather, etc.) and still provide a reliable density estimate. In other words, the density estimate produced by stratifying the data by habitat, observer, weather, etc., should be approximately the same as the estimate produced from the combined data.

Shape Criterion As mentioned previously, the potential for detecting animals should not drop off abruptly at a short distance from the line transect. In other words, $g(x)$ should have a "shoulder" near the line transect (mathematically this means that the derivative $g'(0)$ should be zero). Given this property, spiked functions near zero are excluded from consideration. It is worth noting that histograms of the detection distances often do not reveal the presence or absence of a shoulder, especially if histogram groupings are large.

Efficiency An efficient model is one that has a small variance. Maximum likelihood methods are used as they ensure a minimum variance asymptotically, i.e., as the sample size increases. This characteristic is only useful if the model is robust and when the shape criterion can be met. Otherwise, you may get a very precise estimate that is wrong!

Models of the form $g(x) = \text{key}(x) \times [1 + \text{series}(x)]$, as implemented in the Distance software, have these characteristics, where key(x) is a key function and series(x) is a series expansion. The modeling process involves two steps:

Selection of a key function based on the histogram data (after truncation)
Adjustment of the key function by means of the series expansion

Key functions include the uniform, half-normal, and hazard-rate functions. Series expansions include the cosine series, simple polynomials, and Hermite polynomials. Sometimes a key function without a series adjustment is sufficient.

Model Fit For analyses on exact distances, the Cramér-von Mises and Kolmogorov-Smirnoff goodness-of-fit (gof) tests available in Distance can be used to compare the detection function model to the actual data. In addition, quantile–quantile (qq) plots provided by the Distance software are a graphical means for identifying problems with the data, e.g., rounding to preferred values or systematic departures from the fitted model. These tests avoid arbitrarily grouping exact data into intervals, unlike the χ^2 gof test that can be used to test the fit of the $g(x)$ model to the distance data. The test is based on the grouping of the distance data and compares the observed frequencies C_i (dependent on the groupings selected) to the expected frequencies $\widehat{E}\,(C_i)$ under the model in the usual way $\chi^2 = \sum_{i=1}^{u} \frac{\left[C_i - \widehat{E}(C_i)\right]^2}{\widehat{E}(C_i)}$, which is approximately χ^2 distributed with $u - q - 1$ degrees of freedom if the fitted model is a good approximating model (where u is the number of groups and q is the number of parameters estimated). A defect of the χ^2 gof test is that it has difficulty discriminating between different models at the most critical region near $x = 0$, unless given enough data, and the results are very dependent on the groupings selected. The gof test leads to overfitting, and its power is low, too, and should not be relied on when selecting a model for $g(x)$. The test is useful for highlighting problems with the data.

Model Selection In general, as the number of parameters in the model increases, the bias decreases, but the sampling variance increases. Hence, the number of parameters selected needs to be a compromise between bias and variance. Model selection should only take place once the data have been adequately truncated and various data groupings considered. The fit of the model to the distance data near the line is extremely important (except in the case of heaping at zero). Akaike's information criterion (AIC) is generally used in model selection.

The AIC attempts to find a balance between the number of model parameters q and the model fit and in this way provides a trade-off between variance and bias (more parameters improve model fit and reduce bias, but the cost is an increase in model complexity and variance). For a given data set, the model with the smallest AIC is selected. The AIC is given by $AIC = -2 \cdot \log_e(\mathscr{L}) + 2q$, where $\log_e(\mathscr{L})$ is the log-likelihood function evaluated at the maximum likelihood estimates of the model parameters.

To select a model, fit a small number of key/series adjustment combinations (e.g., uniform + cosine or simple polynomial, half-normal + cosine or Hermite polynomial, and hazard-rate + cosine or simple polynomial). Consider whether detectability might be influenced by group size, observer ability, habitat, or other factors. In that case fit models using MCDS methods checking that the covariates of interest only influence the scale of the detection function. If the covariate influences the shape of the detection function, then if sample size allows one can stratify the detection function by the factor of interest. If the same right-truncation distance or intervals are used, then all of the models can be ranked and compared in terms of their AIC value. Look at the histograms, goodness-of-fit test results, AIC, and summary tables to choose a model. The most important thing to consider is goodness-of-fit test results close to $x = 0$. To improve fit it may be necessary to revert to the exploratory phase. Occasionally it will be difficult to select between models that fit the data well and have similar AIC values (difference between them less than 1). In this case a solution is to resort to multi-model inference. It is worth noting that, for "good" data, with no visible violation of assumptions and a wide shoulder, different detection functions tend to give almost identical results – this is a good situation to be in as it indicates that results are robust. As is more generally the case for any type of model selection, AIC should not replace common sense. For example, in order to reduce the effect of a responsive movement issue, a detection function model might be used that is not the highest ranked model according to AIC (e.g., Williams and Thomas 2007). For a detailed treatment of model selection and multi-model inference, see Buckland et al. (2001), Burnham and Anderson (2002), or Buckland et al. (1997).

6.5.3.3 Final Analysis and Inference

Once a model has been selected for the distance data, then consider (i) options for variance estimation (e.g., bootstrapping to estimate the variance of the estimate), (ii) stratifying some or all of the components of estimation, and (iii) inclusion of covariates in the analysis. Fit the data using the favored model or models and selected options to obtain the estimate of density or abundance. Finally, extract summaries from the analysis and histograms for reporting.

6.5.4 Spatial Modeling

If certain variables are thought to influence density and distribution (based on ecological hypotheses), then a model that incorporates these variables can be fitted to the distance sampling data. Such covariate modeling will also potentially improve precision (Hedley and Buckland 2004; Hedley et al. 2004). The variables can either be collected during the survey as ancillary data, or they can be obtained from other sources, e.g., from a GIS or from other spatially explicit data sources for the study area. These spatially explicit models allow one to investigate factors influencing animal abundance (habitat type, other environmental variables, distance to human settlements or roads, etc.), to extrapolate over the entire area of interest, and to calculate animal abundance for subregions in the study area.

Spatial modeling does not necessarily require data collected from randomly located transects, as there is no need to rely on design-based inference to extrapolate from the surveyed area to the larger area. However, it is preferable to employ a randomized sampling scheme, as both the standard and entirely model-based analysis options are then available. This is a far less risky strategy, as it does not restrict one's options to finding a good approximating model that fits the data well. Another important point here is that a good design produces samples evenly spread through the area – and so forms an ideal start point for a model, where data are available from throughout the spatial region being modeled.

The spatial modeling methods introduced by Hedley and Buckland (2004) are available in the Distance software and in the form of the *dsm* R library, which takes specifically formatted distance sampling data and predicts the spatial distribution of animals in the survey region. The details of density surface modeling (DSM) are provided by Miller et al. (2013). This is one of the several possible two-stage approaches where detectability is first estimated and then subsequently incorporated into a count model using different error distributions (e.g., Poisson or negative binomial) or modeling approaches, such as generalized linear models (McCullagh and Nelder 1989) or generalized additive modeling (Wood 2006). Single-stage approaches move away from counts to spatial point processes that permit the assignment of variable values at the individual level while avoiding the somewhat ad hoc decision of how to split transects into smaller segments for analysis with associated variable values that may only coarsely represent the characteristic across the entire segment (e.g., Johnson et al. 2010).

Further details on one- and two-stage approaches to spatial modeling are given by Buckland et al. (2015) or Miller et al. (2013), for example, a common framework for model-based distance sampling is described by Buckland et al. (2016), which covers nonspatial model-based approaches (e.g., Borchers et al. 2002; Royle and Dorazio 2008; Kéry and Royle 2015), model-based MCDS, model-based MRDS, plot-based models (e.g., Royle et al. 2004), as well as the addition of random effects (Oedekoven et al. 2014, 2015).

If we have the resources to implement a designed experiment rather than simply conduct an observational study (Royle et al. 2004; Buckland et al. 2009; Oedekoven et al. 2013, 2014), then spatial modeling can be used to investigate different treatment effects (e.g., these could correspond to different land management regimes or conservation actions).

6.6 Discussion

Line transect sampling is an efficient and cost-effective method for obtaining estimates of density and abundance for large populations of tiger prey species sparsely distributed over large geographic areas. The method requires that detectability of animal groups decreases with increasing distance from the transect line and that the perpendicular distance can be obtained for each group of animals. The method works particularly well for prey populations at moderate density where populations occur in well-defined groups. Habitat should be sufficiently open to ensure that animal groups on or close to the line can be seen. Aerial surveys should ideally be conducted in habitat with little forest or other cover that might obscure animals from the air; otherwise more costly data collection and analysis options need to be considered to estimate the proportion of animals missed due to availability bias. Foot-based surveys remain the standard for areas that are moderately forested.

In addition, new and novel ways are continually being developed to fit spatial models to distance sampling data to investigate the key drivers of species distribution and other questions of interest. The available options for experimental design using distance sampling also provide tools for the assessment of new or different conservation actions or management regimes.

Standard distance sampling requires a random survey design with a sufficient number of transect lines and total effort to ensure that animal locations are independent of the transect lines and a representative sample is obtained and a sample size of observations large enough to fit a detection function that will yield unbiased estimates. In practice (Chap. 7), systematic placement of transects, but with a random start, works well. With survey design and simulation tools available, it should be possible to determine whether it is feasible to meet these criteria while considering the logistical constraints and field practicalities for a given study area or population of interest.

The remaining assumptions underlying distance sampling may be more challenging to meet, even with rigorous observer training, a good field protocol, and instruments to take accurate and precise measurements. In some cases it is possible to collect additional data to deal with assumption failures. The next chapter will provide details on how best to meet these assumptions to avoid the additional costs and complexity of the associated analyses, where possible.

References

Alldredge MW, Simons TR, Pollock KH (2007a) An experimental evaluation of distance measurement error in avian point count surveys. J Wildl Manag 71:2759–2766

Alldredge MW, Pollock KH, Simons TR (2007b) Multiple species analysis of point count data: a more parsimonious modeling framework. J Appl Ecol 44:281–290

Borchers DL, Langrock R (2015) Double-observer line transect surveys with Markov-modulated Poisson process models for animal availability. Biometrics 71:1060–1069

Borchers DL, Buckland ST, Zucchini W (2002) Estimating animal abundance: closed populations. Springer, London

Borchers DL, Laake JL, Southwell C, Paxton CGM (2006) Accommodating unmodeled heterogeneity in double-observer distance sampling surveys. Biometrics 62:372–378

Borchers DL, Marques TA, Gunnlaugsson T, Jupp P (2010) Estimating distance sampling detection functions when distances are measured with errors. JABES 15:346–361

Borchers DL, Zucchini W, Heide-Jorgensen MP, Canadas A, Langrock R (2013) Using hidden markov models to deal with availability bias on line transect surveys. Biometrics 69:703–713

Buckland ST, Anderson DR, Burnham KP, Laake JL (1993) Distance sampling: estimating abundance of biological populations. Chapman and Hall, London

Buckland ST, Burnham KP, Augustin NH (1997) Model selection: an integral part of inference. Biometrics 53:603–618

Buckland ST, Anderson DR, Burnham KP, Laake JL, Borchers DL, Thomas L (2001) Introduction to distance sampling. Oxford University Press, London

Buckland ST, Anderson DR, Burnham KP, Laake JL, Borchers DL, Thomas L (eds) (2004) Advanced distance sampling. Oxford University Press, Oxford
Buckland ST, Russell RE, Dickson BG, Saab VA, Gorman DG, Block WM (2009) Analysing designed experiments in distance sampling. JABES 14:432–442
Buckland ST, Rexstad EA, Marques TA, Oedekoven CS (2015) Distance sampling: methods and applications. Methods in statistical ecology. Springer, Cham
Buckland ST, Oedekoven CS, Borchers DL (2016) Model-based distance sampling. JABES 21:58–75
Burnham KP, Anderson DR (2002) Model selection and multimodel inference: a practical information-theoretic approach, 2nd edn. Springer, New York
Burnham KP, Anderson DR, Laake JL (1980) Estimation of density from line transect sampling of biological populations. Wildl Monogr 72:1–202
Burnham, KP, Anderson, DR, White, GC, Brownie, C, Pollock, KH (1987) Design and analysis methods for fish survival experiments based on release-recapture. American Fisheries Society, USA
Burt M, Borchers D, Jenkins K, Marques T (2015) Using mark-recapture distance sampling methods on line transect surveys. Methods Ecol Evol 5:1180–1191
Cochran WG (1977) Sampling techniques, 3rd edn. Wiley, Singapore
Durant SM, Craft ME, Hilborn R, Bashir S, Hando J, Thomas L (2011) Long-term trends in carnivore abundance using distance sampling in Serengeti National Park, Tanzania. J Appl Ecol 48:1490–1500
Fewster RM, Pople AR (2008) A comparison of mark–recapture distance-sampling methods applied to aerial surveys of eastern grey kangaroos. Wildl Res 35:320–330
Fewster RM, Buckland ST, Burnham KP, Borchers DL, Jupp PE, Laake JL, Thomas L (2009) Estimating the encounter rate variance in distance sampling. Biometrics 65:225–236
Fiske IJ, Chandler RB (2011) unmarked: an R package for fitting hierarchical models of wildlife occurrence and abundance. J Stat Softw 43:1–23
Gerodette T, Forcada J (2005) Non-recovery of two spotted and spinner dolphin populations in the eastern tropical Pacific Ocean. Mar Ecol Prog Ser 291:1–21
Hedley SL, Buckland ST (2004) Spatial models for line transect sampling. JABES 9:181–199
Hedley SL, Buckland ST, Borchers DL (2004) Spatial distance sampling models. In: Buckland ST, Anderson DR, Burnham KP, Laake JL, Borchers DL, Thomas L (eds) Advanced distance sampling. Oxford University Press, Oxford, pp 48–70
Johnson D, Laake JL, VerHoef J (2010) A model-based approach for making ecological inference from distance sampling data. Biometrics 66:310–318
Karanth KU (1988) Population structure, density and biomass of large herbivores in south Indian tropical forest. MS thesis, University of Florida
Kéry M, Royle J (2015) Applied hierarchical modeling in ecology: analysis of distribution, abundance and species richness in R and BUGS. Volume 1: Prelude and static models, 1st edn. Academic Press, Amsterdam
Laake JL, Borchers DL (2004) Methods for incomplete detection at distance zero. In: Buckland ST, Anderson DR, Burnham KP, Laake JL, Borchers DL, Thomas L (eds) Advanced distance sampling. Oxford University Press, Oxford, pp 108–189
Laake JL, Buckland ST, Anderson DR, Burnham KP (1993) DISTANCE User's Guide V2.0. Colorado Cooperative Fish and Wildlife Unit, Colorado State University, Fort Collins
Marques FFC, Buckland ST (2004a) Incorporating covariates into standard line transect analysis. Biometrics 59:924–935
Marques FFC, Buckland ST (2004b) Covariate models for the detection function. In: Buckland ST, Anderson DR, Burnham KP, Laake JL, Borchers DL, Thomas L (eds) Advanced distance sampling. Oxford University Press, Oxford, pp 31–47
Marques TA, Thomas L, Fancy SG, Buckland ST (2007) Improving estimates of bird density using multiple covariate distance sampling. Auk 127:1229–1243
Marshall L (2016) DSsim: distance sampling simulations. R package version 1.0.8
McCullagh P, Nelder JA (1989) Generalized linear models. Chapman & Hall/CRC, Boca Raton

Miller DL (2013) Distance: a simple way to fit detection functions to distance sampling data and calculate abundance/density for biological populations. R package version 0.7.3

Miller DL, Burt ML, Rexstad EA, Thomas L (2013) Spatial models for distance sampling data: recent developments and future directions. Methods Ecol Evol 4:1001–1010

Oedekoven CS, Buckland ST, Mackenzie ML, Evans KO, Burger LW (2013) Improving distance sampling: accounting for covariates and non-independency between sampled sites. J Appl Ecol 50:786–793

Oedekoven CS, Buckland ST, Mackenzie ML, King R, Evans KO, Burger LW (2014) Bayesian methods for hierarchical distance sampling models. JABES 19:219–239

Oedekoven CS, Laake JL, Skaug HL (2015) Distance sampling with a random scale detection function. Environ Ecol Stat 22:725–737

Oehlert GW (1992) A note on the delta method. Am Stat 46:27–29

Pollard JH, Buckland ST (2004) Adaptive distance sampling surveys. In: Buckland ST, Anderson DR, Burnham KP, Laake JL, Borchers DL, Thomas L (eds) Advanced distance sampling. Oxford University Press, Oxford, pp 229–259

Pollard JH, Palka D, Buckland ST (2002) Adaptive line transect sampling. Biometrics 58:862–870

Powell LA (2007) Approximating variance of demographic parameters using the delta method: a reference for avian biologists. Condor 109:949–954

Royle JA, Dorazio RM (2008) Hierarchical modeling and inference in ecology. Academic Press, London

Royle JA, Dawson DK, Bates S (2004) Modeling abundance effects in distance sampling. Ecology 85:1591–1597

Strindberg S (2001) Optimized automated survey design in wildlife population assessment. Dissertation, University of St Andrews

Strindberg S, Buckland ST (2004) Zigzag survey designs in line transect sampling. JABES 9:443–461

Strindberg S, Buckland ST, Thomas L (2004) Design of distance sampling surveys and Geographic Information Systems. In: Buckland ST, Anderson DR, Burnham KP, Laake JL, Borchers DL, Thomas L (eds) Advanced distance sampling. Oxford University Press, Oxford, pp 190–228

Sturtz S, Ligges U, Gelman A (2005) R2WinBUGS: a package for running WinBUGS from R. J Stat Softw 12:1–16

Thomas L, Buckland ST, Rexstad EA, Laake JL, Strindberg S, Hedley SL, Bishop JRB, Marques TA, Burnham KP (2010) Distance software: design and analysis of distance sampling surveys for estimating population size. J Appl Ecol 47:5–14

Thomas L, Karanth KU (2002) Statistical concepts: estimating absolute densities of prey species using line transect sampling. In: Karanth KU, Nichols JD (eds) Monitoring tigers and their prey: a manual for researchers, managers, and conservationists in tropical Asia. Centre for Wildlife Studies, Bangalore, pp 87–110

Thompson SK (1992) Sampling. Wiley, New York

Thompson SK, Seber GAF (1996) Adaptive sampling. Wiley, New York

Wegge P, Storaas T (2009) Sampling tiger ungulate prey by the distance method: lessons learned in Bardia National Park, Nepal. Anim Conserv 12:78–84

Williams R, Thomas L (2007) Distribution and abundance of marine mammals in the coastal waters of British Columbia, Canada. J Cetacean Res Manag 9:15–28

Wood SN (2006) Generalized additive models: an introduction with R. Chapman & Hall/CRC Press, Boca Raton

Field Practices: Estimating Abundance of Prey Species Using Line Transect Sampling

7

N. Samba Kumar, Abishek Harihar, Hannah J. O'Kelly, and Anak Pattanavibool

7.1 Introduction

Distance sampling using line transects is one of the most frequently used methods to obtain estimates of population density and abundance for tiger prey species. This approach can also be used to investigate factors that influence the spatial distribution of prey across the region of interest, which in turn can further inform management actions. In order to ensure reliable monitoring data are generated, it is important to implement appropriate field practices that adequately reflect the survey design considerations and the methodological framework described in Chap. 6. Ecological, logistical and resource constraints can make it challenging to satisfy the underlying methodological assumptions of distance sampling when planning and actually conducting line transect surveys in the field. This chapter tries to strike

N.S. Kumar (✉)
Wildlife Conservation Society, India Program, Bengaluru, India
e-mail: nrao.skumar@gmail.com

A. Harihar
Tiger Program, Panthera, New York, NY, USA

Nature Conservation Foundation, Mysuru, India
e-mail: aharihar@panthera.org; aharihar@ncf-india.org; https://www.panthera.org/; http://ncf-india.org/

H.J. O'Kelly
Conservation Scientist, Vientiane, Lao People's Democratic Republic
e-mail: okellyhj@gmail.com

A. Pattanavibool
Wildlife Conservation Society (WCS) – Thailand Program, Bangkok, Thailand

Department of Conservation, Kasetsart University, Bangkok, Thailand
e-mail: anakp@wcs.org; https://thailand.wcs.org/; http://conservation.forest.ku.ac.th/

K.U. Karanth, J.D. Nichols (eds.), *Methods For Monitoring Tiger And Prey Populations*, DOI 10.1007/978-981-10-5436-5_7

a balance between fulfilling the theoretical requirements of line transect sampling and also addressing some common practical issues that arise in the field.

7.2 Line Transect Survey Design

7.2.1 Minimum Sample Size Requirements

As with all scientifically robust approaches to animal population estimation, the ability to obtain a sufficient sample size of observations is a critical consideration in the design phase of line transect surveys. If densities of some, or all, prey species are low in the survey area, encounter rates will also be low and the sample size attained (i.e. number of animals or clusters of animals detected) may be small, even with high levels of sampling effort (i.e. many kilometres walked on transects). Small sample sizes will result in poor model fits (Chap. 6) and imprecise density estimates. The minimum sample size required will depend on study objectives and data quality. A practical rule of thumb is about 60–80 observations per prey species (Buckland et al. 2001). There are various analytical procedures available, such as pooling observations across multiple survey periods or borrowing detection functions from other similar studies (see Chap. 6 for details), which can help to deal with the constraints posed by low sample size (also see Chap. 8). Furthermore, new analytical approaches to population estimation are continually being developed, such as combining available data from different sources (Chap. 12).

A practical way of determining what level of line transect walk effort will be required to attain sufficient sample sizes in a specific study is to do a pilot survey. This will consist of test walks on transects during which observations of prey species are recorded to calculate encounter rates (number of animal detections per kilometre walked). Encounter rates from the pilot survey can then be used with the survey design and simulation tools in DISTANCE software (see Chap. 6) to approximately estimate the total effort required. Multiple survey scenarios can be compared to investigate required effort allocation at different levels of resource availability or logistical needs.

Where only some prey species occur at low densities, it may be worth considering if they are really contributing substantially to tiger diet. If they are not, survey design can be optimised primarily keeping principal prey species in mind to determine carrying capacity for tigers.

Line transect surveys remain the recommended approach to reliable density estimation of prey species, at low densities. However, if prey species are at *very* low densities and cannot generate reasonable sample sizes for abundance estimation, then alternative approaches need to be explored. One such alternative may be to monitor changes in habitat occupancy by different prey species by estimating the 'occupancy rates' (see Chaps. 4 and 5). In such a case, the occupancy rate becomes a surrogate measure for relative prey densities, rather than absolute ones.

7.2.2 Appropriate Placement of Line Transects

Appropriate placement of transect lines must come before the conduct of a line transect survey. However, it is in this very first step that the common errors associated with the application of method occur. As described in Chap. 6, one of the basic assumptions of this method is that transects are located randomly with respect to the spatial distribution of target animals. Without this assumption, the distribution of observed detection distances reflects both detection probability and possible gradients in density, leading to biased estimates. Additionally, in the absence of a probabilistic spatial sampling scheme, it may not be reasonable to extrapolate density estimates from transects to the entire area of interest, under traditional design-based line transect surveys.

All too frequently, because of sheer convenience, investigators are tempted to use animal trails, roads, firebreaks, waterways and other linear formations as 'transect lines'. Although this is logistically convenient, particularly in difficult terrain or dense vegetation, it has the potential to induce severe biases in the survey data. Such bias is introduced in two ways: First, these existing linear features are likely to be nonrepresentative, by not sampling existing variations in habitats and human influences that affect densities proportionally. Second, these features may be specific microhabitats for only some prey species. Prey species may be either attracted or repelled by them, again causing the distribution of detection distances to reflect some combination of detection and density variation, rather than only detection.

To illustrate this point, envision a survey design with transects located along existing animal trails in the study area. It is likely that for some or all species, densities along these trails will be higher than off the trails, leading to inflated encounter rates on them. In addition, if more animals occur on or closer to the trails, then the gradient in *detectability* with distance from the transect line is confounded by the natural gradient in *density*. As a result, not only is the sampling becomes nonrepresentative of the study area but seriously undermines the investigators' ability to accurately estimate detection probability via the detection function. Generally, in such situations, such factors will lead to an overestimation of animal density.

A contrasting scenario would be if transects are situated along trails that are heavily used by human hunters. In this case, true animal density along trails is likely to be lower on trails than away from trails, because some species may actively avoid hunters. This scenario will lead to lower encounter rates on the trails and more observations at greater distances than would be the case, if they were proper transects placed randomly. In this scenario, animal densities will be underestimated. To further complicate matters, different species within a given area may exhibit different behaviours, with some species favouring trails and others actively avoiding them.

The only way to avoid these biases is to place line transects using a probabilistic, fully randomised survey design or using a systematic placement with a random start (Chap. 6). Systematic placement of transects with a random starting point is generally the most logistically practical design. Moreover, it also achieves more uniform sample area coverage, as assumed in standard distance sampling analyses (Buckland et al. 2001, Chap. 6). The transect placement process can be visualised as drawing transects on an acetate sheet using a particular survey design and sampler geometry and then placing it on a map of the study area using a randomly chosen starting point for alignment.

In some instances, there may be no other option but to conduct surveys along pre-existing features such as trails. One obvious consideration in such situations is that trails are generally somewhat curved rather than straight, although techniques have been developed to adjust for this during the analysis (Hiby and Krishna 2001). If animals are not orienting their movements with respect to trails, in theory, a valid estimate of animal density can be obtained *for the locations surveyed*, and this may potentially represent a useful metric for trend monitoring purposes. However, there is still no theoretical basis for extrapolating such an estimate to the entire area of interest. The only way to justify such an extrapolation is by demonstrating that average density in the surveyed area (e.g. along trails) is the same as that over the whole area – an extremely difficult task. Some possibilities are (i) to perform strip transect counts of animal dung at right angles to the trails to investigate whether dung density varies with distance from the trail or (ii) to compare encounter rates on trails with those from transects placed at random.

In summary, when all the additional effort that is required to justify the use of existing linear features is considered, it is strongly recommended that wherever possible, investigators should establish transect lines according to some valid sampling design (see Chap. 6).

7.2.3 Requirement for Spatial and Temporal Replication

In general, a larger number of spatially replicated transect lines will reduce the between-transect variance in encounter rates and improve estimates. However, in areas hard to access, there are also logistical problems and costs associated with increasing the number of spatial replicates. Therefore, the number of transects to be established depends on available survey personnel and other resources also. A minimum of 15–20 replicates are recommended and preferably 25 or more (Buckland et al. 1993, 2001; Thomas et al. 2010). If the objective is to obtain one density estimate for the study area, these 15–25 transects can possibly represent the area reasonably well. If, however, separate density estimates for multiple strata (Chap. 6) are required, for example, in different habitat types or management zones, at least 15–20 spatial replicates must be placed in each stratum. In situations where a smaller number of spatial replicates are used, the variance of encounter rates may have to be estimated theoretically rather than empirically (see Chap. 6). Generally, deriving these variance estimates empirically from the survey data on hand is considered a superior option (Buckland et al. 1993, 2001).

If certain prey species are restricted only to specific parts of the study area, separate sets of transect lines may have to be created to target them. More typically, however, the distributions of several prey species will overlap spatially, and once a system of transects is established across the entire area, it should be possible to conduct surveys of several prey species simultaneously.

For multispecies surveys, walking at a speed of approximately 1.5 kilometres per hour is recommended for optimal detection of animals. Assuming that an observer can employ about 2–3 h of concentrated search effort in one session, 3–4 kilometres long can be walked depending on terrain and other ambient conditions. To have 25 spatial replicates, a minimum of 75–100 kilometres of transect lines may need to be established. As discussed in Chap. 6, repeated walks of the same transect lines, called temporal replications, can improve the precision of the estimates of detection probability and of cluster size. Often multiple temporal replications may be necessary to attain sufficient sample size of detections. Generally, in habitats with moderate to high prey densities, 8–10 temporal replicates on each spatial replicates will yield reasonable sample sizes (i.e. 60–80 detection events) to estimate detection function parameters with $CV(\widehat{D}) < 20\%$ for most prey species. However, only a larger number of spatial replicates can improve the precision of encounter rate estimates (Chap. 6).

Repeated temporal use can lead to transect lines becoming de facto animal trails which are not desirable as described in previous sections. Particularly in forested habitats, effort should be made to keep transects as unobtrusive as possible with minimal trimming of vegetation to enable quiet movement of survey personal. If animals are reacting to the transect lines, the standard way of checking is to place shorter (100–200 m) dung count transects perpendicular to the line transects to measure whether dung density increases or decreases with the distance from the line, indicating either attraction or avoidance by prey species.

7.2.4 Sampler Geometry

The geometrical shape of transects (termed 'samplers' in distance sampling literature; Buckland et al. 2001, 2004) can be a straight line, a square or some other shape (Strindberg et al. 2004). A square sampler design offers a number of advantages for line transect surveys of tiger prey species. This layout (i) eliminates the additional effort required to return to the start point after the survey (which is particularly useful in difficult habitat), (ii) increases the uniform sample coverage of the surveyed area, (iii) increases survey efficiency and (iv) eases logistics by facilitating multiple access points to start the survey, which in turn increases the potential options for locating campsites, transportation of personnel, etc.

A set of points or line segments, at chosen spacing and orientation, can be generated by the automated survey design feature in the DISTANCE software (Thomas et al. 2010). These points or line segments subsequently can be converted into square samplers using GIS tools. Square sampler survey designs have been successfully implemented for estimating prey densities in a variety of habitats (O'Kelly et al. 2012; Jathanna et al. 2015; Karanth and Kumar, Unpublished data; Pattanavibool, Unpublished data).

One criticism of square samplers (spatially replicated transects) is that animal detections made at the four corners may be nonindependent events (e.g. detected animals fleeing and flushing others around the corner leading to more detections). However, as explained in Chap. 6, although such dependent detection events may bias analytical estimates of sampling variance, they will have little effect on the density estimates. Furthermore, this problem can be addressed by choosing empirical estimators for variance estimation. Karanth and Kumar (unpublished data) also found that detections close to corners of the square sampler contributed little to the overall estimation of detection function parameters. For example, in a large-scale transect survey data set using a square sampler design collected over the past 15 years in India, 'corner observations' constituted less than 1% of the total detections (Karanth and Kumar, unpublished data).

In dense forests, swamp or rugged terrain, sampler geometry other than the square one may be more appropriate. Zigzag (Strindberg and Buckland 2004) or curvilinear (Hiby and Krishna 2001) transect designs are also such options. The same caveats regarding unrepresentative sampling described in Sect. 7.2.2 apply to these designs also. Moreover, there may also be issues of non-uniform coverage probability, which must be addressed during the analysis (Strindberg and Buckland 2004). Ultimately, the choice of sampler geometry will be somewhat study specific shaped by habitat features, prey population characteristics, logistics and resource considerations.

7.3 Field Protocols for Conducting Line Transect Surveys

7.3.1 Meeting Line Transect Assumptions in the Field

The four key assumptions in line transect sampling (Chap. 6) are as follows:

1. Transect lines are located randomly with respect to the distribution of animals.
2. Animals on the line are detected with certainty.
3. Animals are detected at their initial location.
4. Measurements made are exact.

Field protocols should ensure above assumptions are satisfied as best as possible. In the following section, each field protocol prescription is linked to meeting one or more of these assumptions using above numbers. For example, (1) indicates that the prescription is to meet assumption 1.

The start location and orientation of transects should be decided upon in advance, based on an appropriate survey design (see Chap. 6). From a purely theoretical perspective, survey personnel could go to the transect start point, set a predetermined compass azimuth and simply walk in that direction, searching for animals. However, this is not often feasible because of dense vegetation and other impediments to travel such as rivers or escarpments. Such an approach of traversing and crashing through vegetation would also generate unacceptable levels of disturbance,

Plate 7.1 Transect lines have to be clearly marked, while also minimising cutting of the under growth in order to avoid attracting or repelling prey species to the lines (Image Copyright©: Killivalavan Rayar)

causing systematic evasive movement in response to the survey process itself. Such movement would constitute a serious violation of the assumption 3.

Therefore, transect lines have to be minimally cut and maintained physically, while carefully marking their orientation, and distance traversed, with paint (Plate 7.1). This will keep disturbance from the passage of survey personnel down (3). Lines can be oriented using a sighting compass and/or a GPS unit (1). Survey personnel should be concentrating on searching for animals (2), not searching for their line transect itself! Locations along the line should be clearly marked at regular, closely spaced intervals using brightly coloured paint, metal tags or flagging tape to guide survey personnel.

Observers need not walk 'exactly' on the line at all times (see below), but they must *know exactly where the line is* at all times (2) and record all measurements from the line (4). Transects can be measured using hip chains, GPS units or fixed lengths of rope. Subsections of transect can be marked at every 100 m if covariate or any other spatial data collection is involved in the study (see Sect. 7.3.4). Vegetation along transect should be trimmed minimally as necessary to enable survey personnel to move quietly along the line (3).

In open habitats, it may be possible to maintain lines just by paint marks, without cutting any vegetation. In either case, it is essential that transect marking

Plate 7.2 Two observers move along the marked transect line, recording each visual detection of prey species (Image Copyright©: Eleanor Briggs)

and vegetation trimming are completed fully before conducting the actual survey to avoid all disturbance (1).

It is recommended that a team of two observers should survey each line transect (Plate 7.2). It is essential to pay attention to ensure that animals on the line or close to it are detected without fail (2) (shape criterion; see Chap. 6). Observers must be vigilant to detect animals hiding in the vegetation. The first observer should focus on the line itself (2) and on scanning an arc of 30° on either side of it (shape criterion). The second observer should concentrate on a wider arc, say 30°–90°, on both sides of the transect. The second observer should also be responsible for scanning the canopy above the transect line if primate species are surveyed. In this manner, two observers combined should be able to search an arc of 180° ahead of them adequately (3). Sometimes, animals are detected after observers have already passed them. Data from such 'back sightings' are not a problem only if they are not too frequent and systematic.

The survey objective is for the observers to see the animal before the animal sees them (3). Observers should wear cryptic clothing and footwear that permits silent passage (3). They should move carefully and quietly, refraining from talking or smoking (3). Walking at a uniform pace of 1.5 km per hour helps to minimise *responsive evasive movement before initial detection* (3) and ensures that observers move faster than undisturbed target animals would naturally move (Buckland et al. 2001, 2004, see Chap. 6). In addition, walking at this speed reduces footfall noise and increases detection of auditory cues from animals that often lead to subsequent visual observations. However, auditory detections on their own cannot be used for gathering distance data, without visual confirmation (3).

Both active and passive detections of prey species should be recorded. Active detections occur if observers detect the animal before it becomes aware of them. Passive detections occur if the animal flushes in response to the observers leading to the detection event. In both of these cases, all measurements should be taken to the initial location of the animal or animal cluster (3). Upon sighting an animal or a cluster of animals, the measurements usually taken are true distance to the animal or centre of the cluster, r, compass bearing to the animal or cluster,θ_1, and compass bearing of the line, θ_2. Perpendicular distance, x, can then be calculated during the data entry process, as $x = r\sin(\theta_1 - \theta_2)$.

If animals are first observed close to the line but move away before an accurate measurement of radial distance can be recorded, it is acceptable to walk along the line to a point opposite to their initial location and measure the perpendicular distance directly (3). In this situation, particular care must be taken that one observer continues to scan ahead on the transect (2), while the other keeps sight of the animal's initial location (3). To avoid heaping and rounding errors (see Chap. 6), the distance and angle measurements should be accurately read and exactly recorded to the nearest integer value (4). True distances should be measured with a rangefinder, and compass bearings measured with a sighting compass. Laser rangefinders have become increasingly preferred to optical rangefinders because they are easy to use and provide a clear digital display of the distance. Laser binoculars can record both the distance and compass measurements simultaneously although they are considerably more expensive.

Although it is acceptable to move off the transect line by a few metres, to avoid obstructions, observers must remember the location of the 'true line' and try to stay close to it. All measurements should be made from this true line (4). It is particularly important to record both angles θ_1 and θ_2, unless the animal is dead centre on the line (which is unlikely). Even if a cluster of animals may straddle the line, it is unlikely that exact centre of the cluster is exactly on the line. Once again, both angles θ_1 and θ_2 should be recorded.

For each detection event, distance and angle measurements should be made with reference to the approximate geometric centre of the cluster (4). We emphasise that the animal 'cluster' as defined in line transect surveys refers to an aggregation of individual animals that occurs within the sampled strip. Such a cluster may be smaller than a larger 'social grouping' in gregarious species such as chital, hog deer or gaur, for example. A larger group may sometimes be detected as one or more detections of smaller clusters. Some parts of it may lie beyond the sampled strip and go undetected. As a general practice, considering all animals within a 30 metre radius as one 'cluster' is convenient in surveys of forest ungulates in tiger habitats of southern Asia that we have surveyed.

The number of individuals in a detected cluster (cluster size) should be accurately counted before the observers move on (4). After animals flush and scatter, enabling more animals to be counted, they should also be added to the final cluster size count. Detecting such extra individuals may also change the approximate geometric centre of the cluster, requiring distance and angle measurements to be made again (4). Binoculars can improve the accuracy of counts as well as validate species detected. It is also good practice for both observers to count each cluster and compare counts before recording. However, when one observer is recording data, the second observer must continue to scan the transect line so that no animals on or close to it are missed (2).

Additional information that is commonly recorded with detections includes location of the observer/animals at the point of each detection (e.g. 100 metre subsection, indicated by distance markers) and/or it's habitat type. Such information can potentially be used as covariate data for modelling of spatial distribution or other advanced analytical techniques (see Chaps. 6 and 8). However, spending time and paying attention to all sorts of biological or management data, not related to distance sampling, will seriously undermine the data quality in line transect surveys. Unfortunately, many investigators overreach, ignoring this crucial need for focus on counting prey, and seriously undermine the quality or even validity of the data they collect.

A sample data entry form for use in the field for conducting line transect surveys is shown in Table 7.1. After completion of a field trip, data forms should be checked as soon as possible for data recording errors. Most errors can be corrected easily if such validation is performed immediately after the survey, preferably the same day. Postponing this quality control to after completion of the survey is not advisable at all. Memory of a particular detection event fades rapidly, and even the observer may not be available for such crosschecking later.

Sometimes, a particular temporal replicate walk may have to be abandoned part of the way, because of onset of darkness, bad weather, illness and danger from animals such as elephants or rhinos or even from other humans. In such cases, the location where the survey was abandoned and the distance covered should be recorded to ensure sampling effort is recorded accurately on the data form. The survey may be resumed at a later date if possible.

7.3.2 Season for Conducting Line Transect Surveys

The seasonal timing of line transect surveys depends on several biological factors. If prey species are locally migratory, this factor must be taken into account. The way ungulate species aggregate in groups in and out of rut, presence of antlers or propensity to call in male animals and animals being active or not because of ambient conditions are all ecological factors that do influence the detection process.

Seasonal effect on habitats also influences the process of animal detection from a line transect. Often, visibility is greatly improved when trees and shrubs are seasonally deciduous. Improved visibility increases detectability and adds to encounter rates and ultimately a larger sample size for better estimates of both. Additionally, undercounting cluster size is less likely with greater visibility, reducing size bias (see Chap. 6) in data.

Conversely, deciduousness of the vegetation may increase dry leaf litter along transect, creating excessively noisy underfoot conditions, leading to a more evasive movement of animals. A compromise for forests in South Asia is to conduct line

Table 7.1 Example of field data form used in line transect surveys

Surveyor 1: Surveyor 2 :	Transect No:	Field Site:
Date:	Weather:	Page No:
Start Time:	End Time:	Sample Walk:
Start Location:	End Location:	Total Distance Walked (KM):

Sl. No.	Time	Species	Cluster Size	Sighting Distance (metres) r	Compass Bearings		Habitat Type/ Segment Location
					Animal (degrees) θ_1	Line (degrees) θ_2	

Species Acronym: Sambar: **SBR**; Nilgai: **NLG**; Muntjak: **MJK**; Chital: **CHT**; Chinkara: **CHK**; Blackbuck: **BBK**; Elephant: **ELP**; Wild Pig: **PIG**; Sloth Bear: **BER**; Mouse deer: **CHV**; Common langur: **LGR**; Rhesus macaque: **RHM**; Bonnet macaque: **BNT**; Giant Squirrel: **GSQ**; Cattle: **CTL**; Sheep/Goat: **GOT**

transect surveys towards the end of the dry season but after early monsoon showers (Karanth and Nichols 2002).

Seasonal factors can also seriously impact the logistics of line transect surveys. The high ambient temperature in summer is a major constraint to undertaking arduous fieldwork. On the other hand, access to field may be cut off in the wet season or become dangerous due to poor road conditions or flooded rivers. In addition, the risk of disease to field personnel is often greatly increased in certain seasons affecting their efficiency and safety.

Choice of the season for carrying out line transect surveys is therefore every context dependent and needs to consider the above (and perhaps other) local factors.

Timing of transect walks within the diel cycle is a more straightforward matter. Many ungulate prey species are crepuscular (e.g. sambar), being active around dusk and dawn, and tend to lie up during the day. Others (e.g. chital) are more diurnal but still tend to be active early in the morning and evenings. Therefore, although the underlying ungulate densities do not change, detection probability will differ by time of day. Detection probabilities are higher when prey species are active and moving around and drop dramatically when they are lying down or resting (Karanth and Nichols 2002). We recommend that, in tropical Asia, line transects are surveyed early in the morning and/or late in the afternoon to improve detection probabilities due to increased animal activity. In the hot season, these times are also more comfortable for survey personnel. Furthermore, to even out possible variations in detection rates, even within these windows of survey time, it is recommended that temporal replications along the same transect (1) are split evenly between evenings and mornings and (2) traversed from opposite ends (e.g. for square samplers switching between clockwise and counterclockwise directions). We have found that with a little attention to such 'clever logistics', it is possible to add these refinements to even out variations in encounter rates and improve detection probabilities, thereby increasing the reliability of prey density estimates.

Some other logistical factors that might affect the timing of surveys include the time taken to traverse the transect line, the distance and time involved in reaching starting points of the transect lines and the mode of transport available for doing so.

7.3.3 Training and Human Resource Requirements

Because in an ideal study, over 25–50 spatial replicates of 3–4 km length are to be walked 8–10 times each (temporal replication), within a reasonably short survey period, a large number of skilled and well-trained survey personnel will be required to collect high-quality line transect data. Karanth (1999) demonstrated the efficacy of using highly motivated and well-trained volunteer naturalists or 'citizen scientists' to collect high-quality line transect data. They were trained for a week or so in the use of survey equipment and tested for ability to spot, identify and count target species rapidly. While collecting data, each trainee was paired with an experienced instructor. This model led to collection of large amounts of high-quality field data. Furthermore, a large of number of these citizen scientists continued to be involved in prey monitoring initiatives over the long term making major contributions to tiger conservation (Johnson et al. 2014). This 'social model' of conducting line transect surveys may have relevance in many tiger range countries possessing educated populations of potential volunteers interested in natural history. In alternative scenarios, park management staff or research workers or field assistants can be paid and employed to provide the necessary manpower for prey surveys.

7.3.4 Covariate Data Collection

In addition to generating overall estimates of prey abundance and density in an area, researchers, conservationists and wildlife managers may additionally want to investigate spatial variation in prey density within that area of interest. Such variation may be found because of both ecological and management factors or simply through stochastic processes (Royle and Dorazio 2008). This variation can also be useful to generate model-based predictions of prey density (Chaps. 6 and 8) at unsurveyed locations using covariate data (Chap. 8).

Classical distance sampling covered above cannot achieve these objectives. However, recent advances in standard distance sampling methods (Hedley and Buckland 2004; Hedley et al. 2004) and the development of fully model-based inferential approaches (Royle and Dorazio 2005, 2008; Kumar 2011; Kéry and Royle 2015) have opened new analytical avenues to explore such responses of tiger's prey species to ecological and management factors (e.g. Sillett et al. 2012). These approaches are briefly covered in Chaps. 6 and 8.

The newer analytical extensions to line transect survey methods mentioned above make it necessary to collect additional data on the spatial location of detection events on transects and other relevant covariate information associated with each detection. Some types of covariate data can be extracted from existing GIS layers and/or remotely sensed data sets, while others can be collected during field surveys (e.g. location of observers/animals at a detection event, habitat information, weather conditions, elevation, etc.; e.g. see Kumar 2011, Harihar et al. 2014). It is important to make sure this additional data collection does not distract survey personnel from their primary task and lower the quality of the basic line transect prey survey data. Therefore, we suggest that a separate, one-time survey can be conducted on each transect line to gather these additional data, without trying to record animal detections (e.g. data on vegetation, evidence of fire or other human impacts, etc.; e.g. see Kumar 2011). Such one-time surveys can also be carried out simultaneously with the initial marking, or annual maintenance, of transect lines.

7.4 Implementing Line Transect Surveys Under Challenging Conditions

Tiger habitats outside of Southern Asia pose some additional challenges to implement line transect surveys as we have described above. Protected areas (PAs) in Southeast Asia tend to be far more extensive (1000s of sq. km as opposed to 100s of sq. km) and situated in remote regions with limited infrastructure such as roads and power supply. This makes access to survey areas and movement within them time-consuming, with field teams generally being required to hike and camp for several days to reach transects on foot in difficult terrain. In order to maximise the ratio of survey time to travel time in these situations, it is recommended that transects are temporally surveyed on successive days until the necessary effort (total distance walked) per each spatially replicated transect is attained. For greater efficiency, surveys should be undertaken both in the morning and evening, where possible. There is a small risk that animals may be disturbed as a result of such repeated sampling. Encounter rates should be checked to confirm there is no such systematic decrease over time. Additional precautions should be taken to minimise such disturbance, by ensuring survey teams camp at a sufficient distance away from transects.

Another characteristic of many Southeast Asian PAs is depressed prey densities due to hunting pressure. Low prey densities make it extremely difficult to obtain sufficient detections to estimate species-specific detection function parameters. This issue of small sample sizes is already discussed in Sect. 7.2.1. However, one advantage of distance sampling method is that detection function data can be accumulated and pooled over time (i.e. over multiple years) until a detection function can be reliably modelled using a larger data set. This detection function can then be applied retrospectively to previous surveys. In extreme cases of low density, data can be combined from similar species to model a detection function. Detection functions can also be 'borrowed' from 'similar studies' conducted elsewhere. However, all of these techniques involve implicit assumptions regarding the detectability of the species to be constant across time or space or species. Therefore, they should be applied with due caution. It is worth noting that where densities of all prey species are so low that they preclude all attempts at population estimation, there is probably no sufficient prey base present to support a viable tiger population in the area anyway (Karanth and Stith 1999; O'Kelly et al. 2012).

Another critical constraint that is common in tiger range states is a lack of adequate manpower and limited technical capacity. Outside of Southern Asia, there is typically a lack of volunteers, and even government staff may be reluctant to spend long periods of time surveying at remote field sites. Survey personnel undertaking line transect surveys must be able to use equipment (GPS, map, compass, rangefinder, etc.) and fill out data forms. They should understand the need to meet basic assumptions of distance sampling. Another essential requirement for survey personnel is field skill in spotting, identifying and rapidly counting animals. This is particularly important in forests or where animals are wary. It is vitally important that observers gain sufficient experience to develop a 'search image' for target species, not easy when dealing with low-density or cryptic species. An efficient approach we recommend is to pair one locally recruited observer with only field skills with another more educated observer possessing skills to use instruments and record data (i.e. a high school or university graduate). It is also important that both observers train each other over time. Given the challenges involved in finding and training skilled field personnel, it helps to retain trained survey personnel, for example, by providing employment beyond the duration of the line transect surveys.

Due to all of the above constraints, conducting line transect surveys can be expensive and time-consuming. There are, however, multiple advantages to carrying out annual surveys: increasing the number of data points to better assess prey population trends, retaining skilled staff, gathering auxiliary anecdotal information relevant to management and conservation by being eyes and ears in the tiger habitat being surveyed. However, because of resource constraints, surveys may be feasible only once every 2 or 3 years or only survey different parts of the area of interest each year by rotation. All such resource allocation decisions are context specific and will invariably depend on a wide range of local factors (Chap. 2).

Finally, it must be recognised that in some tiger habitats, transect surveys of prey species are simply not a feasible option because of steep terrain, barriers to movement and even low densities of prey animals or other logistical issues. In these circumstances, investigators do have other methods, perhaps as not well developed as line transects at this point, available for monitoring prey (e.g. Gopalaswamy et al. 2012; Vongkhamheng et al. 2013). Some of these issues and approaches are discussed in Chaps. 2 and 8.

References

Buckland ST, Anderson DR, Burnham KP, Laake JL (1993) Distance sampling: estimating abundance of biological populations. Chapman and Hall, London

Buckland ST, Anderson DR, Burnham KP, Laake JL, Borchers DL, Thomas L (2001) Introduction to distance sampling. Oxford University Press, London

Buckland ST, Anderson DR, Burnham KP, Laake JL, Borchers DL, Thomas L (eds) (2004) Advanced distance sampling. Oxford University Press, Oxford

Gopalaswamy AM, Karanth KU, Kumar NS, Macdonald DW (2012) Estimating tropical forest ungulate densities from sign surveys using abundance models of occupancy. Anim Conserv 15:669–679

Harihar A, Pandav B, MacMillan DC (2014) Identifying realistic recovery targets and conservation actions for tigers in a human-dominated landscape using spatially explicit densities of wild prey and their determinants. Divers Distrib 2014:1–12. doi:10.1111/ddi.12174

Hedley SL, Buckland ST (2004) Spatial models for line transect sampling. J Agric Biol Environ Stat 9:181–199

Hedley SL, Buckland ST, Borchers DL (2004) Spatial distance sampling models. In: Buckland ST, Anderson DR, Burnham KP, Laake JL, Borchers DL, Thomas L (eds) Advanced distance sampling. Oxford University Press, Oxford

Hiby L, Krishna MB (2001) Line transect sampling from a curving path. Biometrics 57:727–731

Johnson MF, Hannah C, Acton L, Popovici R, Karanth KK, Weinthal E (2014) Network environmentalism: citizen scientists as agents for environmental advocacy. Glob Environ Chang 29:235–245

Jathanna D, Karanth KU, Kumar NS, Goswami VR, Vasudev D, Karanth KK (2015) Reliable monitoring of elephant populations in the forests of India: analytical and practical considerations. Biol Conserv 187:212–220

Karanth KU (1999) How volunteer naturalists can monitor tigers and prey. In: Seidensticker J, Christie S, Jackson P (eds) Riding the tiger: tiger conservation in human-dominated landscapes. Cambridge University Press, Cambridge, pp 118–119

Karanth KU, Nichols JD (2002) Monitoring tigers and their prey: a manual for researchers, managers and conservationists in tropical Asia. Centre for Wildlife Studies, Bangalore

Karanth KU, Stith BM (1999) Prey depletion as a critical determinant of tiger population viability. In: Seidensticker J, Christie S, Jackson P (eds) Riding the tiger: tiger conservation in human-dominated landscapes. Cambridge University Press, Cambridge, pp 100–113

Kéry M, Royle JA (2015) Applied hierarchical modeling in ecology: analysis of distribution, abundance and species richness in R and BUGS: Volume 1:Prelude and static models. Elsevier Science, Amsterdam

Kumar NS (2011) Assessment of distribution and abundance of ungulate prey using spatial models in Nagarahole and Bandipur Tiger Reserves of India. A Ph. D. dissertation, Manipal University, India

O'Kelly HJ, Evans TD, Stokes EJ, Clements TJ, Dara A, Gately M, Menghor N, Pollard EHB, Soriyun M, Walston J (2012) Identifying conservation successes, failures and future opportunities; assessing recovery potential of wild ungulates and tigers in Eastern Cambodia. PLoS One 7:e40482. doi:10.1371/journal.pone.0040482

Sillett TS, Chandler RB, Royle JA, Kéry M, Morrison SA (2012) Hierarchical distance-sampling models to estimate population size and habitat-specific abundance of an island endemic. Ecol Appl 22:1997–2006

Royle JA, Dorazio RM (2005) Hierarchical models of animal abundance and occurrence. J Agric Biol Environ Stat 11:249–263

Royle JA, Dorazio RM (2008) Hierarchical modeling and inference in ecology: the analysis of data from populations, metapopulations and communities. Academic, San Diego

Strindberg S, Buckland ST, Thomas L (2004) Design of distance sampling surveys and Geographic Information Systems. In: Buckland ST, Anderson DR, Burnham KP, Laake JL, Borchers DL, Thomas L (eds) Advanced distance sampling. Oxford University Press, Oxford

Strindberg S, Buckland ST (2004) Zigzag survey designs in line transect sampling. J Agric Biol Environ Stat 9:443–461

Thomas L, Buckland ST, Rexstad EA, Laake JL, Strindberg S, Hedley SL, Bishop JRB, Marques TA, Burnham KP (2010) Distance software: design and analysis of distance sampling surveys for estimating population size. J Appl Ecol 47:5–14

Vongkhamheng C, Johnson A, Sunquist ME (2013) A baseline survey of ungulate abundance and distribution in northern Lao: implications for conservation. Oryx 47:544–552

Concepts and Practices: Estimating Abundance of Prey Species Using Hierarchical Model-Based Approaches

8

Robert M. Dorazio, N. Samba Kumar, J. Andrew Royle, and Arjun M. Gopalaswamy

8.1 Introduction

Tigers predominantly prey on large ungulate species, such as sambar (*Cervus unicolor*), red deer (*Cervus elaphus*), gaur (*Bos gaurus*), banteng (*Bos javanicus*), chital (*Axis axis*), muntjac (*Muntiacus muntjak*), wild pig (*Sus scrofa*), and bearded pig (*Sus barbatus*). The density of a tiger population is strongly correlated with the density of such prey species (Karanth et al. 2004). In the absence of direct hunting of tigers, abundance of prey in an area is the key determinant of the "carrying capacity" of that area for tigers (Chap. 2). Accurate estimates of prey abundance are often needed to assess the potential number of tigers a conservation area can support.

Historically, the most accurate estimates of abundance of tiger prey have been obtained using line-transect sampling at randomly selected locations (Buckland et al. 2001, 2004, Chaps. 6 and 7). This approach has the advantage of providing an estimate of prey abundance that is design unbiased for a fixed sample frame

R.M. Dorazio (✉)
Wetland and Aquatic Research Center – United States Geological Survey (USGS), Gainesville, FL, USA
e-mail: bdorazio@usgs.gov

N.S. Kumar
Wildlife Conservation Society, India Program, Bengaluru, India
e-mail: nrao.skumar@gmail.com

J.A. Royle
USGS Patuxent Wildlife Research Center, Laurel, MD, USA
e-mail: aroyle@usgs.gov

A.M. Gopalaswamy
Statistics and Mathematics Unit, Indian Statistical Institute, Bengaluru, India

Department of Zoology, University of Oxford, Oxford, UK
e-mail: arjungswamy@gmail.com; https://www.isibang.ac.in; https://www.zoo.ox.ac.uk/

K.U. Karanth, J.D. Nichols (eds.), *Methods For Monitoring Tiger And Prey Populations*, DOI 10.1007/978-981-10-5436-5_8

(i.e., a finite region of interest) provided the model of detection is correct (or at least approximately correct). The use of randomization in selecting locations ensures that estimates of total prey abundance in the region are unbiased relative to the true, though unknown, total number of prey. This approach is ideally suited to regions that are relatively homogenous in habitat or to regions that can be stratified for sampling into subregions of similar habitats.

In practice, however, spatial heterogeneity in the habitat of tiger prey species is more common, and it is not always possible to conduct surveys at randomly selected locations owing to inaccessibility and other logistical constraints. For example, a randomly selected location may be too steep or heavily forested to be surveyed on foot. The presence of swampy soils or water bodies also may impede survey efforts. These problems have motivated statisticians to develop extensions of conventional line-transect models that account for the spatial variation in density of individuals (Hedley and Buckland 2004; Royle et al. 2004; Johnson et al. 2010; Conn et al. 2012; Buckland et al. 2015). These extensions do not require sample locations to be selected randomly. Instead, the idea is to select locations that span the range of variation in habitat and to estimate the effects of habitat on prey density from the data. This fully *model-based* approach to the analysis of line-transect data may lack the design unbiasedness of conventional methods of analysis, but a model-based approach is more versatile and can be extended to accommodate a variety of sampling protocols that include line-transect surveys. A model-based approach provides much more than a single estimate of population abundance or density. Modeling can be used to learn about how key characteristics of habitat influence the density of prey across a region. Modeling also can be used to estimate the abundance or density of prey within any subregion of interest. In this chapter we describe this fully model-based approach and its relevance to surveys of tiger prey.

The general problem that we consider is the estimation of abundance of a prey species living in a finite region of interest that contains spatial heterogeneity in habitat. Our use of the term habitat is not limited to physical habitat. We mean to include measures of ecological or management-related factors that are known to influence variation in prey abundance over space. We assume that sample locations span the range of variation in habitat within the region and that two or more line-transect surveys are conducted at each of these locations within a brief period of time (e.g., on consecutive days). Repeated sampling at each location is used in surveys of tiger prey to increase the number of detections at each location; therefore, the number of prey animals within range of detection can differ among temporal replicates owing to movements of individuals between surveys. Fortunately, existing models can be extended to account for these differences in abundance when prey animals are assumed to move randomly (i.e., into and out of the region of sampling) between surveys (Chandler et al. 2011, 2014; Kéry and Royle 2016, Section 9.5).

In the following sections, we describe several hierarchical models that may be applicable in the analysis of spatially and temporally replicated surveys of tiger prey. As a continuation of the previous chapters, we begin by describing models of line-transect data recorded by one or two observers and models of counts of these observers. The counts, a by-product of line-transect sampling, are obtained by enumerating individual prey detected by one or two observers during each line-transect survey. We show that these counts alone can be used to estimate prey abundance in surveys where detectability of individuals depends more on habitat (e.g., the presence of concealing vegetation) than on distance from observers. We also show that if prey abundance is relatively low at most survey locations, abundance still can be estimated using only a quantized version of each count (that is, the detection or non-detection of one or more individuals). We provide examples of fitting these models to hypothetical data sets constructed by assuming realistic levels of prey density and prey detection probability and to line-transect surveys of chital in the Nagarahole and Bandipur Tiger Reserves of India.

8.2 Hierarchical Models for Spatially and Temporally Replicated Surveys

In this section we describe a hierarchical, model-based framework for estimating the abundance of a prey species living in a finite region of interest that contains spatial heterogeneity in habitat. One component of the hierarchy is used to specify spatial variation in prey abundance, usually as a function of covariates whose measurements are available throughout the region of interest. These measurements are often accessible using geographical databases and software. The other component of the hierarchy is used to specify a relationship between the data observed at a survey location and the unknown abundance of individuals vulnerable to detection at that location. This hierarchical approach, often called *N-mixture modeling*, was first proposed for the analysis of point counts (Royle 2004); however, N-mixture modeling also has been used to analyze data associated with other sampling protocols (Royle and Dorazio 2006, 2008; Kéry and Royle 2016). We describe several examples of these kinds of models in the following sections.

8.2.1 Sampling Design and Assumptions Common to All Models

We assume that the region of interest may be partitioned into a finite number of disjoint (nonoverlapping) sample units, each of which can potentially be surveyed using a protocol that provides information about the abundance and detectability of individuals. The size of each sample unit should correspond (at least approximately) to the area that can be surveyed. In most cases all units in the region cannot be surveyed, and observations are recorded only for a sample of units. We assume that the sample is selected to span the range of variation in habitat within the region. Ideally, the sample will be representative of the entire region, but representativeness is not a requirement of the models we describe.

We also assume that movements of prey between repeated surveys of the same transect are limited to random excursions into and out of the area surveyed. Such movements are sometimes referred to as a "temporary emigration" or "temporary absence" and occur as individuals move within their home ranges. This assumption is reasonable for most tiger prey because they do not migrate (bearded pigs may be an exception). We assume that the spatial extent of prey movements is sufficiently limited that individuals can only be detected within a single sample unit. This assumption could potentially influence the selection of units to be surveyed, but in practice most regions are large enough to ensure that the fraction of units surveyed is small. Therefore, given our assumptions a sample will likely include sample units that are far apart relative to the spatial extent of prey movements.

8.2.2 Modeling Spatial Variation in Abundance

Suppose a representative sample of K units is selected for the purpose of estimating prey abundance in the region of interest. Within each of these units, assume a survey is repeated on J occasions separated by relatively short time intervals (e.g., consecutive days). For now, we let the survey protocol be unspecified. Later, we will describe models induced by specific kinds of survey protocols. In addition, while the number of sampling occasions can generally vary among units, we assume for simplicity that each unit in the sample is surveyed the same number of times. However, all of the models can be fitted if the number of repeated visits actually differs among sample units.

Let M_k denote the unknown number of prey available to be detected in sample unit k ($k = 1, \ldots, K$), and assume

$$M_k \sim \text{Poisson}(\lambda_k a_k)$$

where a_k is the area of the sample unit and λ_k is the expected density of prey (number of individuals per unit area). To specify a relationship between λ_k and habitat covariates, we assume a log-linear model as follows:

$$\log(\lambda_k) = \boldsymbol{\beta}'\mathbf{v}_k + u_k$$

where $\mathbf{v}_k$ is a vector of predictors of prey density computed using the observed covariates and where $\boldsymbol{\beta}$ is a vector of parameters that corresponds to the effects of the predictors on λ_k. Unobserved sources of variation in prey abundance that may generate overdispersion (relative to the Poisson) can be specified by the parameter u_k. For example, we might assume $\mathbf{u} \sim \text{Normal}(\mathbf{0}, \sigma^2\mathbf{R})$ for the parameter vector $\mathbf{u} = (u_1, \ldots, u_K)'$, wherein the correlation matrix $\mathbf{R}$ is formulated so that abundances in sample units close to one another are more similar than abundances in sample units located farther apart. This correlation model is often used to account for the effects of unobserved, spatially varying covariates (Wikle 2010) and provides a hedge against model misspecification or nonrepresentative sampling, particularly in regions where covariate values change smoothly over space.

The abundance of prey at a sample location can differ among surveys of that location because individuals can move into or out of the range of detection between surveys. To account for these movements, we assume that a fixed proportion ψ of the M_k prey available to be detected is, on average, actually present during each of the J surveys. Specifically, we assume

$$N_{kj} \mid M_k \sim \text{Binomial}(M_k, \psi)$$

where N_{kj} is the number of prey present and within range of detection during the jth survey ($j = 1, \ldots, J$) of transect k.

To estimate the total abundance of prey in the entire sample, we need to estimate the expected number of prey in each unit as follows:

$$\text{E}(N_{kj}) = \text{E}[\text{E}(N_{kj}|M_k)] = \psi\lambda_k a_k$$

using estimated values of ψ and λ_k. This estimator accounts for spatial variation in habitat because each estimate of λ_k depends on spatial predictors $\mathbf{v}_k$ and u_k. An identical calculation is needed to predict the expected number of prey in each *unsampled* unit, so that abundance of prey in the entire region can be estimated by adding these predictions to the estimated abundances of prey in units that were surveyed.

8.2.3 Modeling Different Types of Data Observed in Temporally Replicated Surveys of Each Sample Unit

In this section we describe models of several types of data that can be obtained when J surveys are conducted by one or two observers within each of K sample units. A common characteristic of these models is their explicit conditioning on the unknown number of individuals that are vulnerable to detection. To be specific, each model of the observed data depends explicitly on N_{kj}, the unknown number of individuals that are present and within range of detection during the jth survey of sample unit k. These models are often referred to as "observation models" because their formulation depends only on the sampling process, not on the ecological process that produced the N_{kj} individuals vulnerable to detection.

Our description of each model is primarily conceptual and is focused on the underlying assumptions needed for analyzing different types of data. More technical material, including a derivation of the likelihood function for each model, is provided in the Appendix 1.

8.2.3.1 Single-Observer Surveys

We first describe models for the case where repeated surveys of the same sample unit are conducted by one observer.

Modeling distance of each individual from a transect line: The line-transect sampling protocol (described in Chaps. 6 and 7) yields the perpendicular distance between each individual's location and the transect line. Let x_{kj} denote this distance, and let n_{kj} denote the number of individuals detected during the jth survey of sample unit k.

To estimate N_{kj}, we assume that the probability of detecting an individual during the jth survey of unit k depends on the individual's distance x_{kj} from the transect line. A variety of functions can be used to specify this probability. One commonly used function called a "half-normal" is

$$p(x_{kj}) = \exp\left(-\frac{x_{kj}^2}{2\sigma^2}\right)$$

where σ is a strictly positive parameter to be estimated in conjunction with N_{jk}. Note that $p(0) = 1$, which implies that any individual located on the transect line is assumed to be detected. In contrast, individuals not located on the transect line can be missed, and their detection probability $p(x_{kj})$ is assumed to decrease as distance x_{kj} from the line increases. The magnitude of this decrease is controlled by the magnitude of the scale parameter σ, which has units of distance.

The primary assumptions needed for estimating the model's parameters are:

$$x_{kj} \sim \text{Uniform}(0, B_k) \tag{8.1}$$

$$y_{kj} \mid x_{kj} \sim \text{Bernoulli}(p(x_{kj})) \tag{8.2}$$

where B_k is the maximum perpendicular distance from transect k at which individuals can be detected and where y_{kj} indicates whether an individual is detected ($y_{kj} = 1$) or not detected ($y_{kj} = 0$) during the jth survey of transect k. Therefore, $y_{kj} = 1$ for each of the n_{kj} observed individuals, and $y_{kj} = 0$ for each of the remaining $N_{jk} - n_{jk}$ individuals that were present but undetected.

All that is known of the undetected individuals is that $y_{kj} = 0$ for each individual. Because distance x_{kj} is not observed for these individuals, we need to compute the *marginal* probability that each individual was missed (not detected) during the jth survey of sample unit k. This probability, which we denote by π_k, depends on σ and B_k as follows:

$$\begin{aligned}
\pi_k &= \int_0^{B_k} \Pr(y_{kj} = 0 | x_{kj})(1/B_k)\, \mathrm{d}x_{kj} \\
&= \int_0^{B_k} [1 - p(x_{kj})](1/B_k)\, \mathrm{d}x_{kj} \\
&= 1 - \frac{\sqrt{2\pi}\sigma}{B_k}\left(F(B_k) - \frac{1}{2}\right)
\end{aligned}$$

where $F(\cdot)$ is the cumulative distribution function of a Normal$(0, \sigma^2)$ distribution.

Each of the undetected individuals has probability π_k of being missed; therefore, the sum of their individual Bernoulli responses, which equals $N_{kj} - n_{jk}$, has a binomial distribution with index parameter N_{kj} and probability π_k. Equivalently, we can express the distribution of n_{kj} conditional on N_{kj} as follows:

$$n_{kj}|N_{kj} \sim \text{Binomial}(N_{kj}, 1 - \pi_k)$$

It should be noted that π_k is a *derived* parameter of the model (that is, a function of the formal parameter σ). In addition, we note that the binomial distribution of n_{jk} is a consequence of the assumptions given in Eqs. (8.1) and (8.2); it is not an additional assumption of the model.

It is easily shown that our assumed uniform distribution for distance x_{kj} implies that the locations of individuals are also uniformly distributed over the rectangular region surveyed. This assumption is consistent with our model of spatial variation in abundance (described earlier) *provided* the habitat used to predict abundance is the same within the region surveyed along a transect line. The effects of habitat or observer ability on $p(x_{kj})$ may be specified by formulating σ as a function of habitat or observer (Marques and Buckland 2003).

If habitat varies along the transect, our assumption of uniformity of individual locations may not be tenable. In this case abundance N_{kj} can still be estimated, but the hierarchical model must be formulated using the *locations* of detected individuals, not their perpendicular distances (Hedley and Buckland 2004; Johnson et al. 2010; Kéry and Royle 2016, Section 9.8).

Modeling counts of all individuals: In some situations an individual's location (or distance to transect line) is less informative of its detectability than other factors, such as habitat or observer ability. This is particularly true when individuals are potentially hidden from view due to dense vegetation or other physical structures. For some species of prey (e.g., chital), group size or herd size also can be informative of detectability. Therefore, when distance is uninformative of an individual's detectability, we might choose to estimate abundance using only n_{kj}, the number of individuals detected during the jth survey of sample unit k. In this case each of the N_{kj} individuals present in the sample unit (with or without a transect) is assumed to be detected with the same probability p_k, which leads to the following model of the observed number of individuals:

$$n_{kj} \mid N_{kj} \sim \text{Binomial}(N_{kj}, p_k)$$

Unlike an N-mixture model of repeated point counts (Royle 2004), there is no replication because prey abundance can differ among the J surveys of sample unit k owing to movements of individuals. Some restrictions therefore are needed to make the parameters of this model identifiable (i.e., estimable). In particular, we can prove that

$$n_{kj} \mid M_k \sim \text{Binomial}(M_k, \psi p_k)$$

so the detection probability p_k must be specified as a function of covariates of sample unit k if both ψ and p_k are to be estimated. For example, we might choose to specify differences in p_k using the following logit-linear model: $\text{logit}(p_k) = \boldsymbol{\alpha}'\mathbf{w}_k$, where $\mathbf{w}_k$ is a vector of predictors (e.g., based on habitat) and $\boldsymbol{\alpha}$ is a vector of the effects of those predictors. At least one of the predictors in $\mathbf{w}_k$ must be continuously valued (that is, not categorical) to estimate both ψ and $\boldsymbol{\alpha}$; otherwise, only the product ψp_k is estimable.

Modeling a quantized count of all individuals: For some species of prey, counts of individuals may be unreliable or difficult to obtain. For these species, it is more practical to observe whether one or more individuals appears to be present within a sample unit. This sampling protocol, often called a presence-absence survey, yields a quantized count of the individuals that are present (i.e., a presence or a zero). As with actual counts, any quantized count of individuals is potentially subject to detection errors, and a class of models (known collectively as occupancy models) has been developed to account for these errors (MacKenzie et al. 2006; Royle and Dorazio 2008).

Here we describe an occupancy model to estimate the abundance of individuals under the sampling design described earlier. As in our model of counts, we assume an individual's location is less informative of its detectability than other factors, such as habitat or observer ability. Let z_{kj} indicate whether one or more individuals are detected ($z_{kj} = 1$) or not ($z_{kj} = 0$) during the jth survey of sample unit k. An observed value of z_{jk} depends on the presence or absence of individuals (i.e., on the value of $I(N_{kj} > 0)$) and on the probability q_k of detecting one or more individuals when they are present. Specifically, we assume

$$z_{kj} \mid N_{kj} \sim \text{Bernoulli}(q_k \, I(N_{kj} > 0))$$

following the occupancy model described by Koshkina et al. (2017). Notice that this occupancy model assumes functional independence between q_k and N_{kj}, unlike the occupancy model of Royle and Nichols (2003), which assumes a functional dependence between q_k and N_{kj}. As in our model of counts, some restrictions are needed to identify the parameters of this model. In particular, we can prove that marginalizing N_{jk} yields

$$z_{kj} \mid M_k \sim \text{Bernoulli}(q_k \, (1 - (1 - \psi)^{M_k}))$$

where the term in parentheses equals $\Pr(N_{kj} > 0|M_k)$ under the binomial model of N_{kj}. Therefore, if both ψ and q_k are to be estimated, q_k must be specified as a function of covariates of sample unit k.

This occupancy model is best suited to situations where the abundance M_k in a sample unit is not too high because $\Pr(N_{kj} > 0|M_k) \to 1$ as M_k increases. If M_k is too high, detectability and abundance of individuals cannot both be estimated. The abundance-based occupancy model is therefore useful when individual densities are relatively low throughout the population.

Modeling counts of groups of individuals in distance categories: Some species of prey are detected more conveniently as groups of individuals. We use the term, group, to denote a cluster of individual animals observed together, not a social group of individuals. A group can include one or several individuals, or it can even contain dozens of individuals (e.g., chital). In line-transect surveys of these species, it is difficult to record distances for individual animals, and it is more practical to record observations associated with groups of individuals. Specifically, each group is recorded as being detected in a discrete category of distance. Similarly, group size, the number of individuals per group, is recorded using a discrete category of abundance, particularly if group sizes are difficult to determine accurately during a survey. We assume that the observed data are classified correctly with respect to each category of distance and group size.

To model these group-level data, we need to honor the discrete nature of the observations of distance and group size. We also need to account for differences in detectability associated with these observations. That is, we expect larger groups to be more detectable than groups composed of fewer individuals, and we expect groups farther from the transect line to be less detectable than groups located closer to the transect line.

To simplify our model of group-level data, we ignore the temporal replication of each transect survey and instead concentrate on estimating the total abundance N_k of groups summed over J replicates of sample unit k. Therefore, N_k corresponds to the number of groups present in a hypothetical area of Ja_k, where a_k is the actual area surveyed during each replicate of the kth transect. Given this definition of N_k, the expected density of groups λ_k is a parameter of the following Poisson model: $N_k \sim \text{Poisson}(\lambda_k J a_k)$.

Suppose the observed perpendicular distance from each transect line is divided into the following H_k categories: $[0, b_{k1}], (b_{k1}, b_{k2}], \ldots, (b_{k,H_k-1}, b_{k,H_k}]$, noting that $b_{k,H_k} \equiv B_k$. In addition, suppose the observed number of individuals per group is divided into the following I categories: $[1, c_1], (c_1, c_2], \ldots, (c_{I-1}, c_I]$. The observed data correspond to the number of groups in each of the IH_k combinations of group size and distance categories.

Let $\mathbf{n}_k = (n_{k,11}, n_{k,12}, \ldots, n_{k,IH_k})'$ denote a vector of the IH_k group counts observed in sample unit k. We can relate these counts to group abundance N_k using a multinomial model as follows:

$$\mathbf{n}_k | N_k \sim \text{Multinomial}(N_k, \boldsymbol{\pi}_k)$$

where $\boldsymbol{\pi}_k = (\pi_{k,11}, \pi_{k,12}, \ldots, \pi_{k,IH_k})'$ denotes a vector of IH_k probabilities, one for each combination of group size category i and distance category h. To complete this observation model, we need to formulate an expression for each multinomial probability $\pi_{k,ih}$, which denotes the probability that a group from size category i is present and is detected in distance category h.

Assume $\pi_{k,ih} = \bar{g}_i \bar{p}_{k,ih}$, where $\bar{g}_i$ is the probability that a group is a member of size category i and where $\bar{p}_{k,ih}$ is the probability that a group belonging to size category i is present and detected in distance category h.

Let m denote the number of individuals in a group. We assume that the distribution of group size is the same for all sample units, a zero-truncated Poisson with mean $\mu/(1-\exp(-\mu))$; therefore, the probability that a group contains exactly m individuals is

$$g(m) = \frac{\exp(-\mu)\mu^m}{m!(1-\exp(-\mu))}$$

for $m \in \{1, 2, \ldots\}$. Given this assumption, the probability that a group is a member of size category i is $\bar{g}_i = \sum_{m=c_{i-1}+1}^{c_i} g(m)$, and the average number of individuals in this size category is

$$\bar{m}_i = (1/\bar{g}_i) \sum_{m=c_{i-1}+1}^{c_i} m \times g(m)$$

To calculate $\bar{p}_{k,ih}$, let $x_{ki} \in [0, B_k]$ denote the perpendicular distance at which a group of size category i is detected in sample unit k. We assume

$$\begin{aligned} x_{ki} &\sim \text{Uniform}(0, B_k) \\ y_{ki} \mid x_{ki} &\sim \text{Bernoulli}(p(x_{ki})) \end{aligned}$$

where y_{ki} indicates whether the group is detected ($y_{ki} = 1$) or not ($y_{ki} = 0$) with probability

$$p(x_{ki}) = \exp\left(-\frac{x_{ki}^2}{2\sigma_i^2}\right)$$

This detection probability is functionally dependent on the average number of individuals in size category i because we assume the following log-linear model for the scale parameter σ_i:

$$\log(\sigma_i) = \alpha_0 + \alpha_1(\bar{m}_i - 1)$$

where the parameter α_1 is assumed to be strictly positive to ensure that both σ_i and $p(x_{ki})$ increase monotonically with average group size $\bar{m}_i$. Given these assumptions, we calculate $\bar{p}_{k,ih}$ by noting that

$$\begin{aligned} \bar{p}_{k,ih} &= \Pr(y_{ki} = 1 | x_{ki} \in h)\ \Pr(x_{ki} \in h) \\ &= \Pr(y_{ki} = 1 | x_{ki} \in h)\ (b_{kh} - b_{k,h-1})/B_k \end{aligned} \tag{8.3}$$

where

$$\begin{aligned}\Pr(y_{ki} = 1|x_{ki} \in h) &= \int_{b_{k,h-1}}^{b_{kh}} \Pr(y_{ki} = 1|x_{ki}) \ \Pr(x_{ki}|x_{ki} \in h)\mathrm{d}x_{ki} \\ &= \frac{1}{(b_{kh} - b_{k,h-1})} \int_{b_{k,h-1}}^{b_{kh}} p(x_{ki}) \, \mathrm{d}x_{ki} \\ &= \frac{1}{(b_{kh} - b_{k,h-1})} \int_{b_{k,h-1}}^{b_{kh}} \exp\left(-\frac{x_{ki}^2}{2\sigma_i^2}\right) \mathrm{d}x_{ki}\end{aligned}$$

Substitution of the above expression for $\Pr(y_{ki} = 1|x_{ki} \in h)$ into the right hand side of Eq. 8.3 yields

$$\begin{aligned}\bar{p}_{k,ih} &= \frac{1}{B_k} \int_{b_{k,h-1}}^{b_{kh}} \exp\left(-\frac{x_{ki}^2}{2\sigma_i^2}\right) \mathrm{d}x_{ki} \\ &= \frac{\sqrt{2\pi}\sigma_i}{B_k}(F(b_h) - F(b_{h-1}))\end{aligned}$$

where $F(\cdot)$ is the cumulative distribution function of a Normal$(0, \sigma_i^2)$ distribution.

We conclude this section by noting that the expected number of individual prey in sample unit k may be calculated by multiplying the expected group size, the expected density of groups, and the area surveyed as follows: $\mu\lambda_k a_k/(1-\exp(-\mu))$.

8.2.3.2 Double-Observer Surveys

We now describe models for the case where repeated surveys of the same sample unit are collected by two dependent observers (Cook and Jacobson 1979; Nichols et al. 2000). For example, in a line-transect survey, suppose one observer is responsible for detecting individual prey, while the other observer records the prey's distance from the transect. If the first observer fails to detect an individual that is seen by the second observer, the second observer still records that individual's distance (assuming both observers are equipped with range finders or another measuring device). We assume that the two observers switch roles (as first and second (dependent) observer) during the J surveys of the same transect. Switching roles allows estimation of each observer's ability to detect prey.

Modeling observer-specific distance of each individual from a transect line: Let n_{kj1} and n_{kj2} denote the numbers of individuals detected by the first and second observers, respectively, during the jth survey of sample unit k. Let x_{kj} denote the perpendicular distance between each individual's location and the transect line during the jth survey of sample unit k.

To estimate N_{kj}, we specify the probability of detecting an individual using the half-normal detection function for each observer as follows:

$$p_1(x_{kj}) = \exp\left(-\frac{x_{kj}^2}{2\sigma_1^2}\right)$$

$$p_2(x_{kj}) = \exp\left(-\frac{x_{kj}^2}{2\sigma_2^2}\right)$$

where σ_1 and σ_2 are strictly positive parameters for the first and second observers, respectively. Differences in the values of σ_1 and σ_2 correspond to differences in abilities of the two observers to detect individual prey.

The primary assumptions needed for estimating the model's parameters are:

$$x_{kj} \sim \text{Uniform}(0, B_k) \tag{8.4}$$

$$y_{kj1} \mid x_{kj} \sim \text{Bernoulli}(p_1(x_{kj})) \tag{8.5}$$

$$y_{kj2} \mid y_{kj1}, x_{kj} \sim \text{Bernoulli}((1 - y_{kj1})\, p_2(x_{kj})) \tag{8.6}$$

where B_k is the maximum perpendicular distance from transect k at which individuals can be detected and where y_{kj1} indicates whether an individual is detected ($y_{kj1} = 1$) or not detected ($y_{kj1} = 0$) by the first observer during the jth survey of transect k. The value of y_{kj2} indicates whether the second observer succeeds ($y_{kj2} = 1$) or fails ($y_{kj2} = 0$) to detect an individual that was not detected by the first observer.

Given these definitions, $y_{kj1} = 1$ for each of the n_{kj1} individuals detected by the first observer, and $y_{kj1} = 0$ and $y_{kj2} = 1$ for each of the n_{kj2} individuals detected only by the second observer. All that is known of the $N_{kj} - (n_{kj1} + n_{kj2})$ undetected individuals is that $y_{kj1} = y_{kj2} = 0$ for each individual. Because distance x_{kj} is not observed for these individuals, we need to compute the *marginal* probability that each individual was missed (not detected by either observer) during the jth survey of sample unit k. This probability, which we denote by π_k, depends on σ_1, σ_2, and B_k as follows:

$$\pi_k = \left[1 - \frac{\sqrt{2\pi}\sigma_1}{B_k}\left(F_1(B_k) - \frac{1}{2}\right)\right]\left[1 - \frac{\sqrt{2\pi}\sigma_2}{B_k}\left(F_2(B_k) - \frac{1}{2}\right)\right]$$

where $F_1(\cdot)$ and $F_2(\cdot)$ denote the cumulative distribution functions of Normal$(0, \sigma_1^2)$ and Normal$(0, \sigma_2^2)$ distributions, respectively.

Each of the undetected individuals has probability π_k of being missed; therefore, the sum of their individual Bernoulli responses, which equals $N_{kj} - (n_{kj1} + n_{kj2})$, has a binomial distribution with index parameter N_{kj} and probability π_k. Equivalently, we can express the distribution of the number of observed individuals $(n_{kj1} + n_{kj2})$ conditional on N_{kj} as follows:

$$n_{kj1} + n_{kj2} | N_{kj} \sim \text{Binomial}(N_{kj}, 1 - \pi_k)$$

This binomial distribution is a consequence of the assumptions given in Eqs. (8.4), (8.5), and (8.6); it is not an additional assumption of the model. As with the single-observer model of individual distances, the effects of habitat on $p_1(x_{kj})$ and $p_2(x_{kj})$ may be specified by formulating σ_1 and σ_2 as a function of habitat.

Modeling observer-specific counts of individuals: Suppose the individual distances measured by the two dependent observers are less informative of prey detectability than other factors, such as habitat. In this situation we can estimate prey abundance using only the counts obtained by each observer. A model of these counts conditional on abundance N_{kj} is

$$n_{kj1}, n_{kj2} \mid N_{kj} \sim \text{Multinomial}(N_{kj},\ p_{k1},\ (1 - p_{k1})p_{k2})$$

where p_{k1} and p_{k2} are the detection probabilities of the first and second (dependent) observer, respectively (Shirley et al. 2012). Marginalizing N_{kj} from the model yields the following model of double-observer counts:

$$n_{kj1}, n_{kj2} \mid M_k \sim \text{Multinomial}(M_k,\ \psi p_{k1},\ \psi(1 - p_{k1})p_{k2})$$

As with the model of single-observer counts, the effect of habitat on detection of prey can be included by specifying logit-linear models as follows:

$$\begin{aligned} \text{logit}(p_{k1}) &= \eta_1 + \boldsymbol{\alpha}'\mathbf{w}_k \\ \text{logit}(p_{k2}) &= \eta_2 + \boldsymbol{\alpha}'\mathbf{w}_k \end{aligned}$$

where the parameters η_1 and η_2 are logit-scale effects of the abilities of the first and second observers, respectively, and where $\boldsymbol{\alpha}$ is a vector of the effects of the predictors $\mathbf{w}_k$.

8.3 Analysis of Data Sets

8.3.1 Simulated Data

In this section we analyze six hypothetical data sets constructed by assuming realistic levels of prey density and prey detection probability in a real landscape. To construct the sample frame, we partitioned the Nagarahole Tiger Reserve, which is located in Karnataka, India, into 1358 rectangular sample units, each measuring 1.6×0.4 km. We chose these dimensions to accommodate a transect of length 1.6 km and a maximum detectable distance of 200 m on each side of the transect.

The expected density of individual prey in each unit was modeled as a log-linear function of habitat quality, whose spatial distribution included patches or clusters of high-quality habitat (Fig. 8.1). In real surveys, level of forest cover or understory foliage could be an example of habitat.

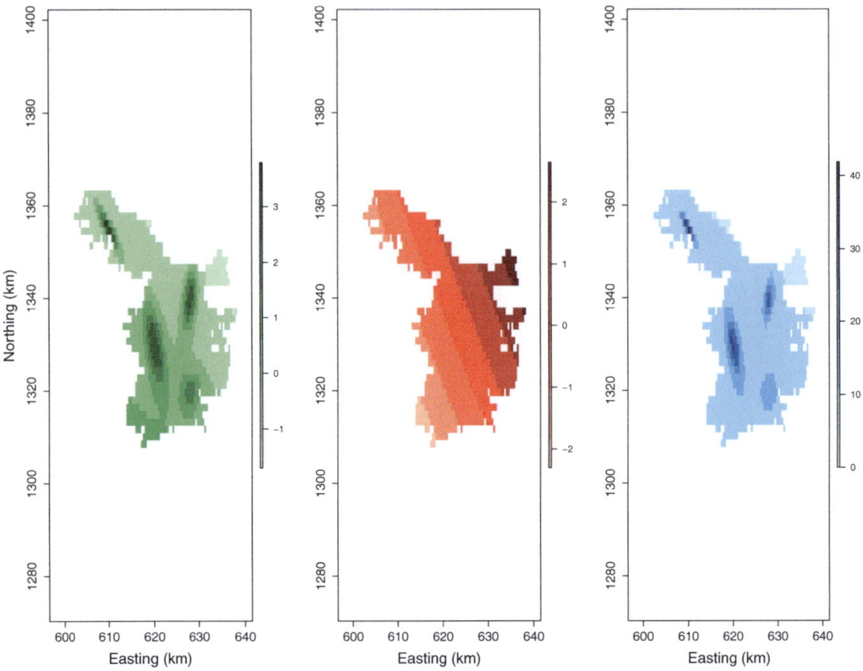

Fig. 8.1 Maps of the spatial distribution of an abundance covariate (*left panel*), a detection covariate (*middle panel*), and the expected density of individuals (*right panel*)

A prey population with medium to high densities of individuals: We simulated one prey population so that the expected density ranged from about 3 to 42 individuals/km^2 (Fig. 8.1); however, in most (85%) of the sample units, the expected density was less than 10 individuals/km^2 owing to the spatial heterogeneity in habitat quality. For purposes of sampling, we selected $K = 200$ sample units randomly and applied two sampling protocols (single observer and double observer) within each unit to generate four types of data (i.e., perpendicular distances and counts of individuals for each two protocols). Each transect was surveyed on $J = 4$ occasions. In the single-observer surveys, the effect of habitat on individual detection probability was specified by modeling σ as a function of a covariate whose value increased from southwest to northeast (Fig. 8.1). The simulated values of σ ranged from 71 to 358 m. A similar approach was used to model the effects of habitat and observer in the double-observer surveys. The simulated values of σ_1 ranged from 71 to 358 m; in contrast, the values of σ_2 ranged from 36 to 242 m, indicating that the second observer was less skilled at detecting individual prey.

A prey population with low densities of individuals: We simulated a second prey population so that the expected density ranged from about 0.2 to 3 individuals/km^2

(Fig. 8.4); however, in most (92%) of the sample units, the expected density was less than 1 individual/km^2 owing to the spatial heterogeneity in habitat quality. We sampled this population intensively by selecting $K = 500$ units randomly and by applying two single-observer sampling protocols (counts of individuals and presence-absence sampling) within each unit during $J = 10$ occasions. The effect of habitat on individual detection probability was specified by modeling σ as a function of a covariate whose value increased from southwest to northeast (Fig. 8.1), as described earlier. The maximum number of counts within a unit ranged from zero to four, and one or more individuals were detected in only 29% of the surveyed units.

8.3.1.1 Results

We estimated the parameters of each model using the method of maximum likelihood and the R software program (R Core Team 2017). Our source code is available upon request. The parameter estimates then were used to compute the expected abundance of individuals in each unit (including units that were not surveyed) as described in Sect. 8.2.2.

A prey population with medium to high densities of individuals: All of the fitted models predicted the spatial distribution of prey density reasonably well (compare the true map in Fig. 8.1 with the estimated maps in Figs. 8.2 and 8.3). Model-based predictions of the expected abundance of prey in all units were similar (5915, 6037,

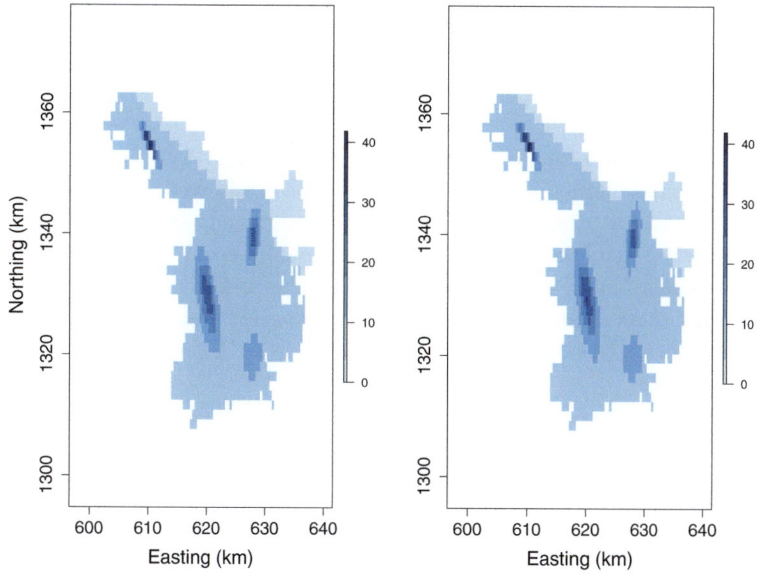

Fig. 8.2 Maps of expected density of individuals predicted from single-observer models of distances (*left panel*) and counts (*right panel*)

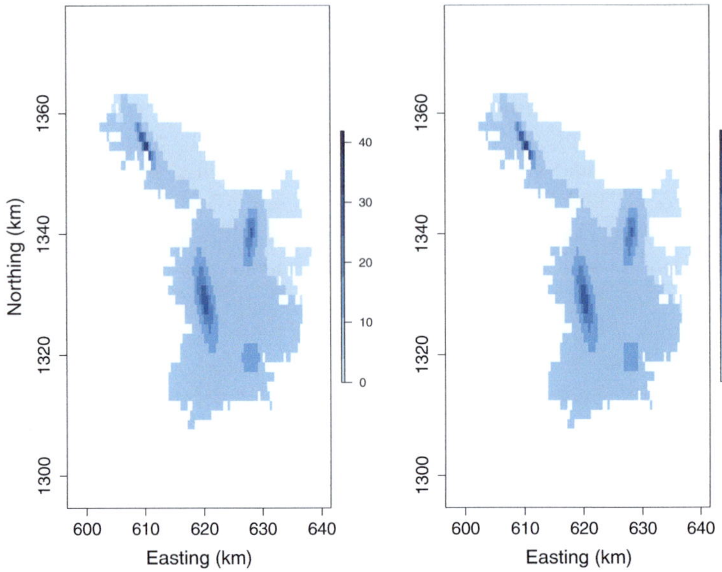

Fig. 8.3 Maps of expected density of individuals predicted from double-observer models of distances (*left panel*) and counts (*right panel*)

5295, and 5238 individuals). Of course, the predictions cannot be expected to equal the true abundance (6302 individuals) exactly because information from only 200 of the 1358 sample units was used to predict the abundance of prey in all 1358 sample units. In addition, the count model (without distances) was a misspecification of the model actually used to simulate the data.

A prey population with low densities of individuals: Both fitted models predicted the spatial distribution of prey density reasonably well (compare panels in Fig. 8.4). Model-based predictions of the expected abundance of prey in all units were similar (475 and 436 individuals) but not exactly equal to the true abundance (450 individuals), even though the sample included information from 500 of the 1358 sample units. These results were not unexpected given that the models of counts and quantized counts were misspecifications of the model actually used to simulate the data.

8.3.2 Line-Transect Surveys of Chital

In this section we analyze counts of groups of chital observed in 2005–2006 while conducting line-transect surveys at locations within the Nagarahole and Bandipur Tiger Reserves of India. These surveys were undertaken to estimate the

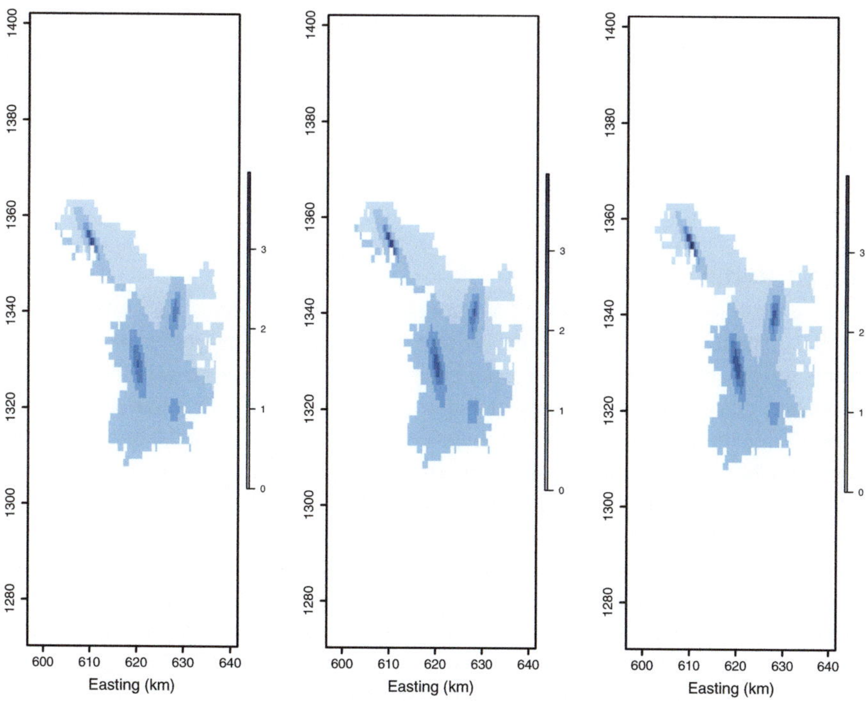

Fig. 8.4 Maps of expected density of individuals predicted from single-observer models of counts (*middle panel*) and presence-absence data (*right panel*). *Left panel* contains a map of the true expected density of individuals

spatial distribution and abundance of chital in the reserves. Kumar (2011) provided a detailed description of these surveys. Briefly, chital were observed along 77 transects that varied in length from 0.8 to 3.2 km. Each transect was surveyed six or seven times within a relatively short period of time (less than 30 days). Categories of group size and perpendicular distance were assigned to each group of chital detected during the surveys. Each group size category included five levels of abundance, and the abundances of 12 group size categories ranged from 1 to 60 individuals (i.e., [1,5], [6,10], . . ., [56,60]). Perpendicular distance categories were obtained by dividing the maximum observable distance (380 m) into 19 categories (i.e., [0,20], (20,40], . . ., (360,380]).

We formulated a log-linear model of the expected abundance of chital groups using a combination of both transect-level and grid-level covariates. A spatial grid of the reserves was obtained by partitioning the study area into 1792 square cells, each of 1 km^2 area. Potential covariates of group abundance at the grid cell level included protection ineffectiveness, ecoclimatic distance, terrain undulation, and distance to water. Potential covariates of group abundance at the transect level included forage quantity and habitat disturbance. A full description of these covariates and of the log-linear model that accounts for spatial misalignment between transect-level and grid-cell-level information was provided by Kumar (2011). Bayesian methods and WinBUGS software (Lunn et al. 2000) were used to fit this model because the marginal likelihood function obtained by integrating out the spatially correlated, random effects cannot be evaluated in closed form (Kumar 2011). Maximization of this likelihood function was therefore not feasible for estimating the model's parameters.

8.3.2.1 Results

During the line-transect surveys, 559 groups of chital were detected while walking a total distance of 1404 km. The estimated detection probabilities of relatively small groups declined with perpendicular distance much more rapidly than the estimated detection probabilities of larger groups (Fig. 8.5). In addition, because most (94%) of the observed chital groups contained ≤ 10 individuals, uncertainty in the estimates of detection probability of groups containing more than 10 individuals was considerably higher than the uncertainty of detection probabilities estimated for groups with fewer individuals.

Our estimate of the expected group size was 6.38 individuals (95% interval, 6.14–6.62), which is slightly higher than the average size of all observed groups (5.2 individuals). Chital group abundance was positively associated with forage quantity and negatively associated with ecoclimatic distance (Table 8.1). The effects of all other covariates of chital group abundance were not significant.

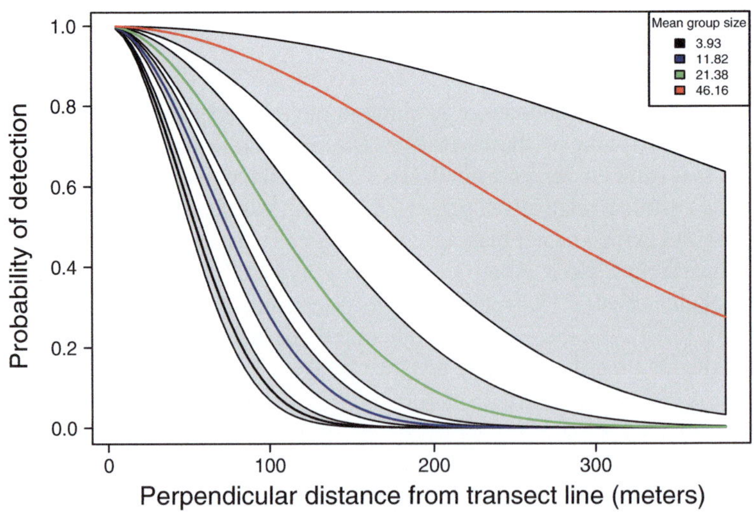

Fig. 8.5 Estimated relationship between the probability of chital group detection and the perpendicular distance from the transect line for each of four categories of chital group size. *Shaded region* indicates the 95% credible interval for the probability of chital group detection

Table 8.1 Estimated posterior means and 95% credible limits for the effects of transect-level and grid-cell-level covariates on expected abundance of chital groups. Monte Carlo standard errors in parentheses

Covariate	Mean		2.5%		97.5%	
Forage quantity	0.30	(0.0043)	0.03	(0.0047)	0.60	(0.0057)
Habitat disturbance	0.10	(0.0028)	−0.16	(0.0037)	0.36	(0.0039)
Terrain undulation	−0.82	(0.0238)	−2.37	(0.0334)	0.28	(0.0181)
Distance to water	−0.17	(0.0034)	−0.47	(0.0043)	0.14	(0.0047)
Ecoclimatic distance	−0.92	(0.0142)	−1.79	(0.0176)	−0.08	(0.0149)
Protection ineffectiveness	−0.21	(0.0049)	−0.57	(0.0063)	0.14	(0.0061)

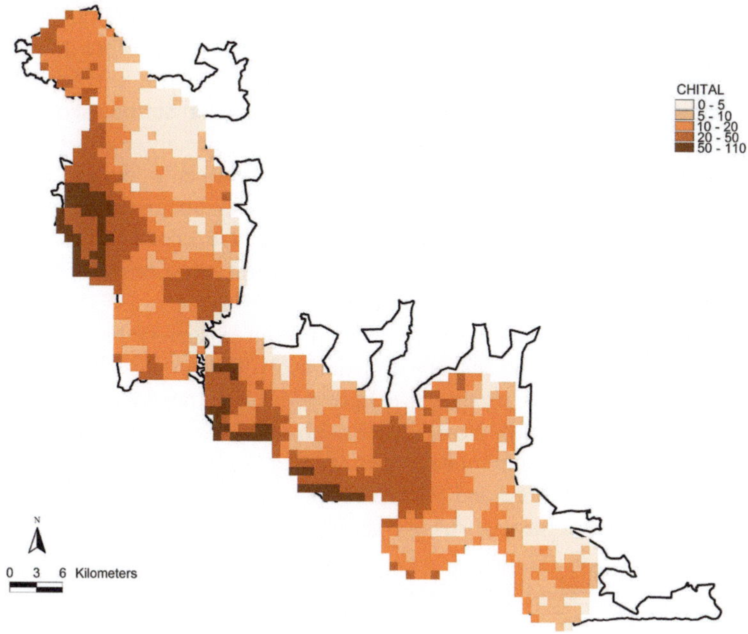

Fig. 8.6 Spatial distribution of chital abundance in the Nagarahole and Bandipur Tiger Reserves of India. Resolution is 1 km^2

Spatial variation was clearly evident in our estimates of the expected abundance of chital in different parts of the tiger reserves (Fig. 8.6). Grid cells from the westernmost regions of the reserves generally contained higher abundances of chital than grid cells lying to the east of these regions. The western-most parts of the reserve are the wettest forest patches (where the degree of deciduous or the ecoclimatic distance is lower) with relatively flat terrain and lower protection ineffectiveness, all of which lend support to higher levels of chital abundance.

8.4 Discussion

In this chapter we described hierarchical models for the analysis of data observed in temporally and spatially replicated surveys. Some of the models require the locations of individuals detected during each survey. Other models require only counts of the individuals detected by either one or two observers. The abundance-based occupancy model, which only requires observations from presence-absence surveys, can even estimate prey abundances when density of individuals is sufficiently low. That said, relatively high sample sizes may be needed to estimate prey abundances accurately from presence-absence data.

An important aspect of all of these models is that they can account for spatial variation in abundance of prey induced by differences in habitat at sample locations. These fully model-based estimators of prey density are therefore much more versatile than their conventional, design-unbiased counterparts. A fully model-based analysis of spatially and temporally replicated survey data can be used to predict a map of the spatial distribution of prey abundance or density, thus offering new opportunities to examine patterns of spatial overlap between tigers and their prey. A fully model-based analysis also provides inferences about the effects of the physical habitat or management-related factors that are known to influence the spatial distribution of prey, thereby permitting central questions of prey species ecology and conservation to be considered in the analysis.

In this chapter we have not attempted to include an exhaustive list of the various approaches developed for the analysis of data observed in temporally and spatially replicated wildlife surveys. Rather, we have used our knowledge of tiger prey sampling methods to develop probability-based models that can be used to solve a variety of estimation and inference problems involving tiger prey populations. We believe that this kind of direct and principled approach to statistical model building has obvious benefits for the analysis of tiger prey surveys.

Appendix 1

In this appendix we derive the likelihood function for estimating the parameters of each model described in the text. We use bracket notation (Gelfand and Smith 1990) to specify probability density functions; thus, $[x, y]$ denotes the joint density of random variables X and Y, $[x|y]$ denotes the conditional density of X given $Y = y$, and $[x]$ denotes the unconditional (marginal) density of X.

A.1 Single-Observer Surveys

Modeling distance of each individual from a transect line: The conditional probability density of data observed during the jth survey of the kth sample unit is

$$[\mathbf{x}_{kj}, n_{kj}|N_{kj}] = \binom{N_{kj}}{n_{kj}} \pi_k^{N_{kj}-n_{kj}} (1/B_k)^{n_{kj}} \prod_{i=1}^{n_{kj}} p(x_{kj,i})$$

where $\mathbf{x}_{kj} = (x_{kj,1}, \ldots, x_{kj,n_{kj}})'$ is a vector that contains the perpendicular distances from the transect line of the n_{kj} individuals observed during the survey. We can marginalize N_{kj} from the joint density $[\mathbf{x}_{kj}, n_{kj}, N_{kj}]$ to obtain the probability density of the data conditional on M_k as follows:

$$\begin{aligned}[\mathbf{x}_{kj}, n_{kj}|M_k] &= \sum_{N_{kj}=n_{kj}}^{M_k} [N_{kj}|M_k][\mathbf{x}_{kj}, n_{kj}|N_{kj}] \\ &= \binom{M_k}{n_{kj}} \psi^{n_{kj}} (\psi\pi_k + 1 - \psi)^{M_k - n_{kj}} (1/B_k)^{n_{kj}} \prod_{i=1}^{n_{kj}} p(x_{ikj})\end{aligned}$$

(derivation omitted for brevity).

To obtain the likelihood function for estimating ψ, σ, and the parameters that specify λ_k ($\boldsymbol{\beta}$), we marginalize M_k from the joint density $[\{\mathbf{x}_{kj}, n_{kj}\}_{j=1}^J,\ M_k]$ as follows:

$$\begin{aligned}[\{\mathbf{x}_{kj}, n_{kj}\}_{j=1}^J] &= \sum_{M_k=max(\{n_{kj}\})}^{\infty} [M_k] \prod_{j=1}^{J} [\mathbf{x}_{kj}, n_{kj}|M_k] \\ &= \sum_{M_k=max(\{n_{kj}\})}^{\infty} \frac{\exp(-\lambda_k a_k)(\lambda_k a_k)^{M_k}}{M_k!} \prod_{j=1}^{J} [\mathbf{x}_{kj}, n_{kj}|M_k]\end{aligned}$$

In practice, this marginalization is done numerically by replacing the infinite limit of summation with a sufficiently high abundance. Assuming independence among sample units, the likelihood function for the entire data set is

$$L(\psi, \sigma, \boldsymbol{\beta}) = \prod_{k=1}^{K} [\{\mathbf{x}_{kj}, n_{kj}\}_{j=1}^J]$$

Modeling counts of all individuals: The conditional probability of the number of individuals n_{kj} observed during the jth survey of the kth sample unit is

$$[n_{kj}|M_k] = \binom{M_k}{n_{kj}} (\psi p_k)^{n_{kj}} (1 - \psi p_k)^{M_k - n_{kj}}$$

where $\text{logit}(p_k) = \boldsymbol{\alpha}'\mathbf{w}_k$.

To obtain the likelihood function for estimating ψ, $\boldsymbol{\alpha}$, and the parameters that specify λ_k ($\boldsymbol{\beta}$), we marginalize M_k from the joint density $[\{n_{kj}\}_{j=1}^J, M_k]$ as follows:

$$[\{n_{kj}\}_{j=1}^J] = \sum_{M_k=max(\{n_{kj}\})}^{\infty} [M_k] \prod_{j=1}^{J} [n_{kj}|M_k]$$

$$= \sum_{M_k=max(\{n_{kj}\})}^{\infty} \frac{\exp(-\lambda_k a_k)(\lambda_k a_k)^{M_k}}{M_k!} \prod_{j=1}^{J} [n_{kj}|M_k]$$

As described earlier, this marginalization is done numerically by replacing the infinite limit of summation with a sufficiently high abundance. Assuming independence among sample units, the likelihood function for the entire data set is

$$L(\psi, \boldsymbol{\alpha}, \boldsymbol{\beta}) = \prod_{k=1}^{K} [\{n_{kj}\}_{j=1}^J]$$

Modeling a quantized count of all individuals: The conditional probability of observing the quantized count z_{kj} during the jth survey of the kth sample unit is

$$[z_{kj}|M_k] = \{q_k\,(1-(1-\psi)^{M_k})\}^{z_{kj}}\{1-q_k\,(1-(1-\psi)^{M_k})\}^{1-z_{kj}}$$

where $\text{logit}(q_k) = \boldsymbol{\alpha}'\mathbf{w}_k$.

To obtain the likelihood function for estimating ψ, $\boldsymbol{\alpha}$, and the parameters that specify λ_k ($\boldsymbol{\beta}$), we marginalize M_k from the joint density $[\{z_{kj}\}_{j=1}^J, M_k]$ as follows:

$$[\{z_{kj}\}_{j=1}^J] = \sum_{M_k=max(\{z_{kj}\})}^{\infty} [M_k] \prod_{j=1}^{J} [z_{kj}|M_k]$$

$$= \sum_{M_k=max(\{z_{kj}\})}^{\infty} \frac{\exp(-\lambda_k a_k)(\lambda_k a_k)^{M_k}}{M_k!} \prod_{j=1}^{J} [z_{kj}|M_k]$$

As described earlier, this marginalization is done numerically by replacing the infinite limit of summation with a sufficiently high abundance. Assuming independence among sample units, the likelihood function for the entire data set is

$$L(\psi, \boldsymbol{\alpha}, \boldsymbol{\beta}) = \prod_{k=1}^{K} [\{z_{kj}\}_{j=1}^J]$$

Modeling counts of groups of individuals in distance categories: The conditional probability of the vector of group counts $\mathbf{n}_k$ observed in the kth sample unit is

$$[\mathbf{n}_k|N_k] = \frac{N_k!}{n_{k,11}! \cdots n_{k,IH_k}!(N_k - n_{k\cdot})!} \pi_{k,11}^{n_{k,11}} \cdots \pi_{k,IH_k}^{n_{k,IH_k}} (1 - \pi_{k\cdot})^{N_k - n_{k\cdot}}$$

where $n_{k\cdot} = \sum_{ih} n_{k,ih}$ is the total number of groups observed in sample unit k and where $\pi_{k\cdot} = \sum_{ih} \pi_{k,ih}$.

To obtain the likelihood function for estimating μ, $\boldsymbol{\alpha}$, and the parameters that specify λ_k ($\boldsymbol{\beta}$), we note that the *unconditional* probability of the vector of group counts can be computed analytically as a product of Poisson probabilities as follows:

$$[\mathbf{n}_k] = \prod_{ih} \exp(-\lambda_k J a_k \pi_{k,ih})(\lambda_k J a_k \pi_{k,ih})^{n_{k,ih}} / n_{k,ih}!$$

Therefore, assuming independence among sample units, the likelihood function for the entire data set is

$$L(\mu, \boldsymbol{\alpha}, \boldsymbol{\beta}) = \prod_{k=1}^{K} [\mathbf{n}_k]$$

A.2 Double-Observer Surveys

Modeling distance of each individual from a transect line: The conditional probability density of data observed during the jth survey of the kth sample unit is

$$[\mathbf{x}_{kj1}, n_{kj1}, \mathbf{x}_{kj2}, n_{kj2}|N_{kj}] = \binom{N_{kj}}{n_{kj}} \pi_k^{N_{kj}-n_{kj}} (1/B_k)^{n_{kj}} \left[\prod_{i=1}^{n_{kj1}} p_1(x_{kj1,i})\right] \times \left[\prod_{l=1}^{n_{kj2}} (1 - p_1(x_{kj2,l}))\, p_2(x_{kj2,l})\right]$$

where $\mathbf{x}_{kj1}$ and $\mathbf{x}_{kj2}$ are vectors that contain the perpendicular distances from the transect line of individuals detected by observers 1 and 2, respectively, and where $n_{kj} = n_{kj1} + n_{kj2}$ is the total number of individuals detected by observers 1 and 2 during the jth survey of sample unit k. We can marginalize N_{kj} from the joint density $[\mathbf{x}_{kj1}, n_{kj1}, \mathbf{x}_{kj2}, n_{kj2}, N_{kj}]$ to obtain the probability density of the data conditional on M_k as follows:

$$
\begin{aligned}
[\mathbf{x}_{kj1}, n_{kj1}, \mathbf{x}_{kj2}, n_{kj2} | M_k] &= \sum_{N_{kj}=n_{kj}}^{M_k} [N_{kj}|M_k][\mathbf{x}_{kj1}, n_{kj1}, \mathbf{x}_{kj2}, n_{kj2}|N_{kj}] \\
&= \binom{M_k}{n_{kj}} \psi^{n_{kj}} (\psi \pi_k + 1 - \psi)^{M_k - n_{kj}} (1/B_k)^{n_{kj}} \\
&\quad \times \left[\prod_{i=1}^{n_{kj1}} p_1(x_{kj1,i}) \right] \times \left[\prod_{l=1}^{n_{kj2}} (1 - p_1(x_{kj2,l}))\, p_2(x_{kj2,l}) \right]
\end{aligned}
$$

(derivation omitted for brevity).

To obtain the likelihood function for estimating ψ, σ_1, σ_2, and the parameters that specify λ_k ($\boldsymbol{\beta}$), we marginalize M_k from the joint density $[\{\mathbf{x}_{kj1}, n_{kj1}, \mathbf{x}_{kj2}, n_{kj2}\}_{j=1}^{J}, M_k]$ as follows:

$$
\begin{aligned}
[\{\mathbf{x}_{kj1}, n_{kj1}, \mathbf{x}_{kj2}, n_{kj2}\}_{j=1}^{J}] &= \sum_{M_k = max(\{n_{kj}\})}^{\infty} [M_k] \prod_{j=1}^{J} [\mathbf{x}_{kj1}, n_{kj1}, \mathbf{x}_{kj2}, n_{kj2} | M_k] \\
&= \sum_{M_k = max(\{n_{kj}\})}^{\infty} \frac{\exp(-\lambda_k a_k)(\lambda_k a_k)^{M_k}}{M_k!} \prod_{j=1}^{J} [\mathbf{x}_{kj1}, n_{kj1}, \mathbf{x}_{kj2}, n_{kj2} | M_k]
\end{aligned}
$$

In practice, this marginalization is done numerically by replacing the infinite limit of summation with a sufficiently high abundance. Assuming independence among sample units, the likelihood function for the entire data set is

$$
L(\psi, \sigma_1, \sigma_2, \boldsymbol{\beta}) = \prod_{k=1}^{K} [\{\mathbf{x}_{kj1}, n_{kj1}, \mathbf{x}_{kj2}, n_{kj2}\}_{j=1}^{J}]
$$

Modeling counts of all individuals: The conditional probability of the numbers of individuals detected by observers 1 and 2 during the jth survey of the kth sample unit is

$$
\begin{aligned}
[n_{kj1}, n_{kj2} | M_k] &= \frac{M_k}{n_{kj1}! n_{kj2}! (M_k - n_{kj})!} (\psi p_{k1})^{n_{kj1}} (\psi (1 - p_{k1}) p_{k2})^{n_{kj2}} \\
&\quad \times (\psi (1 - p_{k1})(1 - p_{k2}) + 1 - \psi)^{M_k - n_{kj}}
\end{aligned}
$$

where $n_{kj} = n_{kj1} + n_{kj2}$ is the total number of individuals detected by observers 1 and 2 during the jth survey of sample unit k and where

$$
\begin{aligned}
\text{logit}(p_{k1}) &= \eta_1 + \boldsymbol{\alpha}' \mathbf{w}_k \\
\text{logit}(p_{k2}) &= \eta_2 + \boldsymbol{\alpha}' \mathbf{w}_k
\end{aligned}
$$

as described earlier.

To obtain the likelihood function for estimating ψ, η_1, η_2, $\boldsymbol{\alpha}$, and the parameters that specify $\lambda_k(\boldsymbol{\beta})$, we marginalize M_k from the joint density $[\{n_{kj1}, n_{kj2}\}_{j=1}^{J}, M_k]$ as follows:

$$[\{n_{kj1}, n_{kj2}\}_{j=1}^{J}] = \sum_{M_k=max(\{n_{kj}\})}^{\infty} [M_k] \prod_{j=1}^{J} [n_{kj1}, n_{kj2} | M_k]$$

$$= \sum_{M_k=max(\{n_{kj}\})}^{\infty} \frac{\exp(-\lambda_k a_k)(\lambda_k a_k)^{M_k}}{M_k!} \prod_{j=1}^{J} [n_{kj1}, n_{kj2} | M_k]$$

As described earlier, this marginalization is done numerically by replacing the infinite limit of summation with a sufficiently high abundance. Assuming independence among sample units, the likelihood function for the entire data set is

$$L(\psi, \eta_1, \eta_2, \boldsymbol{\alpha}, \boldsymbol{\beta}) = \prod_{k=1}^{K} [\{n_{kj1}, n_{kj2}\}_{j=1}^{J}]$$

References

Buckland ST, Anderson DR, Burnham KP, Laake JL, Borchers DL, Thomas L (2001) Introduction to distance sampling. Oxford University Press, Oxford

Buckland ST, Anderson DR, Burnham KP, Laake JL, Borchers DL, Thomas L (2004) Advanced distance sampling. Oxford University Press, Oxford

Buckland ST, Oedekoven CS, Borchers DL (2015) Model-based distance sampling. J Agric Biol Environ Stat 21:58–75

Chandler RB, Royle JA, King DI (2011) Inference about density and temporary emigration in unmarked populations. Ecology 92:1429–1435

Chandler RB, Royle JA, King DI (2014) Corrections to "Inference about density and temporary emigration in unmarked populations". Ecology 94:794

Conn PB, Laake JL, Johnson DS (2012) A hierarchical modeling framework for multiple observer transect surveys. PLoS One 7:e42294

Cook RD, Jacobson JO (1979) A design for estimating visibility bias in aerial surveys. Biometrics 35:735–742

Gelfand AE, Smith AFM (1990) Sampling-based approaches to calculating marginal densities. J Am Stat Assoc 85:398–409

Hedley S, Buckland ST (2004) Spatial models for line transect sampling. J Agric Biol Environ Stat 9:181–199

Johnson DS, Laake JL, Ver Hoef JM (2010) A model-based approach for making ecological inference from distance sampling data. Biometrics 66:310–318

Karanth KU, Nichols JD, Kumar NS, Link WA, Hines JE (2004) Tigers and their prey: predicting carnivore densities from prey abundance. Proc Natl Acad Sci 101:4854–4858

Kéry M, Royle JA (2016) Applied hierarchical modeling in ecology. Academic Press, London

Koshkina V, Wang Y, Gordon A, Dorazio RM, White M, Stone L (2017) Integrated species distribution models: combining presence-background data and site-occupancy data with imperfect detection. Methods Ecol Evol 8:420–430

Kumar NS (2011) Assessment of distribution and abundance of ungulate prey using spatial models in Nagarahole and Bandipur tiger reserves of India. Ph.D. dissertation, Manipal University

Lunn DJ, Thomas A, Best N, Spiegelhalter D (2000) WinBUGS – a Bayesian modelling framework: concepts, structure, and extensibility. Stat Comput 10:325–337

MacKenzie DI, Nichols JD, Royle JA, Pollock KH, Bailey LL, Hines JE (2006) Occupancy estimation and modeling. Elsevier, Amsterdam

Marques FFC, Buckland ST (2003) Incorporating covariates into standard line transect analyses. Biometrics 59:924–935

Nichols JD, Hines JE, Sauer JR, Fallon FW, Fallon JE, Heglund PJ (2000) A double-observer approach for estimating detection probability and abundance from point counts. Auk 117:393–408

R Core Team (2017) R: a language and environment for statistical computing. R Foundation for Statistical Computing, Vienna. http://www.R-project.org/

Royle JA (2004) N-mixture models for estimating population size from spatially replicated counts. Biometrics 60:108–115

Royle JA, Dorazio RM (2006) Hierarchical models of animal abundance and occurrence. J Agric Biol Environ Stat 11:249–263

Royle JA, Dorazio RM (2008) Hierarchical modeling and inference in ecology. Academic Press, Amsterdam

Royle JA, Nichols JD (2003) Estimating abundance from repeated presence-absence data or point counts. Ecology 84:777–790

Royle JA, Dawson DK, Bates S (2004) Modeling abundance effects in distance sampling. Ecology 85:1591–1597

Shirley MH, Dorazio RM, Abassery E, Elhady A, Mekki MS, Asran HH (2012) A sampling design and model for estimating abundance of Nile crocodiles while accounting for heterogeneity of detectability of multiple observers. J Wildl Manag 76:966–975

Wikle CK (2010) Hierarchical modeling with spatial data. In: Gelfand AE, Diggle PJ, Fuentes M, Guttorp P (eds) Handbook of spatial statistics, Chapman & Hall/CRC, Boca Raton, pp 89–106

Concepts: Assessing Tiger Population Dynamics Using Capture–Recapture Sampling

9

J. Andrew Royle, Arjun M. Gopalaswamy, Robert M. Dorazio, James D. Nichols, Devcharan Jathanna, Ravishankar Parameshwaran, and K. Ullas Karanth

9.1 Introduction

Capture-recapture can be viewed as an animal survey method in which the count statistic is the total number of animals caught, and the associated detection probability is the probability of capture. This probability is estimated using the pattern of captures and recaptures of animals over the survey period. For example, a typical design might involve a small mammal trapping grid. Traps are set, and animals trapped on the first sampling occasion are individually marked (tagged) to permit identification in subsequent sampling periods. On the second sampling occasion (e.g., the next day), animals that represent recaptures (they were caught at sampling

J.A. Royle (✉)
USGS Patuxent Wildlife Research Center, Laurel, MD, USA
e-mail: aroyle@usgs.gov

A.M. Gopalaswamy
Statistics and Mathematics Unit, Indian Statistical Institute, Bengaluru, India

Department of Zoology, University of Oxford, Oxford, UK
e-mail: arjungswamy@gmail.com; https://www.isibang.ac.in; https://www.zoo.ox.ac.uk/

R.M. Dorazio
Wetland and Aquatic Research Center – United States Geological Survey (USGS), Gainesville, FL, USA
e-mail: bdorazio@usgs.gov

J.D. Nichols
Crofton, MD, USA
e-mail: jamesdnichols2@gmail.com

D. Jathanna • R. Parameshwaran
Wildlife Conservation Society, India Program, Bengaluru, India

Centre for Wildlife Studies, Bengaluru, India
e-mail: devcharan@gmail.com; ravishankar.cws@gmail.com

K.U. Karanth, J.D. Nichols (eds.), *Methods For Monitoring Tiger And Prey Populations*, DOI 10.1007/978-981-10-5436-5_9

occasion 1) are recorded, and newly captured animals are recorded and given individual marks. The process is repeated for the duration of the study (perhaps five consecutive days). The data resulting from the study then consist of the total number of animals that were caught and the record of *when* (i.e., on what sampling occasions) each animal was caught. The temporal sequence of encounter for each individual is referred to as the *individual encounter history*. The total number of animals caught is the count statistic, and the detection or capture probability is estimated from the records of the individual capture histories (Otis et al. 1978).

Although capture-recapture methods were originally developed for such cases where animals are physically caught and marked with artificial tags, the underlying concepts are valid even when animals are identified using natural markings. In the case of tigers (and other patterned animals such as leopard, chital, etc.), we can rely on their distinctive markings in the form of stripes or spots. In addition, identification of individuals based on DNA obtained from hair or scat has become widely adopted in surveys of carnivore populations (Long et al. 2008). As long as individuals can be identified reliably and sampling methods provide records of detection and non-detection over time (sampling periods) or space (traps or areas searched), then the resulting data can be used in a capture-recapture framework to estimate animal abundance.

As these new technologies for obtaining individual identity have been adopted, they have facilitated the study of species that could not have been effectively studied two decades ago and, at the same time, have produced vast quantities of *spatially explicit* encounter information. Using classical capture-recapture methods (Otis et al. 1978), such spatial information is not directly used (as data in the likelihood) in making inferences about population size or density. To make use of this spatial information, new classes of capture-recapture models have been developed. These are called spatial capture-recapture (SCR) or spatially explicit capture-recapture (SECR) methods (Efford 2004; Royle et al. 2014). They are spatially explicit in the sense that they involve explicit spatial models for both the sampling process and how individuals interact with the sampling process. A benefit of SCR is that they provide a means of direct estimation of population density. However, SCR is far more than simply an extension of technique meant to resolve certain technical deficiencies of ordinary capture-recapture or for improved density estimation. Rather, SCR is a general framework for the study of spatial population ecology, which allows ecologists to characterize wildlife populations and address many aspects of spatial

K.U. Karanth
Wildlife Conservation Society (WCS), New York, NY, USA

Centre for Wildlife Studies, Bengaluru, India

Wildlife Conservation Society, India Program, Bengaluru, India

National Centre for Biological Sciences-TIFR, Bengaluru, India
e-mail: ukaranth@wcs.org; https://www.wcs.org/; http://cwsindia.org/; http://wcsindia.org/home/; https://www.ncbs.res.in/

population ecology from individual encounter history data, including such things as resource selection, movement, landscape connectivity, and factors that influence density (Royle et al. 2017).

9.2 Photo-Trapping Tigers

Camera trapping of tigers for the purpose of estimating abundance is described in some detail by Karanth (1995) and Karanth and Nichols (1998, 2000). The field has grown rapidly in the subsequent years, seeing publication of numerous syntheses and review articles on camera trapping including Long et al. (2008) and O'Connell et al. (2010).

The practical aspects of camera trapping tigers are explained in Chap. 5. The basic methodology involves setting out camera traps within some area of interest. The configuration of the traps can vary with the situation. As camera trapping methodology was being developed in the 1980s and 1990s, one of the main concerns in implementing field studies was to ensure that an area was covered fairly uniformly with cameras (Nichols and Karanth 2002; Chap. 11) in the sense that it would be difficult for a tiger in the sampled area to travel about and not encounter at least one camera trap. That is, traps should be set in a manner to prevent having "holes" in the sampled area in which tigers could move without any chance of being "captured" (photographed). The rationale for this "no holes" objective was to ensure that all individuals in a well-defined geographic area were catchable by at least one camera. Presumably the existence of holes implies $p = 0$ individuals which are not accounted for in the classical nonspatial capture-recapture models that were used. By assuming that no such individuals exist, then the estimate of N obtained by capture-recapture methods could be asserted to apply to a prescribed area containing the trap array. With spatial capture-recapture (SCR) models, there are no such restrictions necessary on the spatial arrangement of camera traps. This is because the population size is defined by the specification of the state space (see below), and the model explicitly accommodates $p = 0$ individuals, which arise as a result of individuals that are not in close proximity to any traps (Royle et al. 2014, Chap. 10).

In general, camera traps are set for several consecutive days. Each day is often considered as a sample period for defining encounter histories of individuals, although it could be sensible in some cases to group days together. In an ideal situation, there are sufficient camera traps to cover the entire area of interest with the available traps. However, if the area to be sampled is large, then various possible schemes for rotating traps to cover the entire area are possible. In these cases, the entire area is covered once every 2 or 3 days, perhaps. In this sampling situation, the sampling period is defined by the number of days necessary to cover the entire area. With the use of SCR methods, when the region being sampled is very large relative to the number of available traps, the use of cluster designs has proven to be very practical and efficient (Efford and Fewster 2013; Sun et al. 2014). Additional discussion of details associated with sampling designs will be presented in Sect. 7 below (also see Royle et al. 2014, Chap. 10).

The data obtained from camera trapping are summarized as capture history data, which is a record of the occasions and traps of capture for each individual. Each animal captured at least once during the survey has a capture history. While ordinary capture-recapture capture histories are simply a row vector of 1s and 0s indicating on what periods an animal was caught, the spatial encounter history for each individual consists of a two-dimensional matrix with J (number of traps) rows and K (occasions) columns where element $(j,k) = 1$ if an individual was encountered in trap j during occasion k. The programs such as SECR (Efford 2016) and SPACECAP (Gopalaswamy et al. 2012) make use of a compact encounter data file (EDF) and a trap deployment file (TDF) which describes the occasions in which each trap was operated. We note that capture histories need not be based on binary detection/non-detection data. Models can be developed based on total encounter frequencies per individual (Efford et al. 2009; Royle et al. 2009), and also recent advances in continuous time models have been made (Borchers et al. 2014; Dorazio and Karanth 2017).

In the remaining sections of this chapter, we elaborate on the basic modeling concepts that allow inference about encounter histories from camera trapping arrays. We cover classical capture-recapture methods and also relatively recent spatial capture-recapture methods. While the essential concepts are the same for both classes of methods, spatial capture-recapture models make use of spatial information on encounter and, in doing so, resolve some important technical limitations of classical capture-recapture and also allow researchers to study elements of spatial population ecology using individual encounter history data (Royle et al. 2017).

9.3 Closed Capture-Recapture

Capture-recapture models can be viewed as probabilistic expressions describing the processes that give rise to encounter history data (Nichols 1992, Williams et al. 2002). Capture-recapture models are frequently classified according to requisite assumptions about population *closure*. Closed population models are used when no gains to, or losses from, the population occur between sampling occasions. Because of this assumption of no population change, closed models are generally applied to studies conducted over relatively short time periods. Closed population models permit estimation of abundance or density. In contrast, open population models are used when there are gains, losses, or both occurring between sampling periods. Longer time intervals typically separate sampling occasions. Such models permit estimation of abundance and also local survival rate and number of recruits. Both closed and open models have been used with tiger camera trap data. For applications of closed population models, see Karanth (1995) and Karanth and Nichols (1998). Camera trap data from long-term studies at Nagarahole reserve, India have also been analyzed extensively to understand tiger population dynamics using open models (Karanth et al. 2006; Gardner et al. 2017).

9.3.1 Classical Closed Population Capture-Recapture Models

We suppose that the population of N individuals is subjected to repeated sampling for a specified number of occasions, say K (e.g., nights of a camera trapping study), where, in the first sampling occasion, all captured individuals are marked and released, and then at each subsequent sampling occasion, the detection of marked individuals is recorded and new unmarked individuals are marked or identified. This repeated sampling produces *individual encounter histories* that describe whether or not individuals were detected in each of the K occasions. For example, in a $K = 5$ occasion capture-recapture study, an individual with an encounter $y_i = (01010)$ was encountered two times; first on occasion 2, and then again on occasion 4, and it was not encountered in occasions one, three, or five. Estimating abundance using encounter history data collected using the general sampling scheme described above can be thought of as the process of estimating how many individuals were *missed* by the sampling, i.e., how many individuals have encounter history $y_i = (00000)$. Under the closure assumption, an individual detected at least once during the study was alive and present for the entire study, and therefore, failure to detect that individual in any occasion was due to imperfect detection alone. This information therefore allows for the estimation of the frequency of all-zero encounter histories.

The basic idea of all closed population capture-recapture methods is that the pattern of detections (the encounter histories) of individuals observed at least once provides information about detection probability, p, which, in turn, can be used to estimate the number of individuals that were not encountered. The underlying concept can be understood by recognizing that, under the assumption that $n \sim$ Binomial $(N, \tilde{p})$, the observed number of individuals n is related to the total population size N by the expression:

$$E(n) = N\tilde{p}$$

where $E()$ denotes statistical expectation, and $\tilde{p}$ is the probability that an individual is captured *at least once* during the study. Thus, the canonical estimator of N (also sometimes called the heuristic estimator) is obtained by plugging in an estimator of $\tilde{p}$ into this expression and rearranging to yield $\hat{N} = n/\hat{\tilde{p}}$.

The simplest closed population capture-recapture model is often called "model M_0," or the null model. Under model M_0, the encounter probability "per sample" p_i for each individual $i = 1, 2, \ldots, N$, is assumed to be the same for all individuals in the population, i.e., $p_i = p$. That is, there are no individual or temporal covariates that affect p. Then, whether or not we encounter an individual $i = 1, 2, \ldots, N$ during sampling occasion k, y_{ik}, is a Bernoulli trial (a "coin flip") with constant probability p. In a study of K survey occasions, the probability of being captured at least once, $\tilde{p}$, is directly related to the "per sample" probability of detection parameter p by the formula

$$\tilde{p} = 1 - (1 - p)^K$$

The expression relating p to $\tilde{p}$ is different depending on the specific capture-recapture model being considered. The parameter p can be estimated from the observed encounter histories, and, in turn, this is used to estimate $\tilde{p}$, and then finally we estimate N using the canonical estimator introduced previously. Alternatively, one may obtain the MLE of N directly by maximizing what is usually referred to as the "full likelihood," which is an explicit function of both p and N (Sanathanan 1972).

The assumption of constant encounter probability p is usually not satisfied in practice. Otis et al. (1978) described a family of models that can be used to deal with most sources of variation in individual encounter probabilities:

M_{o} – The null model. Capture probability is constant.

M_{t} – Time effects model. Capture probability is the same for all individuals but varies among sampling occasions. (Note that we use k for the time index here.)

M_{b} – The behavioral response model. Capture probabilities vary depending on whether or not individuals have been captured previously.

M_{h} – The individual heterogeneity model. Capture probabilities vary among individuals.

Variations of these different models exist, and the effects in some cases can be included together in a model. For example, the usual application of model M_t involves occasion-specific parameters p_k, but we can also consider systematic variation in detection probability that results from explicit covariates such as related to environmental conditions or systematic variation in time. For example, in classical small mammal trapping using catch traps, traps may be checked in the morning and evening, and we might expect encounter probability to vary over time (e.g., by sampling occasion). In camera trapping studies which usually involve a daily sample period then we might have systematic variation in encounter probability that varies by "day" according to a quadratic polynomial in Julian day, J_k, such as:

$$\text{logit}\,(p_k) = \alpha_0 + \alpha_1 J_k + \alpha_2 {J_k}^2$$

where the parameters α_0, α_1, and α_2 are estimated. The behavioral response model is usually parameterized as a permanent change in p for individuals *subsequent* to their initial capture (i.e., p_{pre} and p_{post}, for capture probability before and after first capture, respectively). This could be the result of "trap happiness" due to using bait or lures at traps, or it could be the result of "trap shyness" due to aversion.

Model M_h has been an important model in capture-recapture because a number of factors usually induce variation in p among individuals, and it has long been recognized that the existence of individual heterogeneity in capture probability will lead to underestimation of N when it is not accounted for (Otis et al. 1978). Thus much attention has been focused on developing more flexible classes of model M_h (Dorazio and Royle 2003). Norris and Pollock (1996) formulated the model in terms of a finite mixture or latent class model in which each individual in the population belongs to a finite (and small) number of classes represented by distinct values of p (see Pledger 2000), but membership in these classes is not known. This finite mixture model is probably the most widely used version of model M_h.

There are a number of basic technical or conceptual problems with the use of classical closed population capture-recapture models in camera trapping studies. For example, classical CR methods do not allow modeling of trap-level covariates, which may include type of trap (e.g., model of camera), effort (or days of operation of each camera), behavioral response which in many cases should depend on the specific trap, and local habitat conditions which might affect encounter probability through differential resource selection (Royle et al. 2013a). Another problem that has been widely recognized is that the spatial nature of camera trap arrays induces heterogeneity in encounter probability (Karanth and Nichols 1998) due to the juxtaposition of individual home ranges with the trap array (Fig. 9.1 below). Individuals with home ranges on the edge of an array will have lower probabilities than individuals with home ranges having a more interior location. This induced individual heterogeneity has traditionally been addressed by fitting individual heterogeneity models (i.e., "model M_h"), although this approach has technical limitations (Link 2003). Alternatively, it may be preferable to explain heterogeneity using explicit covariates that are the cause, such as spatial proximity of individuals relative to traps. This was the idea behind Boulanger and McLellan (2001) who proposed using Huggins-Alho type models with a "distance to the edge" covariate, where distance to edge is computed as the distance from the average capture location to the boundary of a convex hull containing the trap locations. SCR models can be seen as an extension and formalization of this approach (Royle et al. 2014) by accommodating uncertainty in the definition of average capture location and making use of trap-level encounter data. However, one of the main deficiencies with classical closed population models is that they do not permit direct estimation of animal density because, in almost all practical field applications, it is not possible to precisely define the area sampled by a set of trapping devices.

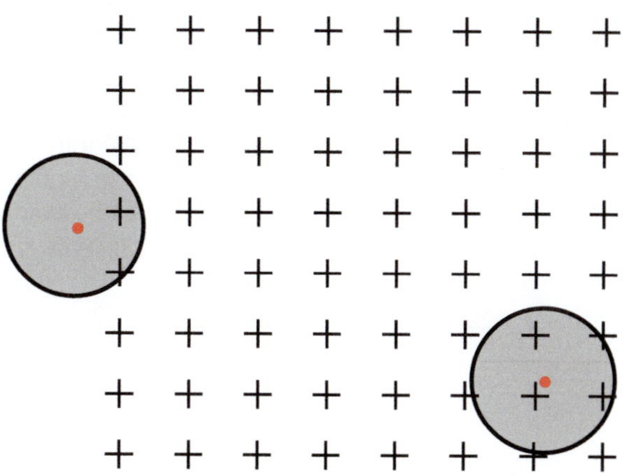

Fig. 9.1 Two home ranges of individuals (gray circles) juxtaposed with a spatial sampling grid of traps showing the variable exposure to trapping based on home range location

This is because individuals being captured move about space and can be captured without the biologists knowing whence those individuals originated or how much space they are using. Historically this has been accommodated in closed capture-recapture models by the use of buffer strips defined by the mean maximum distance moved (Wilson and Anderson 1985). Inference based on this method is discussed by Nichols and Karanth (2002; Chap. 11 of the first edition p. 129). Spatial capture-recapture models provide a formal technical framework for dealing with density estimation (Efford 2004; Borchers and Efford 2008; Royle and Young 2008; Royle et al. 2014) by prescribing an explicit "state space," effectively a prior distribution on potential home-range locations, and estimating the number of points (individual activity centers) contained on that state space.

9.4 Closed Spatial Capture-Recapture

The sampling scheme for a spatial capture-recapture analysis is the same as described above, i.e., there is a population of N individuals, and we consider each individual having an individual covariate which is their activity center ($\boldsymbol{s}_i=[s_{i,X}, s_{i,Y}]$), a spatial coordinate. This is regarded as a latent variable (i.e., unobserved). Now the goal is to estimate the number of individuals (or activity centers) within a region of interest which is referred to as the *state space*, or S. The basic ("null") SCR model assumes that individual activity centers are distributed uniformly throughout the state space:

$$\mathbf{s}_i \sim \text{Uniform}\ (S).$$

In SCR models, the realized density is the number of activity centers in the state space divided by the area of the state space: $D = N\ /\ ||S||$, where $||S||$ is the area of S. Moreover, the realized density for any subset of the state space, say $D(R)$ can be estimated. For example, using a fine partition of the state space allows explicit density maps to be produced. In some formulations of SCR, a prior distribution is imposed on N, e.g., $N \sim Poisson(\lambda||\mathrm{S}||)$ (e.g., Borchers and Efford 2008), and instead of the realized density noted previously, the expected density λ is estimated.

As before, the population is subjected to sampling using camera traps. However, we explicitly acknowledge both how many traps there are, $j=1,\ldots,J$ traps, and the location of each of the traps, which we denote as $\boldsymbol{x}_j$. The acknowledgment of the spatial structure of the traps means observations can be spatially indexed so encounter histories describe *which individual* (i), *when* (k), and importantly *where* (j) individuals were encountered, i.e., $y_{i,j,k}$. Typically, these observations are assumed to be Bernoulli outcomes:

$$y_{i,j,k} \sim \text{Bernoulli}\left(p_{i,j,k}\right),$$

where $p_{i,j,k}$ is the probability of encountering individual i in trap j and occasion k, which depends on the distance between the trap location ($\boldsymbol{x}_j$) and the individuals activity center (s_i) as follows:

$$p_{i,j,k} = p_0 \times e^{-\left(1/2\sigma^2\right)d\left(\boldsymbol{x}_j, s_i\right)^2}.$$

It may also depend on sample occasion k in some manner and trap or time-varying covariates, as well as other individual covariates. This is referred to as the half-normal encounter model where the parameter logit(p_0)=α_0 is the baseline encounter probability, corresponding to the maximum probability of encountering an individual, which occurs when an individual's activity center coincides with a trap location. The parameter σ describes the rate at which detection probability declines as a function of distance, and $d(\mathbf{x}_j,\mathbf{s}_i)$ is the Euclidean distance between trap j and the activity center of individual i. In a spatial capture-recapture analysis, the parameters to be estimated are α_0 and σ in addition to population size N. As in model M_h, the additional parameter σ accommodates individual heterogeneity in p, but, unlike model M_h, the parameter represents an explicit source of heterogeneity that is due to distance between individual activity or home-range centers and trap locations.

Thus, SCR models involve two very prominent modifications compared to ordinary capture-recapture models: (1) the encounter model describes trap-specific encounters and (2) the trap-level encounter probabilities are a function of a latent individual covariate, describing the activity center of individuals. These two modifications provide solutions to many of the technical and conceptual deficiencies of ordinary capture-recapture models. SCR provides an explicit model of heterogeneity in encounter probability induced by spatial sampling, and SCR models address the density estimation problem directly by specification of the model for activity centers and an explicit state space S. The inference problem then reduces to estimating the number of such activity centers in any well-defined subset of S.

SCR models are extremely flexible from the standpoint of modeling covariates on detection probability and also density. For example, occasion or trap-specific covariates (or even individual covariates such as individual's sex) on baseline detection probability p_0 can be modeled using logistic models of the form:

$$\text{logit}\left(p_{0,ijk}\right) = \alpha_0 + \alpha_1 \times \text{Sex}_i + \alpha_2 \times \text{Covariate}_{jk}$$

Covariates can also be modeled on the parameter(s) of the encounter probability model. For example, on the σ parameter of the half-normal model:

$$\log\left(\sigma_{ik}\right) = \gamma_0 + \gamma_1 \times \text{Sex}_i + \gamma_t \times \text{Covariate}_k.$$

The assumption that individual activity centers are uniformly distributed in space (i.e., constant density) is often regarded as being overly simplistic. However, other very general models describing the distribution of activity centers are possible. For example, when spatially referenced covariates, say $z(\mathbf{s})$, can be identified which affect the density of individuals (Borchers and Efford 2008, Dorazio 2013), then we can formulate a point process model in which the intensity parameter of the point process is a function of such covariates:

$$\log(\lambda(\mathbf{s})) = \beta_0 + \beta_1 z(\mathbf{s})$$

where the parameter β_1, to be estimated, allows density to depend on $z(\mathbf{s})$.

9.5 Example: Analysis of the Nagarahole Data

We provide a brief example here using tiger camera trapping data collected in 2006 from the Nagarahole reserve, India. The data have been analyzed using spatial capture-recapture models by Royle et al. (2009) and Gopalaswamy et al. (2012) (the data are available as part of the R package SPACECAP). Both of those studies used Bayesian methods, but here we obtain a density estimate under two models using maximum likelihood in the R package oSCR (Sutherland et al. 2016). The study used 120 camera trap sites with 30 cameras rotated every 12 days ($K = 12$). The data set contains encounter histories of 44 individuals captured up to five times each. The state space was defined by a grid with 1000 m spacing and buffering the trap array by 3000 m (Fig. 9.2). We fit the null SCR model having constant p, σ, and D which produced the following summary results:

```
> print(out0)
 Model:    D ~ 1 p0 ~ 1 sig ~ 1
 Run time:    21.92167  minutes
 AIC:    901.432

Summary table:
                       Estimate     SE        z  P(>|z|)
p0.(Intercept)     -5.515 0.269 -20.478        0
sig.(Intercept)     7.567 0.108  69.882        0
d0.(Intercept)     -2.035 0.194 -10.494        0
*Density intercept is log(individuals per pixel)
  Nhat(state-space) = exp(d0.)*nrow(ssDF)
  (caution is warranted when model contains density covariates)
```

Because the default parameterization involves transformations of the basic parameters, we have to back-transform them to their natural scale for interpretation. Estimated density in the raw output is "per pixel" of the input state space (1000 m × 1000 m for the state space used here), on the log scale, and this has to be back-transformed and multiplied by 100 to get tigers per 100 km^2:

```
> 100*(exp(-2.035))
[1] 13.06805
```

(Note that results differ modestly from Royle et al. (2009) due to variations in the definition of the state space and parameterization used here.) The baseline detection

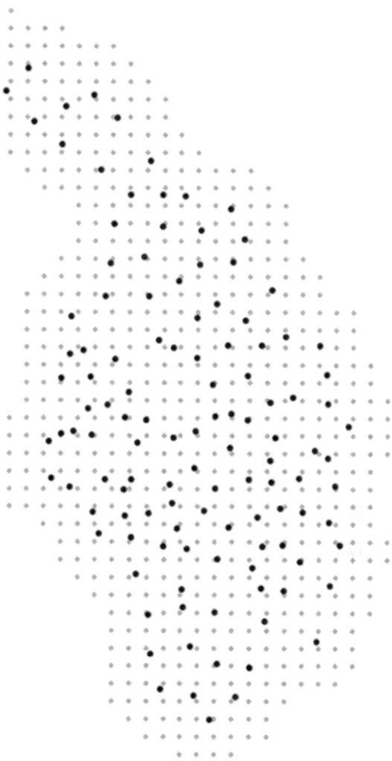

Fig. 9.2 120 camera trap locations (black dots) in Nagarahole reserve, India and the state-space grid (gray dots) used to fit the null SCR model

probability is computed by taking the inverse logit transformation of the intercept parameter:

```
> plogis(-5.515)
[1] 0.004009784
```

The estimated σ parameter, in meters, requires back-transforming the "sig.int" parameter which is estimated on the log scale:

```
> exp(7.567)
[1] 1933.332
```

And finally we produce a realized density map using the predict.oSCR function and plotting the output shown in Fig. 9.3. This density map can be very useful in the context of conservation and management because it depicts where the population is distributed, and thus estimate of total population for subregions, such as management units, can be obtained directly.

9.6 Open Model Capture-Recapture (Conventional and Spatial)

9.6.1 Classical Open Models

As discussed above, capture-recapture models for closed populations are used when a marked population is sampled over a relatively short time span such that it is not

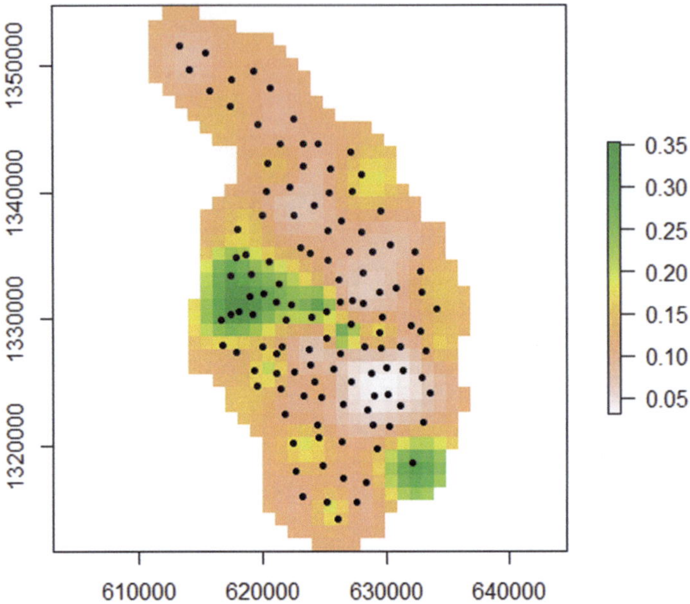

Fig. 9.3 Estimated tiger density expressed in units of "individuals per state-space pixel" (individual activity centers per km^2) of the Nagarahole reserve, India. *Black dots* are camera trap locations

expected to exhibit substantial changes from births, deaths, or movements in and out of the population. The quantities of primary interest are usually abundance (number of animals in the sampled area) or density (number of animal activity centers per unit area) of animals during this short sampling period. Encounter probabilities (ps) and (for SCR models) parameters describing the decrease in encounter probabilities associated with increasing distance between activity center and trap (σ) are also estimated.

Capture-recapture models for open populations shift the emphasis from statics to dynamics. Open populations are defined as those to which changes are expected to occur between sampling occasions. Those changes include births, deaths, and movement in and out of the sampled population. The very first capture-recapture models for open populations focused on losses to the population between sampling occasions (Cormack 1964; also see Jolly 1965; Seber 1965). We refer to these as Cormack-Jolly-Seber (CJS) models. The data for these models are encounter histories, the same as for classical closed population models, but the modeling differs in allowing for losses from the population between sampling occasions. The CJS model describes probabilities of these observable encounter histories in terms of detection probability and survival probability parameters.

We define the following CJS parameters used to model encounter history data:

p_t = probability that an animal in the sampled population at time/occasion t is encountered during sampling efforts

ϕ_t = probability that a member of the sampled population at occasion t is alive and in the population at occasion $t+1$

Consider the detection history, 01010, arising during a $K = 5$ occasion study. Using the above parameters, the CJS probability associated with this history can be written as:

$$\Pr\left(01010|\text{release at occasion } 2\right) = \phi_2\left(1 - p_3\right)\phi_3 p_4\left(1 - \phi_4 p_5\right)$$

The animal was initially encountered in occasion 2, so we condition on this encounter (i.e., we don't model this initial encounter, but use it as a starting point, effectively focusing interest on the fate of the marked population). We know the animal survived until at least occasion 4, because we encountered the animal on that occasion, and the probability associated with surviving from occasion 2 to 4 is $\phi_2\phi_3$. Despite being alive, the animal was not encountered at occasion 3 (1-p_3) but was encountered at occasion 4 (p_4). Uncertainty characterizes occasion 5, as the animal may have been lost (death or movement) from the population or may have survived but simply not been encountered. The probability associated with these two events can be written as the complement of the probability of surviving and being detected: (1-$\phi_4 p_5$). Note that this CJS model incorporates both the observation process (the p_t) and the dynamical ecological process (ϕ_t) within the same model structure.

Every animal detected in a capture-recapture study has an associated encounter history, and each such history has an associated probability, as above. The product of these probabilities for all of the animals encountered during a study forms the likelihood and can be used to estimate the survival and detection/encounter parameters. The CJS model is very general, with many time-specific parameters to estimate in long-term studies. It is possible to constrain survival or encounter parameters to be constant over time, or they can be modeled as linear-logistic functions of time-specific covariates (e.g., weather variables) to yield more parsimonious models. Software is available to make computations relatively easy for the full CJS model, constrained models, and models with time-specific covariate relationships (e.g., MARK; White and Burnham 1999).

Although Cormack's (1964) original model focused on survival, Jolly (1965) and Seber (1965) recognized that encounter data also contained information about population size and recruitment to the population. Under the Jolly-Seber (JS) model, abundance and recruitment are estimated as derived parameters. That is, survival and encounter probabilities are estimated directly as parameters of the model structure and then used to estimate these other quantities with an additional step. The two additional quantities in the JS modeling approach both concern ecological processes:

N_t = abundance, number of animals in the population at sampling occasion t

B_t = number of new individuals joining the population between occasions t and $t + 1$ and still in the population at time $t + 1$

Abundance is estimated by counting all of the individuals encountered at each occasion (denote n_t) and then dividing this statistic by the estimate of encounter probability:

$$\widehat{N} = \frac{n_t}{\hat{p}_t}. \tag{5.1}$$

Note that the above estimator represents another application of the canonical estimator of Chap. 3. Recruitment can then be estimated as the difference between the estimated abundance at occasion $t + 1$ and the expected survivors from occasion t:

$$B_t = \hat{N}_{t+1} - \hat{N}_t \phi_t. \tag{5.2}$$

The basic JS approach for modeling open populations thus provides estimates of the time-specific state variable, $\widehat{N_t}$; the number of new recruits, $\hat{B}_t$; the rate of loss, $1 - \hat{\phi}_t$; and the number of losses, $\hat{N}_t \left(1 - \hat{\phi}_t\right)$. These estimates provide a remarkably good description of population dynamics, based only on a set of encounter histories, i.e., a set of vectors populated by 0s and 1s. Other open population modeling approaches have been developed more recently that include parameters associated with recruitment and change in abundance directly in the likelihoods (Crosbie and Manly 1985, Pradel 1996, Schwarz and Arnason 1996). The logic of these approaches is identical to that of Jolly (1965) and Seber (1965), however, and selection of a modeling approach should depend on the objectives of the analysis.

Several basic assumptions underlie these estimators (also see Williams et al. 2002). First, time-specific encounter and survival probabilities are assumed to be the same for all animals in the population. This assumption can be relaxed in numerous ways. For example, animals may be stratified by any variable relevant to either observation or ecological processes. Some variables remain constant throughout the study (e.g., sex), and the most general modeling approach essentially uses separate JS models for each stratum. Age and prior (to time t) encounter history are dynamic state variables for which specific models have been developed (e.g., Pollock 1975, Pollock 1981, Brownie and Robson 1983, Brownie et al. 1986, Williams et al. 2002). Stochastically dynamic variables that characterize animals (e.g., size, location, reproductive state) and may influence survival and/or detection can be dealt with using multistate models (Arnason 1972, 1973; Brownie et al. 1993; Lebreton et al. 2009). In addition to survival and encounter probabilities, these models also estimate probabilities of moving from one state to another between sampling occasions. These so-called multistate models are discussed more fully in Chap. 12.

Marks permitting individual identification are assumed to remain with animals (no loss of marks) and to be read accurately and not overlooked. Sampling occasions are assumed to be short relative to the periods between occasions, and mortality during sampling occasions is assumed to be negligible. Animals are assumed to behave independently with respect to probabilities of survival and/or encounter. Finally, all emigration from the sampled population is assumed to be permanent. This assumption can be relaxed using either the robust design (Kendall et al. 1997; see below) or open population models with certain constraints (Kendall and Nichols 2002).

The above assumptions are needed for CJS and JS modeling. One assumption required by JS and not CJS models is that previously encountered individuals have the same encounter probability as individuals not previously encountered. In CJS modeling, the p_t parameters only apply to previously encountered animals. However, under the JS model, these encounter probabilities are assumed to apply to animals not previously encountered (new animals) as well.

9.6.2 Robust Design

Pollock (1982) recognized that many sampling programs include groups of sampling occasions that are close together in time (e.g., a few days apart) but separated from the next group of occasions by a long time period (e.g., approximately 1 year). He noted that encounter data within a season (referred to as secondary period data) could be analyzed using models developed for closed populations, whereas encounters from one season to the next (primary period data) could be analyzed using models for open populations. Advantages of this approach included robust estimators and the ability to estimate quantities that might not be estimable using only open or only closed models (Pollock 1982, Kendall and Pollock 1992, Kendall et al. 1997, Williams et al. 2002).

Pollock's (1982) initial proposal was to simply use open models (e.g., CJS) to estimate survival probabilities, to use classical closed models to estimate abundances, and then combine the two kinds of estimators to estimate recruitment using the basic structure of Eq. 5.2. More recent work has incorporated the closed and open model components into a single likelihood (Kendall et al. 1995), but historic work with tiger populations has thus far relied on Pollock's (1982) initial ad hoc approach of combining estimates from closed and open population models. Note that a distinct advantage of this ad hoc approach is the ability to use SCR models for closed populations to estimate abundances (Duangchantrasiri et al. 2016). A clear advantage provided by SCR models is an explicit mechanism for dealing with an important source of heterogeneity in encounter probabilities, the spacing of animal activity centers relative to camera trap positions. Another advantage provided by SCR models is the ability to estimate time-specific density, $\hat{D}_t$, directly. Studies using capture-recapture models for open populations typically span multiple years (e.g., Karanth et al. 2006, Duangchantrasiri et al. 2016). A common occurrence in such studies is variation over time in the amount of area sampled. It is common to begin a camera trap study, for example, by covering a relatively small area. Additional funding often allows investigators to buy more cameras and increase areas surveyed. One consequence of such temporal changes in area surveyed is that traditional estimates of population change, $\hat{\lambda}_t = \hat{N}_{t+1}/\hat{N}_t$, reflect changes in both population dynamics *and* sampled area. However, we typically want to estimate λ_t in order to draw inferences about ecology, not sampling design changes. One way to deal with this issue is to base ecological inferences on ratios of density estimates, i.e., $\hat{\lambda}_t = \hat{D}_{t+1}/\hat{D}_t$. As long as the habitat characteristics of the added area sampled are similar to those of the original study area, then these density-based estimates of population change should pertain largely to ecological and not observational processes. A similar approach can be used for recruitment, replacing $\hat{N}_t$ in Eq. 5.2 with $\hat{D}_t$. The resulting $\hat{B}_t$ is now interpreted as new recruits *per unit area*, but their use restricts inference to ecological processes rather than to changes in study area.

The robust design has been used to draw inferences about tiger population dynamics in Nagarahole reserve, Karnataka state, India (Karanth et al. 2006), and in Huai Kha Khaeng Wildlife Sanctuary in the Western Forest Complex of Thailand (Duangchantrasiri et al. 2016). Both studies used CJS-type models to estimate annual survival rates. The Karanth et al. (2006) study used classical models for closed populations to estimate tiger abundance and density, as SECR models were not well developed at the time of their analyses. Duangchantrasiri et al. (2016) used SECR models to estimate densities and abundances over time.

The original references should be consulted for methodological details, but Table 9.1 shows some of the population dynamic parameters that can be estimated using the robust design (from Karanth et al. 2006). The increases in estimated abundances that accompanied increases in study area size emphasize the need

Table 9.1 Estimated abundance, annualized (not period to period) survival rate, area sampled by camera traps, population density, and rate of change in density for primary sampling periods, for the tiger population in Nagarahole, India, 1991–2000

Primary period t	Date	Abundance $\hat{N}_t$	Survival rate $\hat{\phi}_t$	Area sampled $\hat{A}_t$	Density $\hat{D}_t$	Density change $\hat{\lambda}_t$
1	May 1991	9 (0.0)	0.77(0.051)	41.4 (3.3)	21.73 (1.7)	0.78 (0.30)
2	Dec 1991	7 (2.6)	0.77(0.051)	41.4 (3.3)	16.91 (2.6)	0.64 (0.40)
3	Apr 1992	11 (5.5)	0.77(0.051)	101.5 (5.2)	10.84 (5.4)	1.91 (1.01)
4	Jan 1994	21 (3.2)	0.77(0.051)	101.5 (5.2)	20.69 (3.3)	0.57 (0.10)
5	Jan 1995	12 (0.0)	0.77(0.051)	101.5 (5.2)	11.82 (0.6)	0.99 (0.08)
6	Mar 1996	27 (1.4)	0.77(0.051)	231.8 (7.8)	11.65 (0.7)	0.74 (0.13)
7	Jun 1997	20 (3.2)	0.77(0.051)	231.8 (7.8)	8.62 (1.4)	0.85 (0.17)
8	Jan 1998	17 (1.7)	0.77(0.051)	231.8 (7.8)	7.33 (0.8)	1.35 (0.18)
9	Mar 1999	23 (1.7)	0.77(0.051)	231.8 (7.8)	9.92 (0.8)	1.30 (0.15)
10	May 2000	30 (2.1)	–	231.8 (7.8)	12.94 (1.0)	–

Estimated standard errors are in parentheses. From Karanth et al. (2006)[a]

[a]Estimation of number of new recruits between primary sampling occasions required estimates of survival from one occasion to the next (computed as annual survival raised to the power corresponding to the interval between sampling periods). Estimated recruits (*SE*) for the last four intervals (no change in study area size) are $\hat{B}_6 = 3(3.2), \hat{B}_7 = 0(3.0), \hat{B}_8 = 11(2.8), \hat{B}_9 = 14(2.9)$

to account for such changes when making ecological inferences. Because of the changing study area size, rate of population change, λ_t, was estimated using density estimates, and recruitment was estimated for the largest number of primary sampling occasions with unchanged area sampled. Annual survival rates were best estimated using a time constant model. Results of these analyses from India (Karanth et al. 2006) and Thailand (Duangchantrasiri et al. 2016) provided evidence of dynamic stability, with fluctuations in abundance and turnover of individuals, yet no sustained increases or decreases.

9.6.3 Open SCR Models

Spatial capture-recapture models allow for a much richer class of open population models compared to classical Jolly-Seber and CJS models. Not only can explicit population dynamics (survival, recruitment) be modeled but also spatial dynamics such as dispersal and transience can be modeled by allowing for the point process model of activity centers to be temporally dynamic (Ergon and Gardner 2013; Schaub and Royle 2014; Royle et al. 2016; Gardner et al. in review). We describe basic ideas of open SCR models here. We note that so far all applications of open SCR models have been based on Bayesian analyses using MCMC because of the complexity of the latent variable structure in these models. That said, we imagine that such models are amenable to likelihood analysis using techniques of hidden Markov models.

One class of open SCR model which does not contain explicit dynamics is the multi-session models. These models assume that year-specific population sizes N_t are Poisson random variables that are independent from 1 year to the next: $N_t \sim Poisson(\lambda_t)$. This allows for variation among years to be modeled on the expected population size parameter λ_t such as $\log(\lambda_t) = \beta_0 + \beta_1 \times \text{year}_t$ for a simple trend model. Under this Poisson assumption for annual population size, it is possible to obtain MLEs of model parameters using marginal likelihoods (see Royle et al. 2014, ch. 6). These models are usually called multi-session models in the context of SCR (e.g., as implemented in the software package "SECR," Efford 2016). Multi-session models provide a flexible framework for combining data from multiyear studies into a single analysis framework while at the same time providing year-specific density or population size estimates and also estimates of basic parameters that describe inter-annual variation in density or population size (recall that with SCR models, density is related directly to N_t and the area of the prescribed state space, as discussed above). Multi-session models do not preserve individual identity across years and so information about survival and recruitment is lost. Nevertheless the approach is straightforward to apply and versatile because the populations need not be strictly temporal but can be demographic groups, or different species, or different locations (such as different tiger reserves). However, as noted in the previous section, it is often important to be able to obtain explicit estimates of population dynamics (survival and recruitment, state-transition probabilities, growth rate, etc.). This is especially true in many camera trapping studies which are often focused on extremely rare carnivores for which little is known due to the difficulty in conducting population ecological studies of such species.

It is possible to incorporate explicit population dynamics into spatial capture-recapture models. So far, all applications of "open SCR" models have been based on a Bayesian formulation of the models using data augmentation (Royle and Dorazio 2008, ch. 10; Royle and Dorazio 2012; Kery and Schaub 2012, ch. 10; Royle et al. 2014, ch. 16). Bayesian analysis of these models has been done because SCR models contain many latent variables (i.e., the individual activity centers), and no one has worked out likelihood formulations yet, although in principle we imagine this can be done.

To specify a spatially explicit version of the Jolly-Seber model, let $y_{i,j,k,t}$ be the encounter history for individual *i*, at trap *j*, occasion *k*, and during primary occasion *t*. We think of *t* here indexing biological seasons or years, or "primary periods" in the terminology of the robust design, such that dynamics occurs across the *t* index of the variables. We note that there may or may not be replication within primary periods (i.e., k=1 is allowable). The observation model is specified conditional on a latent state variable $z_{i,t}$ which we define as the "alive state" of individual *i* in year *t*. That is, $z_{i,t}$=1 if individual *i* is alive in year *t* and $z_{i,t}$=0 if not. Thus, the observation model has the form:

$$y_{i,j,k,t} \mid z_{i,t} \sim \text{Bernoulli}\left(p\left(\boldsymbol{x}_j, \boldsymbol{s}_i\right) z_{i,t}\right)$$

where $p(\boldsymbol{x}_j, \boldsymbol{s}_i)$ is an ordinary "closed" SCR encounter probability model such as the half-normal model defined above. Population dynamics are parameterized in a model for the individual state variable $z_{i,t}$ (Royle and Dorazio 2008). The initial state is assumed to be a Bernoulli random variable: $z_{i,1} \sim \text{Bernoulli}(\gamma_1)$, where the interpretation of γ_1 is related to the initial population size. Under the data augmentation scheme (Royle and Dorazio 2012) where the maximum size of the super-population is set at M, then $E(N_1)=\gamma_1 M$. The latent state variable $z_{i,t}$ is assumed to be Markovian so that values for t=2,...,T depend on the previous states. For $t \geq 2$, $z_{i,t} \sim \text{Bernoulli}(\phi^* z_{i,t-1} + \gamma^* r_{i,t})$; where $r_{i,t}$ is an indicator of whether an individual is available to be recruited or not (set to 0 if that individual has ever been previously recruited and to 1 otherwise to ensure that individuals are recruited at most once). Thus if $z_{i,t-1}$=1, then the individual can survive with probability ϕ, and if $r_{i,t} = 1$, then an individual can be recruited with probability γ. In this parameterization of the Jolly-Seber model (Schwarz and Arnason 1996), the recruitment parameters are interpreted as "conditional entrance probabilities," not per capita recruitment. Under this formulation, the number of recruits during interval t, R_t, is a derived parameter, being the sum of all values of $z_{i,t}$ which transition is from $z_{i,t} = 0$ to $z_{i,t} = 1$. Further, per capita recruitment can be derived by dividing the number of recruits at time t, R_t by N_{t-1}. We note that the model can also be formulated directly in terms of per capita recruitment (Chandler and Clark 2014; Gardner et al. in review).

An application of the spatial Jolly-Seber type model was first given by Gardner et al. (2010) for the simplest case where T=2 and the individual activity centers

s_i were static from one year to the next. In general, however, it is possible to allow for explicit population dynamics (survival and recruitment) and also spatial dynamics in which individual activity centers are not static. Ergon and Gardner (2013) consider this in the context of modeling dispersal of voles using spatial capture-recapture data, while also modeling survival and recruitment. Schaub and Royle (2014) consider a Cormack-Jolly-Seber version of this type of system in which the spatially dynamic model allows for the explicit separation of dispersal and true survival from capture-recapture data. Recently Gardner et al. (in review) apply a fully open SCR model to Nagarahole camera trapping data in which the activity centers are modeled according to Gaussian random walk such that

$$\boldsymbol{s}_{i,t} \sim \text{Normal}\left(\boldsymbol{s}_{i,t-1}, \sigma_s^2 \boldsymbol{I}\right),$$

where I is the 2×2 identity matrix. Here, σ_s^2 is the variance of the random walk (quantifying the degree of movement in activity centers between primary occasions), which is different than σ^2, the scale parameter in the encounter probability model defined above (quantifying within primary occasion movement about an individual's activity center).

9.6.4 Spatially Dynamic Models, Without Population Dynamics

SCR models allow for both population dynamics (survival, recruitment) and also spatial dynamics (space-time dynamics of the point process model for activity centers). These two types of dynamics can be represented together in the same model (as in Gardner et al. in review), but, in addition, models can be developed that contain only one or the other dynamic components. For example, the basic open SCR model developed by Gardner et al. (2010) accommodates population dynamics but not spatial dynamics. In addition, a special kind of "open" population model which is only possible in the context of SCR is a model in which there are no population dynamics, but the individual activity centers are spatially dynamic (Royle et al. 2016). This type of model might be relevant when, during the study, some portion of the population disperses or when some portion of the population exhibits transient space usage. Royle et al. (2016) considered Markovian movement or dispersal models in demographically closed systems and showed that while parameters of the movement model could be estimated from repeated detections of individuals subject to demographic closure, ignoring movement dynamics did not produce biased estimates of density. Thus if estimating density is the main objective of a study then a misspecified encounter probability model that neglects movement is not necessarily detrimental to inferences.

We imagine intermediate types of models too in which some reduced population dynamics are paired with spatially dynamic activity centers. For example, we might allow for recruitment, but not survival, and spatially dynamic activity centers.

9.7 Survey Design Considerations

An overview of the sampling associated with camera trapping was provided above. Here we provide general considerations about important aspects of sampling when the objective is focused on estimating population size or density within a study area.

The basic consideration in all camera trapping studies is this: you need to sample "enough" (time and space) to accumulate a sufficient sample size of unique individuals and a sufficient sample size of recaptures, whether or not you intend to use ordinary CR or SCR models. If you are using SCR models then, in addition, you need to obtain sufficient "spatial recaptures" (i.e., captures of individuals in more than one trap). Unfortunately, what constitutes sufficient depends on just about everything: the actual unknown population size, the actual unknown detection probability and any other parameter of the model, and the extent and geometry of the space being sampled. Importantly, however, it depends on the length of time or number of occasions that camera traps are operational and also the number of traps and configuration of traps within the study area. As such, it is almost impossible, in general, to give prescriptive sample design guidance. As a practical matter, the best approach to make sensible choices of sampling design is to use Monte Carlo simulation. That is, propose a number of realistic sampling designs, and evaluate them by simulating populations and the sampling process using a priori guesses of the parameter values (or values taken from the literature). This should always be done, for any field study, before deployment of camera traps.

Using simulation we will usually seek to find a design to meet a certain precision target (e.g., CV of 10% for estimating N), or we will seek to evaluate a handful of logistically feasible designs and use the best one in terms of CV or perhaps also considering economic cost of implementing the design. All things being equal, we find for given true values of the parameters p and N (or other parameters of the model), you can always reduce the CV of an estimator by increasing the amount of effort (number of sample occasions, number of traps), because this produces larger sample sizes. On the other hand, this comes at increased economic cost. Thus the design problem always comes down to trading off a cost constraint with a statistical precision objective. As such, many study design problems are formulated in terms of optimizing precision for a fixed number of sample locations. This is roughly the same in practical terms as having a fixed project budget because "number of sites" is easy to translate into financial cost (cost of cameras, effort to establish sites, check cameras). In these situations, the design question boils down to "where do we allocate those traps?" Then various designs can be defined, and simulations run to produce a sequence of results (estimates). Then, looking at the results, you have to ask the question: is the realized statistical precision sufficient for management needs? If not, you increase the effort (number of traps, number of days operational) and repeat the exercise.

As in traditional capture-recapture models, most important design elements represent aspects of sampling in space and time: how many spatial samples (and where, and what spacing) and how long should sampling occur? We discuss some specific aspects of space and time here.

Study duration. A general objective in camera trapping is to keep sampling duration short relative to tiger population turnover so that demographically closed models can be used. However, the definition of short will be situation dependent and will likely be determined by the logistics of each sampling situation. It would be excellent, but likely impossible, to complete sampling within a 2-week period. Periods of 4–6 weeks or even longer will more likely be required. Longer time intervals increase the likelihood of models for open populations being needed. Closed models are preferred for reasons of precision and estimator robustness, but if closed models cannot be used, then open models are available (Seber 1982; Pollock et al. 1990) and can be used to accommodate non-closure. This general consideration relating to the duration of the study is relevant to both SCR and ordinary capture-recapture methods. From the standpoint of estimating density, it would be preferable to focus on simpler (demographically closed) models, but if it is necessary to increase the length of time in order to obtain sufficient sample sizes, the open models can be used to accommodate non-closure. An economical way to deal with non-closure that avoids having to model complex population dynamics is the use of multi-session models as described previously. These could involve one additional structural parameter per session and yet still accommodate temporal variation in density but "share" the parameters of the detection probability model among the sessions.

Trap spacing. General trap spacing is an important element of SCR models. Effective estimation requires that spatial recaptures are obtained, that is, recaptures of individuals in multiple traps. Intuitively, designing a study to have traps close together should maximize the number of spatial recaptures. On the other hand, the total sample size of unique individuals should be maximized if the traps are placed uniformly far apart in any prescribed region of interest. Thus, practical SCR design represents a compromise between obtaining sufficient spatial recaptures and a large sample size of individuals (Royle et al. 2014, ch. 10), and the two objectives contradict one another in the sense that close trap spacing is optimal for the former objective, while distant trap spacing is optimal for the latter. Simulation studies (Sollmann et al. 2013; Efford and Fewster 2013; Royle et al. 2014, ch. 10; Augustine unpubl. Results: https://groups.google.com/forum/#!topic/spatialcapturerecapture/SnswEKOwFb0) indicate that trap spacing of 1.5–2.5 times σ produces optimal designs for estimating population size or density under the half-normal encounter probability model with parameter σ. Thus if sufficient traps exist to achieve this spacing with uniform coverage of the area, then this should be done. Note that this approach (uniform coverage) is the same as the "Design 1" recommendation of Nichols and Karanth (2002) who recommended to "...spread [traps] throughout the area once and then checked each day for say 5-30 consecutive days". This approach yields standard (spatial) capture history data amenable to analysis by the classes of models described in this chapter. On the other hand, if the study area is too large to cover uniformly given the number of available camera traps or effort to deploy them, then there are three additional design options available:

- (Design 2) Cover a central area of the study uniformly to achieve close to the optimal spacing.
- (Design 3) Use a type of cluster design where small clusters achieve optimal within-cluster spacing in order to generate spatial recaptures, but then clusters are more widely spaced to cover the area uniformly, expose a larger portion of the population to sampling, and obtain a higher sample size of encountered individuals.
- (Design 4) Cover a subregion uniformly and then rotate the traps so that a larger area is covered.

Design 2 is logistically appealing because it saves on the labor and cost of moving traps. On the other hand, by moving traps you can obtain a larger sample size of observed individuals and possibly more spatial recaptures. If there is a strong local behavioral response, then there is probably some statistical benefit to moving traps since the recaptures provide less information. When implementing this design, the area sampled should be representative of the area as a whole. If there is a strong gradient in density, then this approach is not necessarily robust to density models, and this should be studied by simulation for the specific landscape under consideration. The use of cluster designs (Design 3) has proved to be an effective and logistically practical way of carrying out studies for density estimation (Efford and Fewster 2013; Sun et al. 2014; Fuller et al. 2016). Spacing of within and between clusters can be analyzed by Monte Carlo simulation. As in the first edition of this manual, designs which involve sampling an area and then moving traps (rotating blocks of traps, Design 4) can be viable in some situations. The advice from Edition 1 is probably still good advice, as long as we keep the basic trap spacing considerations of SCR in mind. In that regard, we probably should define grid cells to be about 1.5–2 times σ and then rotate the traps to achieve uniform coverage of the grid cells over time but making sure that sufficient neighboring cells contain traps so as to produce the needed spatial recaptures.

9.8 Future Directions

The field of capture-recapture has undergone rapid and profound evolution over the last few years. This has been driven in large part by technological innovations such as camera trapping and DNA sampling which allow for capture-recapture studies of species such as tigers that could not be effectively studied using historical methods requiring physical capture and marking. Such methods produce huge quantities of spatially explicit encounter information that are informative about all aspects of spatial population ecology, not just estimation of density and population size. The development of spatial capture-recapture methods has been driven to some extent by the widespread adoption of these new technologies for sampling which produce rich spatial encounter data. Spatial capture-recapture methods have proved useful for modeling spatial variation in density, resource selection, movement, and population dynamics (Royle et al. 2017). We believe that SCR methods will continue to grow in importance, and there are many unresolved and important aspects related to the study of animal populations by capture-recapture that have received little or no attention in the literature so far but which we feel have enormous practical relevance for the study of tiger and other populations. We identify a few of these here.

Uncertain Identity The advent of new technologies for noninvasive sampling, including camera trapping and DNA sampling, has increased the frequency of samples with uncertain identity. When a sample is obtained but cannot be identified to individual, observed encounter histories may not be the true encounter history of an individual. Accommodating uncertain identity is extremely important for unbiased population size estimation when, for example, it can produce "ghost" individuals which have the effect of inflating the population size estimation (Lukacs and Burnham 2005) or if behavioral response is present and not properly modeled in the presence of uncertainty (Augustine et al. 2014). Uncertain identity is especially relevant in camera trapping where, historically, obtaining photos of both sides of an individual has been necessary in order to identify an individual. However, a single-side photograph can be viewed as a sample having uncertain identity, and this uncertainty can be accommodated in capture-recapture models. Augustine et al. (2016) showed that spatial location of encounter is informative about sample identity and used this information in the development of SCR models to allow for single-side samples in camera trapping. Their method can be applied even to whole arrays of single cameras, in which case it may not ever be possible to obtain both sides of an individual.

Landscape Connectivity SCR provides an empirical basis for estimating not only density but also connectivity of the landscape. For example, explicit models of connectivity, such as least-cost path, can be integrated directly in the SCR likelihood, and parameters that describe the resistance of the landscape to individual space usage can be estimated (Royle et al. 2013b; Sutherland et al. 2015; Fuller et al. 2016). Furthermore, SCR provides an empirical framework for making explicit landscape management decisions using population density and connectivity information. For example, Morin et al. (2017) evaluated the potential of camera trapping studies to estimate density-weighted connectivity (DWC) and proposed that landscape management (corridor, reserve design) should be based on maximizing DWC of a landscape or minimizing the loss of DWC. We imagine that these ideas might be useful in the development of connected landscapes for tigers and other large cats based on camera trapping studies.

Biologically Realistic Point Process Models Most applications of SCR models to date are based on the simplistic homogeneous point process model in which individual activity centers are distributed uniformly and independent of one another. There is an obvious need for more biologically realistic point process models which accommodate important biological features of populations such as territorial behavior and territory overlap of individuals and among species. General classes of point process such as Markov point process models accommodate dependence among points (Illian et al. 2008) might be realistic for modeling such biological phenomenon. Reich and Gardner (2014) proposed an SCR model for territorial species based on a Strauss process model which accommodates point processes that are more regular than complete spatial randomness. To the best of our knowledge, no other examples of non-independent point process models in the context of capture-recapture modeling exist.

Continuous Time SCR Models Both classical and spatial capture-recapture models accommodate the modeling of temporal variation in the form of time effects or behavioral response. In the vast majority of applications to date, these effects are modeled in discrete time, that is, assuming that sample occasions are registered in discrete time intervals $k=1,2,...,K$. Recent efforts toward developing continuous time models (Borchers et al. 2014; Dorazio and Karanth 2017) show promise in improving the modeling of biological processes such as a behavioral response (due to transient space use of an individual's home range) and make more efficient use of detection data.

References

Arnason NA (1972) Parameter estimates from mark-recapture experiments on two populations subject to migration and death. Res Popul Ecol 13(2):97–113

Arnason NA (1973) The estimation of population size, migration rates and survival in a stratified population. Res Popul Ecol 15(2):1–8

Augustine BC, Tredick CA, Bonner SJ (2014) Accounting for behavioural response to capture when estimating population size from hair snare studies with missing data. Methods Ecol Evol 5(11):1154–1161

Augustine BC, Royle JA et al (2016) Spatial capture-recapture with partial identity: an application to camera traps. bioRxiv, p.056804

Borchers DL, Efford MG (2008) Spatially explicit maximum likelihood methods for capture-recapture studies. Biometrics 64:377–385

Borchers DL, Distiller G, Foster R, Harmsen B, Milazzo L (2014) Continuous-time spatially explicit capture–recapture models, with an application to a jaguar camera-trap survey. Methods Ecol Evol 5:656–665

Boulanger J, McLellan BN (2001) Closure violation in DNA-based mark-recapture estimation of grizzly bear populations. Can J Zool 79(4):642–651

Brownie C, Robson DS (1983) Estimation of time-specific survival rates from tag-resighting samples: a generalization of the Jolly-Seber model. Biometrics 39(2):437–453

Brownie C, Hines JE, Nichols JD (1986) Constant-parameter capture-recapture models. Biometrics 42(3):561–574

Brownie C, Bowman DT, Burton JW (1993) Estimating spatial variation in analysis of data from yield trials: a comparison of methods. Agric J 85(6):1244–1253

Chandler RB, Clark JD (2014) Spatially explicit integrated population models. Methods Ecol Evol 5:1351–1360

Cormack RM (1964) Estimates of survival from the sighting of marked animals. Biometrika 51(3/4):429–438

Crosbie SF, Manly BFJ (1985) Parsimonious modelling of capture-mark-recapture studies. Biometrics 41(2):385–398

Dorazio RM (2013) Bayes and empirical bayes estimators of abundance and density from spatial capture-recapture data. PLoS One 8(12):e84017

Dorazio RM, Karanth KU (2017) A hierarchical model for estimating spatial distribution and abundance of animals detected by continuous-time recorders. Plos One 12:e0176966

Dorazio RM, Royle JA (2003) Mixture models for estimating the size of a closed population when capture rates vary among individuals. Biometrics 59:351–364

Duangchantrasiri S, Umponjan M, Simcharoen S, Pattanavibool A, Chaiwattana S, Maneerat S, Kumar N, Jathanna D, Srivathsa A, Karanth KU (2016) Dynamics of a low-density tiger population in Southeast Asia in the context of improved law enforcement. Conserv Biol 30(3):639–648

Efford MG (2004) Density estimation in live-trapping studies. Oikos 106:598–610

Efford MG (2016) secr: Spatially explicit capture-recapture models. R package version 2.10.3. http://CRAN.R-project.org/package=secr

Efford MG, Fewster RM (2013) Estimating population size by spatially explicit capture–recapture. Oikos 122(6):918–928

Efford MG, Dawson DK, Borchers DL (2009) Population density estimated from locations of individuals on a passive detector array. Ecology 90:2676–2682

Ergon T, Gardner B (2013) Separating mortality and emigration: modelling space use, dispersal and survival with robust-design spatial capture–recapture data. Methods Ecol Evol 5: 1327–1336

Fuller AK, Sutherland CS, Royle JA, Hare MP (2016) Estimating population density and connectivity of american mink using spatial capture-recapture. Ecol Appl 26:1125–1135

Gardner B, Reppucci J, Lucherini M, Royle JA (2010) Spatially explicit inference for open populations: estimating demographic parameters from camera-trap studies. Ecology 91:3376–3383

Gardner B, Sollmann R, Karanth KU, Kumar S (2017) Open population spatial capture-recapture models and their potential for long-term wildlife population monitoring. In review

Gopalaswamy AM, Royle JA, Hines JE et al (2012) Program SPACECAP: software for estimating animal density using spatially explicit capture–recapture models. Methods Ecol Evol 3:1067–1072

Illian J, Penttinen A, Stoyan H, Stoyan D (2008) Statistical analysis and modelling of spatial point patterns. Wiley, New York

Jolly GM (1965) Explicit estimates from capture-recapture data with both death and immigration-stochastic model. Biometrika 52(1/2):225–247

Karanth KU (1995) Estimating tiger Panthera tigris populations from camera-trap data using capture—recapture models. Biol Conserv 71(3):333–338

Karanth KU, Nichols JD (1998) Estimation of tiger densities in India using photographic captures and recaptures. Ecology 79:2852–2862

Karanth KU, Nichols JD (2000) Ecological status and conservation of tigers in India. Final technical report to the US fish and wildlife service (division of international conservation), Washington, DC, and Wildlife Conservation Society, New York. Centre for Wildlife Studies, Bangalore

Karanth KU, Nichols JD, Kumar N, Hines JE (2006) Assessing tiger population dynamics using photographic capture–recapture sampling. Ecology 87:2925–2937

Kendall WL, Nichols JD (2002) Estimating state-transition probabilities for unobservable states using capture–recapture/resighting data. Ecology 83(12):3276–3284

Kendall WL, Pollock KH (1992) The robust design in capture-recapture studies: a review and evaluation by Monte Carlo simulation. In: Wildlife 2001: populations. Springer, Netherlands, pp 31–43

Kendall WL, Pollock KH, Brownie C (1995) A likelihood-based approach to capture-recapture estimation of demographic parameters under the robust design. Biometrics 51(1):293–308

Kendall WL, Nichols JD, Hines JE (1997) Estimating temporary emigration using capture–recapture data with Pollock's robust design. Ecology 78(2):563–578

Kery M, Schaub M (2012) Bayesian population analysis using WinBUGS: a hierarchical perspective. Academic Press, Waltham

Lebreton JD, Nichols JD, Barker RJ et al (2009) Modeling individual animal histories with multistate capture–recapture models. Adv Ecol Res 41:87–173

Link WA (2003) Nonidentifiability of population size from capture-recapture data with heterogeneous detection probabilities. Biometrics 59(4):1123–1130

Long RA, MacKay P, Zielinski WJ, Ray JC (2008) Noninvasive survey methods for carnivores. Island Press, Washington, DC

Lukacs PM, Burnham KP (2005) Review of capture–recapture methods applicable to noninvasive genetic sampling. Mol Ecol 14(13):3909–3919

Morin DJ, Fuller AK, Royle JA, Sutherland C (2017) Model-based estimators of density and connectivity to inform conservation of spatially-structured populations. Ecosphere 8(1)

Nichols JD (1992) Capture-recapture models. Bioscience 42:94–102

Nichols JD, Karanth KU (2002) Statistical concepts: estimating absolute densities of tigers using capture-recapture sampling. In: Karanth KU, Nichols JD (eds) Monitoring tigers and their prey: a manual for researchers, managers, and conservationists in tropical Asia. Centre for Wildlife Studies, Bangalore

Norris IIIJL, Pollock KH (1996) Nonparametric MLE under two closed capture-recapture models with heterogeneity. Biometrics 52(2):639–649

O'Connell AF, Nichols JD, Karanth KU (eds) (2010) Camera traps in animal ecology: methods and analyses. Springer, New York

Otis DL, Burnham KP, White GC, Anderson DR (1978) Statistical inference from capture data on closed animal populations. Wildl Monogr 62:3–135

Pledger S (2000) Unified maximum likelihood estimates for closed capture–recapture models using mixtures. Biometrics 56(2):434–442

Pollock KH (1975) A K-sample tag-recapture model allowing for unequal survival and catchability. Biometrika 62(3):577–583

Pollock KH (1981) Capture-recapture models allowing for age-dependent survival and capture rates. Biometrics 37(3):521–529

Pollock KH (1982) A capture-recapture design robust to unequal probability of capture. J Wildl Manag 46(3):752–757

Pollock KH, Nichols JD, Brownie C, Hines JE (1990) Statistical inference for capture-recapture experiments. Wildl Monogr:3–97

Pradel R (1996) Utilization of capture-mark-recapture for the study of recruitment and population growth rate. Biometrics 52(2):703–709

Reich BJ, Gardner B (2014) A spatial capture-recapture model for territorial species. Environmetrics 25:630–637

Royle JA, Dorazio RM (2008) Hierarchical modeling and inference in ecology: the analysis of data from populations, metapopulations and communities. Academic Press, Waltham

Royle JA, Dorazio RM (2012) Parameter-expanded data augmentation for bayesian analysis of capture–recapture models. J Ornithol 152(2):521–537

Royle JA, Young KV (2008) A hierarchical model for spatial capture-recapture data. Ecology 89:2281–2289

Royle JA, Karanth KU, Gopalaswamy AM, Kumar N (2009) Bayesian inference in camera trapping studies for a class of spatial capture–recapture models. Ecology 90(11):3233–3244

Royle JA, Chandler RB, Sun CC, Fuller AK (2013a) Integrating resource selection information with spatial capture-recapture. Methods Ecol Evol 3:545–554

Royle JA, Chandler RB, Gazenski KD, Graves TA (2013b) Spatial capture-recapture models for jointly estimating population density and landscape connectivity. Ecology 94:287–294

Royle JA, Chandler RB, Sollmann R, Gardner B (2014) Spatial capture-recapture. Academic Press, Waltham

Royle JA, Fuller AK, Sutherland C (2016) Spatial capture–recapture models allowing markovian transience or dispersal. Popul Ecol 58:53–62

Royle JA, Fuller AK, Sutherland C (2017) Unifying population and landscape ecology with spatial capture-recapture. Ecography (in press). doi: https://doi.org/10.1111/ecog.03170

Sanathanan L (1972) Estimating the size of a multinomial population. Ann Math Stat 43(1):142–152
Schaub M, Royle JA (2014) Estimating true instead of apparent survival using spatial cormack-Jolly–Seber models. Methods Ecol Evol 5:1316–1326
Schwarz CJ, Arnason AN (1996) A general methodology for the analysis of capture-recapture experiments in open populations. Biometrics 52(3):860–873
Seber GA (1965) A note on the multiple-recapture census. Biometrika 52(1/2):249–259
Seber GAF (1982) The estimation of animal abundance and related parameters. 2nd Ed. Macmillan, New York
Sollmann R, Gardner B, Parsons AW, Stocking JJ, McClintock BT, Simons TR, O'Connell AF (2013) A spatial mark-resight model augmented with telemetry data. Ecology 94:553–559
Sun CC, Fuller AK, Royle JA (2014) Trap configuration and spacing influences parameter estimates in spatial capture-recapture models. PLoS One 9:e88025
Sutherland C, Fuller AK, Royle JA (2015) Modelling non-euclidean movement and landscape connectivity in highly structured ecological networks. Methods Ecol Evol 6:69–177
Sutherland C, Royle JA, Linden D (2016) oSCR: multi-session sex-structured spatial capture-recapture models. R package version 0.25.0
White GC, Burnham KP (1999) Program MARK: survival estimation from populations of marked animals. Bird Study 46(sup1):S120–S139
Williams BK, Nichols JD, Conroy MJ (2002) Analysis and management of animal populations. Academic Press, Waltham
Wilson KR, Anderson DR (1985) Evaluation of two density estimators of small mammal population size. J Mammal 66(1):13–21

Field Practices: Assessing Tiger Population Dynamics Using Photographic Captures

10

K. Ullas Karanth, James D. Nichols, Abishek Harihar, Dale G. Miquelle, N. Samba Kumar, and Robert M. Dorazio

10.1 Introduction

In Chap. 9, sampling animal populations by repeatedly "catching" identifiable individuals to generate capture histories was described. From these histories, capture frequency statistics and estimates of capture probabilities can be derived. Estimates of capture probabilities permit us to estimate the abundance and density of animals

K.U. Karanth (✉)
Wildlife Conservation Society (WCS), New York, NY, USA

Centre for Wildlife Studies, Bengaluru, India

Wildlife Conservation Society, India Program, Bengaluru, India

National Centre for Biological Sciences-TIFR, Bengaluru, India
e-mail: ukaranth@wcs.org; https://www.wcs.org/; http://cwsindia.org/; http://wcsindia.org/home/; https://www.ncbs.res.in/

J.D. Nichols
Crofton, MD, USA
e-mail: jamesdnichols2@gmail.com

A. Harihar
Tiger Program, Panthera, New York, NY, USA

Nature Conservation Foundation, Mysuru, India
e-mail: aharihar@panthera.org; aharihar@ncf-india.org; https://www.panthera.org/; http://ncf-india.org/

D.G. Miquelle
Wildlife Conservation Society (WCS), New York, NY, USA

Department of Ecology, Far Eastern Federal University, Vladivostok, Russia
e-mail: dmiquelle@wcs.org

K.U. Karanth, J.D. Nichols (eds.), *Methods For Monitoring Tiger And Prey Populations*, DOI 10.1007/978-981-10-5436-5_10

in the surveyed area, after accounting for imperfect detection, without having to catch all individuals in the population.

In traditional "capture–recapture" or "mark-recapture" studies of smaller animals such as rodents, birds, or fish, individuals are physically caught "invasively" and then "recognizably marked" using artificial tags, branding, or mutilation. Such physical captures, marking and recaptures of tigers, at the scale needed, are not practical because of logistical and ecological constraints.

However, the fact that tigers are "naturally marked" confers a major advantage. Because unambiguous individual identification of tigers from their stripe patterns is possible (Plate 10.1), statistical sampling of wild populations through "photographic

Plate 10.1 Tigers can be individually identified from differences in their stripe patterns. However, stripe patterns on the two sides of the same animal differ. Camera trap photos above show two distinct individual tigers photo-captured in Malenad landscape of India. They are labelled BRT-105 (*top*) and BRT-103 (*bottom*) (Image Copyright©: Ullas Karanth/WCS)

N.S. Kumar
Wildlife Conservation Society, India Program, Bengaluru, India
e-mail: nrao.skumar@gmail.com

R.M. Dorazio
Wetland and Aquatic Research Center – United States Geological Survey (USGS), Gainesville, FL, USA
e-mail: bdorazio@usgs.gov

captures" of individuals become feasible. Such sampling can generate data appropriate for capture–recapture (hereafter CR) analyses (Chap. 9) for estimating demographic parameters. The volume by O'Connell et al. (2011) provides a general overview of the use of camera traps in animal ecology.

Currently, two reliable, noninvasive methods for "marking" tigers exist: capture and identification of individual tigers from photographs (this chapter) and from DNA extracted from their scats or hair (Chap. 11, Bhagavatula and Singh 2006, Mondol et al. 2009, Gopalaswamy et al. 2012a). Although individual identifications based on other methods, such as recognition of tigers from their unique scent using trained dogs (Kerley and Salkina 2007) or from voice recordings may be feasible, these have not been rigorously proven for practical population sampling.

Royle et al. (2014; pp. 381–399) present various options to statistically describe different ways in which a "detection device" and the "target object" potentially intersect resulting in a "capture event" (fixed detectors, areal searches, route searches), while Efford et al. (2009) describe different types of fixed detection devices (i.e., proximity detectors, multi-catch or single-catch) and "encounter process" models (Chap. 11) corresponding to each. The process of a tiger naturally moving around its area of activity and being photographed in an array of multiple, stationary camera traps matches the "proximity detector" encounter process (Poisson) model. This process yields counts of captures of an individual at a trap in a sampling occasion but can also be approximated by the binomial model (captured/not captured) if counts (disaggregated by individual, trap, and occasion) of >1 are rare, as typical in camera trap studies of tigers (Royle et al. 2009, Dorazio 2013). On the other hand, survey personnel moving and detecting stationary tiger scat samples requires the use of models for search encounter data (Royle et al. 2014, Chap. 11 of this volume).

In the following sections, we present possible designs for field surveys as well as protocols for camera trap data collection, storage, management, and analyses to fully exploit the analytic power of methods described in Chap. 9.

10.2 Design of Field Surveys Using Camera Traps

Choosing the right kind of equipment, selecting the best spots to set traps, and designing an array of such camera traps to efficiently sample an area, to "catch" individuals from the tiger population exposed to the array, are key factors to be considered even before data collection can begin. These practical aspects are covered in this section.

10.2.1 Some General Considerations

Implementing a survey design is not a mere ad hoc step. We emphasize that the details of the field monitoring effort should be inherited from the ultimate objectives of the program for tiger conservation and/or science that the survey is expected to inform (Chap. 14).

Design of camera trap surveys should also be guided by site-specific aspects of the ecology of the tiger population being sampled, as well as environmental and logistical factors as described in Chap. 2. In this monitoring context, prior knowledge of ecological variables such as expected tiger home range sizes, numbers, travel routes, as well as available investigator skills and other resources will shape the quality of results.

Standard capture–recapture analyses (Chap. 9) require unambiguous identification of individual tigers to construct accurate capture histories. The stripe patterns on the left and right sides of the same tiger differ. Therefore, to identify any particular individual with certainty, both sides of each tiger must be simultaneously photographed. This should happen at least once during a survey to definitively link the two profiles of the same animal. In practical terms, this means that every camera trap location should have two cameras, with one positioned on either side of the expected path of the tiger. This critical need greatly increases the resources required to conduct camera trap surveys, placing constraints on the extent of area to be sampled.

Because not all tigers present are likely to be photo-captured, any unmatched left or right profiles constitute uncertain data with lower value. Therefore, the practice has been to generate reliable individual identifications using two cameras per trap. However, recent analytical advances (Augustine et al. 2016, Chap. 9) may permit more efficient use of "single-flank only" data by accounting for the associated uncertainties in conjunction with data on capture locations. However, even these new models perform best if at least some unambiguous (two-flank) detections are available. Use of single-flank data in spatial capture–recapture (SCR) models is a new and rapidly developing methodological research area. Within the next few years, we expect to be able to provide some general recommendations about what fraction of survey camera trap stations should contain two traps. These new hybrid designs (some two-trap stations and some one-trap stations) should permit major gains in analytical and cost efficiencies. However, many tiger populations are small and under serious threat (Chap. 1), and reliably identifying individuals also has applications in law enforcement or tracking "problem tigers"; for these reasons using two cameras per trap to get definite identifications of as many individuals as possible has merit. In this chapter, we recommend two cameras per trap as justifiable "best practice" at this point in time, while recognizing that this recommendation may be modified in the future.

10.2.2 Choice of Camera Trap Equipment

It is beyond the scope of this chapter to provide precise guidance on the exact type of camera trap equipment suited for every ecological, logistical, and resource availability context. Based on several decades of camera-trapping tigers across a range of survey conditions (see also O'Connell et al. 2011), we provide some general guidelines below.

Camera traps used in tiger population surveys have two components: the cameras and the tripping device that fires them.

10.2.2.1 The Camera

Nowadays, digital cameras are used in practically all camera trap models. The final image quality is important because poor images make it difficult to identify individual tigers. Specifically, when pattern matching software is used for rapid identification of tigers (Sect. 10.4.1), grainy images may not meet quality standards necessary for automated pattern matching.

Some camera traps also offer an additional video feature. This may help when several individual tigers pass through together or when additional information from video footage may improve the ability to identify individuals or to classify them by age and sex. However, video footage usually produces poorer-quality photos for any given frame in a sequence, rendering identification difficult. Reviewing video footage is also much more time-consuming than reviewing still photos. Use of video also requires the extra step of identifying the best frame of a video and converting the video into a still frame for use in database management and identification software programs. Unless video footage can produce images of sufficient quality, we generally recommend relying on still images of higher quality.

Camera traps with white flash can take color pictures of tigers at night, which is better for individual identification compared to monochromatic infrared flash images, particularly because infrared images also tend to be more "grainy." However, visible white flash may increase a "trap-shy" behavioral response in tigers (Chap. 9, Karanth and Nichols 1998, Wegge et al. 2004) increasing analytical complexity. White flash may also attract thieves or vandals more readily to the camera trap.

The cameras used should be capable of shooting several flash photos in sequence rapidly without exhausting their batteries over several days. They should withstand extreme heat or cold, moisture, and rough field use as anticipated.

To conserve battery power, most cameras are designed to go to "sleep" when the shutter cover is left open for some time. With such "idiot-proof" cameras, the tripping device should be capable of periodically "waking the camera up" so that no animal is missed.

Some camera traps have moisture-resistant containers. In humid areas, this is an advantage. Sometimes, other species of large animals such as elephants or bears may damage cameras. More commonly, people may steal or vandalize them. In such situations, a steel shell may be needed to protect the cameras. Design for such an elephant-proof steel shell is provided in the Karanth and Nichols (2002; Appendix 5) manual.

10.2.2.2 The Tripping Device

Electronic tripping devices are of two types. The "active" type has a transmitter that emits an infrared beam, which impinges on a receiver placed opposite. When the tiger walks between the two, the beam is interrupted, triggering the cameras. Most cameras on the market today rely on a "passive" infrared monitor that detects the combination of body heat and motion in the area it is monitoring. The area of sensitivity forms a wedge radiating outward in front of the monitor.

Generally, passive infrared units tend to be less prone to false tripping by moisture condensation, insects, or vibrations compared to active infrared units. In areas where wind and rain are problems, this may be an important factor favoring passive infrared units. However, passive infrared units are also prone to taking multiple exposures in response to changes in ambient light and movement of non-target animal species.

10.2.2.3 Important Considerations in Camera Trap Choice

In recent years, technological advancements and increased demand have improved camera trap quality and availability. Continued advancements in camera design will quickly make any recommendations on specific cameras obsolete. Therefore, we do not attempt to provide precise guidance on exact type of camera traps best suited for every ecological, logistical, and resource availability context. Based on several decades of camera-trapping tigers (see also O'Connell et al. 2011), we provide some general guidelines below. These highlight the need for camera traps to meet six major design/functionality criteria.

1. *Fast trigger speed:* It is critical under most field situations that the delay between the tripping of the sensor and the triggering of camera is minimal. In general, a trigger speed of about 250 ms (1/4 sec) is considered optimal.
2. *Long battery life:* During field surveys, camera traps are often required to be set in remote locations for long periods of time. Hence, having long battery life helps reduce maintenance costs. Batteries that do not require changing for about 100 days and are resilient to changes in temperature would be ideal.
3. *Low weight:* Deploying camera traps in the field often involves teams carrying several traps on foot to remote locations; hence, having lightweight units reduces the burden of transportation.
4. *Durability:* Some camera trap manufacturers enclose their units in moisture-resistant containers. In wet or humid areas, this is an important advantage. Sometimes, other large animals such as bears or elephants damage cameras, or people may steal or vandalize them. In such situations, a rugged protective shell may need to be deployed to protect the cameras. Appendix 5 in Karanth and Nichols (2002; pp. 184–186) shows the design of the *Javaji* steel shell that has been used to house many camera trap brands.
5. *Highly configurable:* Refining the camera's functionality (e.g., trigger speed, photo and video mode) to suit local conditions is important. Having models that are easy to configure in the field is critical.
6. *Costs:* Finally, depending on the field study to be conducted, effective surveys often require hundreds of camera trap units. With tight budgets, typical of most conservation programs, cost of individual units becomes a primary concern.

Several commercial camera traps are available in the market with models changing rapidly. We advise users to check relevant literature, websites, and list servers for current details (http://www.crowsystems.com/cameras.htm, http://www.trailmaster.com/thttp://www.camtrakker.com, http://www.trailsenseengineering.com/); for reviews and comparisons of cameras, see http://www.chasingame.com/ and http://www.jesseshunting.com.

10.2.3 Field Surveys: Camera Trap Locations and Setup

Because tigers live at low population densities of 10–15 animals/100 km^2 even in the best habitats (Chap. 2), getting a "photographic capture" of a tiger is a highly uncertain, rare event even with many camera traps in an array.

Therefore, if estimation of demographic parameters is the monitoring goal, camera trap sites must be chosen to maximize tiger capture probabilities: they *should not be* randomly selected spots in the sampled area (Karanth and Nichols 2010, Karanth et al. 2011). Tigers use favored travel routes (Chap. 2) and do not move randomly through space. If traps are set randomly, most will have near-zero probabilities of catching a tiger. Therefore, practitioners should avoid randomly selected GPS coordinates for setting camera traps, if the study objectives focus on tiger population parameters.

Probabilities of tigers encountering camera traps are inherently low. In order to maximize capture probabilities (Chap. 9), traps can be placed along tiger travel routes identified based on natural history skills (Plate 10.2) or radio-tracking data if available. More generally, trap site selection depends on the ability to "think like a tiger": convergence of game trails and presence of salt licks and water holes that attract prey animals may indicate good sites. Usually, trails with soft underfoot conditions, sandy streambeds, and forest logging roads make good "trap lines."

Although we do not routinely recommend scents or lures to "hold tigers" in front of the cameras, these could be employed if found useful. Capture–recapture models can account for any potential "trap-happy" response induced by baits or lures, if they are used (Chap. 9).

Plate 10.2 Camera trap locations should be non-random and selected carefully to maximise tiger photo captures (Image Copyright©: Eleanor Briggs)

Once a trap site is chosen, the goal is to frame the picture so that the tiger's flanks are clearly photographed. Although stripes on any part of the tiger's body can be used for identification, pictures of flanks are the easiest for comparing and identifying individuals.

The trail should not be modified drastically, and traps should be placed as unobtrusively as possible. Often, a simple trick such as throwing a few twigs around a trap can prevent it from being conspicuous. Masking tape of a dull color should be used on bright shiny parts. All loose cables should remain hidden.

At each trap site, cameras should be positioned about 3–4 meters away on both sides of the path. In active infrared-triggered units, the electronic beam should be set 45 cm above the ground and at a trigger sensitivity-level sufficient to catch even tiger cubs.

It may be necessary to experiment with cameras to determine the optimal distance from traps to the tiger's likely path. For cameras with easily accessible memory cards, it is best to experiment and review photographs to ensure the best positioning of the camera. Having an extra camera or tablet computer to view photographs from the memory card is useful for this task if the camera trap does not have a built-in screen. Even if the tiger walks on the far edge of the path, usable pictures should be obtainable at this distance.

10.2.4 Survey Design: Spatial Sampling of the Tiger Population

Achieving "randomness" is sometimes offered as the reason for placing traps in some regular geometrical pattern, such as a square grid. Although the grid itself is certainly not random, the distances between unknown animal home range centers and the camera traps (a determinant of capture probability) can be viewed as random. It should be noted that, as far as possible, sampling should produce similar capture probabilities for all animals in the sampled area, but employing a square grid (or any other shape) is not necessarily advised. What should be avoided is the choice of inferior locations for setting traps to fulfill some vague pursuit of randomness while ignoring superior ones.

Although similar capture probabilities among individual animals are desirable for classical CR methods, we also note that the SCR models (Chap. 9) reduce this need, in the sense that variation among individuals in the distances between home range centers and camera traps is explicitly incorporated into the modeling (also see Royle et al. 2014).

Placement of traps in areas less likely to be visited is not helpful for increasing capture probabilities or making them similar among individual tigers. Not using the best available trap sites will often dramatically lower capture probabilities (fewer animals caught and fewer recaptures). Tigers are rare and difficult to detect in traps, and good trap sites are difficult to find, particularly in low tiger density areas. The aim is to get more photos from more individual tigers exposed to the trap array. Thus, choosing the best possible trap sites for each individual in the sampled area should be the primary consideration.

With the advent of SCR models (Borchers and Efford 2008, Royle et al. 2009, 2014), conventional capture–recapture analyses have become much less important. Therefore, our general prescriptions on spatial survey design below are made keeping SCR analyses in mind. Whether the subsequent analyses are spatial or nonspatial, all closed model capture–recapture surveys assume that "samples" of a few individual tigers are drawn multiple times within a "short period." The population is assumed to remain unchanged by births, immigration, deaths, or emigration because the survey duration is short.

Although not all tigers need to be "detected" for capture–recapture methods to work (Chaps. 3 and 9), every individual exposed to the sampling effort must have non-zero *probability* of being detected by trapping, in order for conventional closed models to work. For this reason, large "holes" without any traps in the array within which a tiger can potentially remain unexposed to cameras are best avoided. Therefore, expected home range sizes of breeding females, which are typically smallest (Chap. 2), should critically influence the trap spacing. This recommendation of "no holes" is not so important for SCR models, because these explicitly deal with the fact that capture probability is a function of distance between range center and individual traps, thus permitting animals with range centers far from any traps to have capture probabilities that approach zero.

With a limited number of traps available, which is most often the case, any survey design faces an unavoidable trade-off. A closer trap spacing increases capture probabilities for each tiger exposed to trapping. However, close spacing reduces the number of individual tigers exposed to cameras. Placing traps farther apart (increasing trap array size) will expose more individuals to capture if the sampled population occupies a larger area. Because sample size for capture–recapture analyses is dependent on the number of individuals caught (n), larger sample sizes lead to stronger inferences. Balancing these competing needs is a key to efficient spatial design of a tiger survey (see Table 7.2 in Karanth et al. 2011 for an example).

In absolute terms what should be the spacing between traps? It should be lower at smaller expected minimum tiger home range sizes. If the expected home ranges are larger, traps can be set farther apart. From previous studies of tiger social organization (Chap. 2, Smith 1993, Kerley et al. 2003, Goodrich and Miquelle 2010, Hernandez-Blanco et al. 2015) within a population, resident tigresses that breed have the smallest home ranges. Therefore, if at least 2–3 camera traps are placed within an expected female home range (Karanth and Nichols 1998), individual tigers in other age–sex classes will automatically be exposed to an equal (juveniles, cubs) or larger (breeding males, transients) number of traps in their home ranges.

Home ranges of breeding tigresses in prey-rich areas can be as small as 10–15 km^2 (Sunquist 1981; Smith 1993; Sunquist et al. 1999, Karanth and Sunquist 2000, Chap. 2). In such high-density areas, camera traps can be set at about 2–3 km apart to potentially expose a resident tigress to 2–3 traps. At the other extreme, female ranges can be as large 200–600 km^2 (Goodrich et al. 2010, Hernandez-Blanco et al. 2015, See Chap. 2). In such areas, camera traps can be deployed 5–10 km apart. We suggest the above figures only as broad guidelines: specific local knowledge about likely tiger range size is a key factor in designing the trap array.

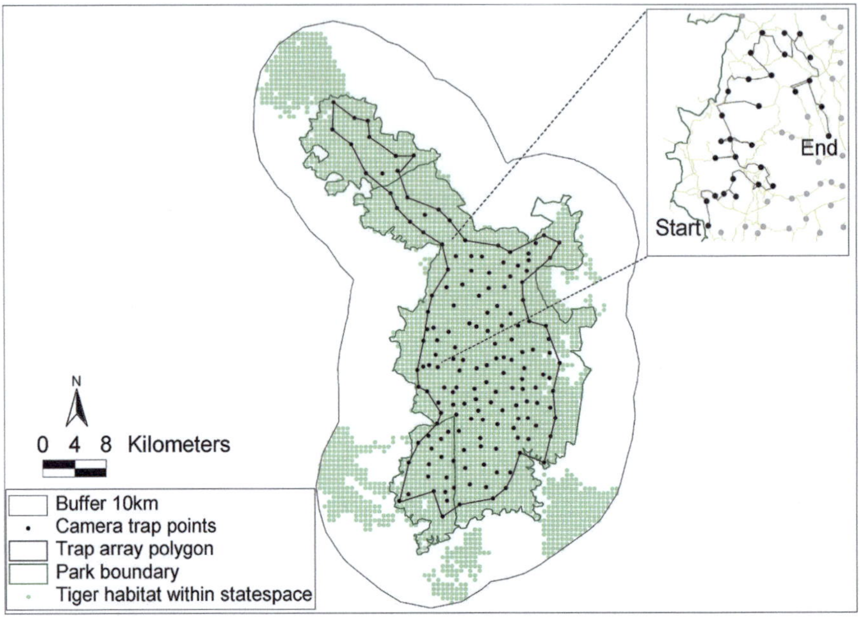

Fig. 10.1 Camera trap survey design, showing the array of camera trap points and tiger habitat in Nagarahole reserve, India, within the defined state space. Inset shows one of eight "trap lines" used by field teams to regularly check camera traps

Within each expected home range area, it is better to place "more traps" than "fewer traps" depending on the availability of traps.

However, the need is also to catch *as many different individual tigers* as possible. Potentially more individual tigers can be caught by increasing trap spacing. Therefore, trap spacing becomes a compromise between these two competing needs as already noted.

The area to be sampled should be initially explored by skilled naturalists to locate and map most of the potential camera trap sites. Their number should be 2–3 times more than the final number required by the spatial design employed. While deciding the final trap array shape and size, logistical factors, such as topography, barriers to movement of personnel, and potential theft of traps, also have to be considered. Final choice of camera trap sites is made after all these factors are balanced.

Outer boundaries of camera traps—final location of the trap array—typically form an irregular "trap polygon" (Fig. 10.1). Although not all factors that lead to a specific design are controllable, solid, compact shapes for a trap array—say circles, squares, ovals—are preferable to shapes resembling "donuts," "hub and spokes," or "narrow strips" from the analytical perspective.

Furthermore, it is necessary to delineate "real habitat" in which tigers are resident or move through frequently. These habitats may have adjacent "non-habitat" areas, such as human settlements, some types of agricultural land, or large water bodies. Such "non-habitat" areas, which are not routinely used by tigers, should be excluded from the effectively sampled area during analysis. In this context, more efficient sampled area coverage with a given number of traps can sometimes be achieved by placing the outermost traps at some reasonable distance inside, rather than right on the edge of known and mapped tiger habitats (Fig. 10.1).

SCR models of different "flavors" are available now (Efford 2004, Borchers and Efford 2008, Royle et al. 2009, Dorazio 2013, Dorazio and Karanth 2017). All these approaches require adding a "buffer" or "mask" around the trap array to delineate the area sampled in the survey. The purpose of adding such a buffer is to identify a "catchment" area, beyond the trap array, from which captured tigers could have come. The linear distance from the edge of the trap array to the outer periphery of the buffer (mask) depends on how far we expect tigers to routinely move during the survey period.

Because the sampled tiger population may include post-dispersal transients, which have the largest home ranges (Chap. 2; Smith 1993), we suggest a buffer distance of 3–4 times greater than the radius of expected resident female range. For example, in high tiger density sites, the expected female resident home range radius of 2.5 km leads to a buffer distance of 10 km. As in the case of defining the trap array, "non-tiger habitat" is excluded from this buffer area also. Thereafter, the area effectively sampled, including the trap array and buffer and excluding non-tiger habitat, can be identified and mapped. An example is shown in Fig. 10.1.

10.2.5 Survey Design: Temporal Sampling of the Tiger Population

Capture probabilities are estimated from temporally repeated "samples" drawn from among all tigers exposed to trapping. These temporal replicates can be thought of as multiple "sweeps" over the area. The camera trap survey therefore extends over the total number of days required to sweep *the entire sampled area* successively, multiple times. Each successive "sweep" is considered a sample ("sampling occasion" in conventional capture–recapture terminology). SCR permits greater flexibility in this regard, permitting each temporal sweep to cover the sampled area either fully or partially. However, the former approach will provide better quality data, particularly if temporal changes in capture rates can be expected due to changing weather or some other factor.

Camera trap monitoring conducted once in each year, or season, is called a primary sample. It may include multiple subsamples, designated as "secondary samples" (see Chap. 9). The length of each primary sampling period is guided by theoretical and practical considerations. Recall that the critical parameter we are trying to estimate under closed models is the number of tigers in the surveyed area (absolute abundance). This tiger population is assumed to be "demographically closed" (see Chap. 9) through the primary period. Therefore, ideally the survey duration should be as "short" as possible to reduce chances of violating the population closure assumption. We recommend durations of fewer than 45 days, preferably 15–30 days. As this primary sample duration increases, violations of the closure assumption may occur, which can lead to biased estimates (Williams et al. 2002, Chaps. 2 and 9).

Generally, if there are more secondary samples, it is possible to use analytic models that can deal better with potential sources of variation in capture probabilities such as individual heterogeneity, trap response, or time-related variation (Otis et al. 1978, White et al. 1982). Having 5–15 secondary temporal samples is better than having just two. However, extending the survey duration (the primary sample period) increases the risk of violating the demographic closure assumption.

For illustration, assume that 100 trap locations must be sampled in a 28-day closed model survey. If the investigator possesses 100 camera traps, each day's trapping can be considered a secondary sample. Thus, capture data from 28 days of trapping will cover the primary period. However, investigators may not have enough camera traps to cover all trap locations simultaneously. Therefore, they may spatially segregate the trap array into logistically convenient "blocks" or "trap lines" (Fig. 10.1). These blocks are each sampled for a few successive days, one after another. In the above case, if we had only 25 pairs of camera traps, the 100 locations would be partitioned into four blocks, and each block is trapped continuously for 7 days, in sequence, to complete the entire survey in 28 days.

The key point is that the capture data for each primary period, as well as for each secondary period, must come from across the entire sampled area and not from parts of it as is sometimes erroneously done.

For example, in conventional CR analyses, seven sampling occasions can be artificially constructed from the above example data. Data from day 1 in each block are combined as sample 1, data from day 2 data as sample 2, and so on (See Karanth and Nichols 2002, page 133 for details). In SCR analyses, capture data from such a "block" trapping scheme can be readily handled, because the data structure can specify which traps are active during each sample.

Frequent checking of traps can inform which traps were operational and which ones were not. If this is not the case, nonfunctional traps may go into the SCR input data structure considered as "active traps." However, practically, if traps are not checked frequently, even if analyses use SCR models, our preference is for sampling the entire area of interest in each sweep, rather than using the block trapping approach.

We provide a cautionary note: camera trapping an area continuously over months, as some surveys reported by Jhala et al. (2015) appear to have done, is likely to seriously violate the assumption of population closure. It is incorrect to apply closed capture–recapture models to such data from what are really open populations of tigers. Although open CR models (Sect. 10.4.4) can be used in such cases, their primary use is for estimation of survival and recruitment rates rather than abundance or density as already seen (Chap. 9).

Of course, if there are no constraints on available camera trap numbers or on logistics, the area can be camera trapped continuously over the year. In such a situation, data from successive "short periods," in which closure can be assumed, can be separately analyzed and yield an "average abundance" for the overall period (see Jedrezejewsky et al. 2017, for such an analysis of jaguar capture data). However, continuous trapping will frequently not represent an efficient use of resources and should prompt the question "Is some sort of year-round average abundance/density preferable in any way to abundance/density in a specific season?"

It is also pertinent to note that such an approach is not readily extendable to the estimation of vital rates using open models under the robust design (Chap. 9, Karanth et al. 2006). For example, it is desirable that time intervals between successive primary sampling occasions are long relative to the duration of each primary occasion. Artificial discretization of continuous sampling data is an ad hoc approach that can be used but is not recommended by us. SCR models for closed populations now exist for data collected continuously in time, as is really the case with camera trap surveys (Borchers et al. 2014, Dorazio and Karanth 2017).

The central point is that temporal scale of camera trap sampling, as with all other aspects of study design, should be based on survey objectives and assumptions of models to be used in analysis. We can think of no reason to design studies that do not correspond to any of the numerous currently available analytic modeling approaches.

Sometimes, the tiger survey objective is deliberately scaled down to determine the "minimum number of tigers alive" at any point in time (e.g., Phase IV, Section 2, Part-D in NTCA 2012 protocol). However, because of high population turnover rates (Chap. 2), if surveys are extended, violating the closure assumption, accumulated number of animals photo-captured can potentially exceed the total number of tigers alive at any point during the survey! Given the major investments made in camera trap surveys of tigers, we see no logic in setting up such inadequate objectives or in applying poor methods to achieve them. The real tiger conservation need is for reliable demographic estimates obtained using the most rigorous methods available.

10.3 Conduct of Camera Trap Surveys in the Field

10.3.1 Setting Up, Checking, and Troubleshooting Camera Traps

Meticulous field protocols will prevent inaccuracies creeping into tiger individual identifications and records of capture location and time. In this regard, we make the following recommendations:

1. All camera trap equipment should be maintained as per the manufacturer's manual. Fresh battery cells should be used in the cameras and tripping devices to ensure that data are not lost. Camera traps should be checked on a daily basis—or as frequently as possible—to record data. Particular attention should be paid to battery status, amount of remaining memory, and proper mounting of the camera and tripping device. It is important to make sure that the date, time, time zone set for capture, camera delay, and expected target size (period of beam breakage) are all set correctly. The date and time of each exposure as read by the camera's data-back (metadata) should be carefully double-checked with data recorded by the tripping device as well as manual records. Spatial capture–recapture analyses also require records of "working cameras" in the trap array for each sampling occasion.
2. Standard data forms (e.g., Tables 10.1 and 10.2) should be used to record information from camera traps. Unique identification numbers should be given

Table 10.1 Camera trap field data form for details of date, time, location, equipment, images, and other remarks. Camera trap survey field data form

CAMERA TRAP SURVEY FIELD DATA FORM

Names of Surveyors: Name of Supervisor: Field Site:

Other details: Page No:

Date	Time	Location	Exposure details			Remarks
			Cam No.	USB Pen Drive No.	No. of Images	
Setup Check						
Setup Check						
Setup Check						
Setup Check						
Setup Check						

Cam No. = Camera ID number
USB Pen Drive No. = Unique USB Pen Drive ID number
No. of Images = Should be filled from the log file of each camera

to each camera and tripping device for identifying its location and functioning and for easy troubleshooting in case of malfunction.

3. Memory cards should be reformatted prior to camera deployment to avoid inadvertent additions of photos from earlier surveys. This may particularly occur when setting of dates on the camera is incorrect. It is important to uniquely label each memory card (or film roll) before it is loaded into the camera. Because cameras tend to get moved around in field surveys, this practice is critical for accurately determining the time and location for any tiger capture event, as well as for linking left and right profiles of tigers. Therefore, the record of each identified memory card (or film roll) must be traceable from the time of loading until image data are processed.
4. For most currently available analysis methods, individual tigers must be identified correctly and unambiguously by getting photographs of both sides. Investigators using single camera units can end up with two smaller data sets of left and right profiles, without being able to match the profiles for a certain identification. Analyses of such data, however, are now possible using new models under development (Augustine et al. 2016). However, even these models typically

Table 10.2 Camera trap field data form for equipment maintenance and troubleshooting

Details of trouble shooting

Trap location	Camera 1 working	Camera 2 working	Camera Battery Status	Clock Battery Status	No False Tripping	All OK	Details of Units replaced/Rolls changed/Batteries replaced	
							Camera 1	Camera 2

rely on a subset of animals identified using images of both flanks. We do not recommend studies be designed using single camera traps at all, or even most, trap sites at this point.

10.3.2 Collection and Management of "Raw" Camera Trap Data

We have seen that quite often poor management of raw capture data results in analytical problems because of inaccuracies in individual tiger identifications as well as in photo-capture locations and dates. The field protocols outlined below will help ensure reasonable integrity of survey data:

1. Each "camera trap" consists of two cameras and a tripping system, each labeled with unique ID numbers. Each image storage device (e.g., memory card or film roll) should have a unique ID label, all of which are entered into data forms.
2. While setting up camera traps, date, time, and GPS location of the cameras and tripping device should be fully synchronized. The camera trap setup and checking activities, with all relevant ancillary data, should be recorded on a data form (Tables 10.1 and 10.2).

3. It is important that all field and supervisory personnel record their names on data forms to assign responsibility for the integrity of the survey exercise.
4. Downloading of raw image data from the camera traps in the field and their subsequent storage on computers in a carefully designed folder structure are critical. Details of such a data storage folder structure are provided at http://projecttiger.nic.in/WriteReadData/userfiles/file/Protocol_Phase_IV_Monitoring_r.pdf. These raw data can be stored in a spreadsheet (or some other convenient format). However, they should be extractable using specific criteria, for subsequent construction of capture histories (Chap. 9). Only after a high-quality tiger photo database is established can analyses described in Section 10.4 be undertaken.
5. However, several software packages to manage camera trapping data are now available, from desktop applications such as 'AARDWOLF' (Krishnappa and Turner 2014) and 'Camera Base' (Tobler 2013), to citizen-science oriented web applications, e.g. 'SNAPSHOT SERENGETI' (Swanson et al. 2015) and 'Camera CATalogue' (https://www.zooniverse.org/projects/panthera-research/camera-catalogue) to more project-specific cyber-infrastructures like 'TEAM NETWORK' (Fegraus et al. 2011) or 'WILDLIFE INSIGHTS' (https://www.wildlifeinsights.org). In recent years more flexible open source packages such as CamtrapR (Niedballa et al. 2016) and TRAPPER (Bubnicki et al. 2016) have been developed.

10.4 Capture–Recapture Analyses of Camera Trap Data

10.4.1 Individual Identification of Tigers

After tiger images from a survey are correctly labeled and stored, the next step is to link the left and right profiles of individual tigers to identify and assign them unique identification (ID) numbers.

Tiger images can of course be visually compared and identified entirely manually. However, as the number of images increases over the years, manual comparison of each new image with all previous ones becomes cumbersome. It is necessary to set up a semiautomated system to store, manage, and identify tigers from images.

Although many image storage and management software programs are now available (camtrapR: Niedballa et al. 2016; PhotoWarehouse: Ivan and Newkirk 2015; trapper: Bubnicki et al. 2016), as are animal identification programs (WildID: Bolger et al. 2011 and HotSpotter: Crall et al. 2013), we have found the database and identification approach developed by Hiby and Lovell (1990), implemented in the free software ExtractCompare (Hiby et al. 2009) to be versatile and reliable for tiger identifications. Although not the most user-friendly software available, requiring substantial training before the user can gain proficiency, this very feature ensures data integrity by automatically excluding incorrect data entries that may slip through with other more "user-friendly" software programs.

When comparing tiger stripe patterns, diagnostic features such as uniquely shaped stripes, either singly or in juxtaposition with others, are used for identification (Plate 10.1). The shapes of specific individual stripes as well as their relative positioning on the flanks of tigers are particularly useful. If necessary, additional stripe patterns found on the head, limbs, and the tail can also be used (Plate 10.1).

The following three demographic stages can be assigned to photo-captured tigers based on body size and proportions: cubs <12 months; juveniles 12–24 months; and post-dispersal subadults/adults 24 months or older, when clear photographs are available (Karanth and Stith 1999, Karanth et al. 2006). Often, sex can also be identified for juveniles and adults.

Tiger cubs have lower capture probabilities than older animals due to wariness. They also have higher mortality rates, violating the closure assumption and lowering numbers of recaptures for individuals. As a result, typically fewer cubs are photo-captured and identified. In this sense, tiger cubs are akin to a different species with extremely low capture probabilities. Therefore, we recommend that estimates of tiger density, abundance, and vital rates should be based on data that exclude cubs <12 months old.

The website of India's NTCA (2012) describes a unique identification and labeling system: http://projecttiger.nic.in/WriteReadData/userfiles/file/Protocol_Phase_IV_Monitoring_r.pdf. This system, developed by Karanth et al. (unpublished data), is based on a database of images of >800 individual tigers accumulated from a long-term camera trap study in India (1991–2017).

10.4.2 Construction of Capture Histories of Individual Tigers

After individuals are given unique identification numbers, every capture event for each individual is assigned to a particular secondary sampling occasion (Chap. 9) using the location, date, and time of capture. From this database, the capture history of each individual tiger can be constructed. For conventional CR analyses, only the individual identity and the sampling occasion(s) on which the individual was captured are used. For SCR analyses, data that pertain to location and time of capture, as well as the temporal schedule of operation of each trap, are included.

Spatial CR analyses will also require maps in the form of GIS shape files for the sampled area, the trap array, buffer distance used (mask), and non-tiger habitat, as described in Sect. 4.3.2. Typically, GIS software such as ArcView (ESRI 2011) is employed to create these spatial data layers.

Thereafter, analyses generally proceed as below.

10.4.3 Closed Model CR Analyses: Basic Approaches and Software

We believe conventional CR approaches highlighted in the Karanth and Nichols (2002) manual are obsolete and superseded by spatial CR models for most uses. Details of conventional CR analyses can be found in Otis et al. (1978),

White et al. (1982), and Amstrup et al. (2005). Specifically, with reference to tiger data, readers are referred to Chaps. 11 and 12 in Karanth and Nichols (2002). Typically, estimation of tiger abundance using conventional methods is performed using software programs CAPTURE (White et al. 1982, Rexstad and Burnham 1991) or MARK (White and Burnham 1999). Both of these are free software, available respectively at www.mbr-pwrc.usgs.gov/software.html and whttp://ww.warnercnr.colostate.edu/~gwhite/mark/mark.htm. Their users are also supported by helpful online lists (e.g., http://www.phidot.org/forum).

10.4.3.1 Testing for Demographic Closure

The first step in the analysis of CR data from a "secondary sample" (one snapshot of the tiger population, obtained over 30–45 days, in a single season or year) involves testing for potential violation of the demographic closure assumption, using tests for demographic closure implemented in software such as CAPTURE, MARK, and ClosTest (Stanley and Burnham 1999). White et al. (1982) suggest that the power of these closure tests has not been intensively investigated, and they may be sensitive to certain forms of variation in capture probability, highlighting the importance of making realistic biological assumptions.

If the demographic closure assumption appears reasonable for the sample data, the CR analysis can proceed. If the closure assumption appears to have been violated seriously, the analyst may be compelled to either use one of the open model analyses (Sect. 4.4.) or to restrict analysis to capture data from a reasonable shorter period within the primary sampling period.

10.4.3.2 Spatial Capture–Recapture: Basic Approaches

Statistical concepts of spatial capture–recapture models have been summarized in Chap. 9. These are based on statistical advances that emerged rapidly in the last decade (Efford 2004, Borchers and Efford 2008, Efford et al. 2009, Royle et al. 2009, 2014).

Spatial capture–recapture models have been developed under likelihood-based (Borchers and Efford 2008) and Bayesian (Royle et al. 2009, 2014) approaches. The reader is referred to relevant statistical literature (e.g. Royle and Dorazio 2008, Link and Barker 2010) on comparative merits of these two fundamental inferential approaches. We note that for estimates of tiger density, the "credible intervals" generated under Bayesian approaches constitute direct probability statements, whereas the frequentist confidence intervals have more complicated interpretations (see Link and Barker 2010 pages 7, 29–36 for a lucid explanation of the distinction).

Although the Bayesian spatial CR model formulations require more computing time and power, we believe they provide greater general flexibility. We also believe that the scope for future development of integrated modeling of "closed" and "open" capture–recapture population analyses is greater under the Bayesian framework. The Bayesian approach also has inherent advantages for drawing on prior knowledge and for generating estimates from small samples of data (Royle and Dorazio 2008, Link and Barker 2010, Royle et al. 2014).

Several computer programs are available for analyses of tiger capture–recapture data. Under the likelihood approach, the following programs are popular: DENSITY (Efford et al. 2004), R package *secr* (Efford 2011), and R package *oSCR* (Sutherland et al. 2016). There are several options for performing Bayesian analyses also. They range from software specifically designed for camera trap data, such as R package *SPACECAP* (Gopalaswamy et al. 2012b), to the more general package *SCRbayes* (https://github.com/jaroyle/SCRbayes). Some analysts prefer to develop computer code specifically tailored to individual studies using the R programming language (R Development Core Team 2012) or Bayesian programming environments such as WinBUGS (Gilks et al. 1994) and JAGS (Plummer 2003).

Going into details of all these analytical options is beyond the scope of this chapter. We recommend detailed study of the original literature cited above. Table 10.3 has results of SCR analyses from tiger studies in India and Malaysia We find it comforting from our experience that, if the data are clean and properly handled, these alternative SCR analyses usually generate estimates that are very similar (Table 10.4). The Appendix shows an example of a full analysis performed in the R programming environment, using an SCR model developed by Dorazio (2013). It shows the input files, part of the R code used, and the output, along with definitions and notation employed.

10.4.4 Open CR Models and Analyses: Approaches and Software

The secondary sample surveys are embedded within an annual primary sample as described above that provides a "snapshot" of the tiger population. They generate estimates of population size (abundance, N), density, and other spatial- and detection-related parameters for that particular period.

Such surveys are repeated over some longer time period (e.g. annually), for the overall period (consisting of multiple primary samples, each of which contains secondary samples over which the population is assumed to be closed). We treat the tiger population as being open between primary samples: subject to increases due to births and immigration, and decreases from deaths and permanent emigration. Such a multi-year capture–recapture analysis can generate estimates of demographic parameters such as survival, losses, and gains that drive tiger population dynamics and are critical needs for tiger population management and recovery.

Ideally, the annual surveys should be similar in terms of area sampled, duration, and other aspects of field implementation. However, as it often happens, as more equipment and resources become available over the years, the sampled area can be increased. If such expansion is systematic and takes place from the original "core area" outward, it is still possible to combine data across the years to obtain some parameter estimates over the entire period, while some parameters can only be estimated when the sampled area remains constant (see Karanth et al. 2006, 2011 for an example). However, many practitioners employ additional resources that become

Table 10.3 Estimates of tiger densities at different sites from camera trap surveys using spatial capture–recapture analyses under closed models

Site	Year	Software package used for analyses	Park boundary (km^2)	Trap array area (km^2)	Duration (days)	n	$\widehat{D}\left(\widehat{SE}\right)$ (/100 km^2)	$\widehat{N}$
Nagarahole-Wayanad North, India[a]	2013	SPACECAP	774	503.8	30	82	11.24 (0.9)	94[d]
Bandipur-Wayanad South, India[a]	2013	SPACECAP	1178	647.8	30	99	10.30 (0.81)	132[d]
Bhadra, India[a]	2013	SPACECAP	492	571.9	30	20	2.31 (0.40)	13[d]
Dandeli-Anshi, India[a]	2013	SPACECAP	1306	936.0	81	3	0.20 (0.07)	3[d]
Biligiri Rangaswamy Temple, India[a]	2013	SPACECAP	539.5	351.7	30	52	11.42(1.33)	69[d]
Rajaji National Park, India[b]	2004–2005	WinBUGS	148	NA	45	4	3.31(1.51)	8[e]
	2005–2006	WinBUGS	148	NA	45	5	2.67 (0.97)	13[e]
	2006–2007	WinBUGS	148	NA	45	6	5.17 (1.94)	25[e]
	2007–2008	WinBUGS	148	NA	45	6	5.81 (2.26)	28[e]
The Royal Belum State Park, Malaysia[c]	2009	secr	1175	~400	90	17	1.95 (0.48)	24[f]
Temengor Forest Reserve, Malaysia[c]	2009	secr	1489	~400	90	4	0.61 (0.31)	9[f]

n: number of individuals photo-captured; $\widehat{D}$: estimated density/100 km^2; $\widehat{N}$: estimated abundance
Sources: [a]Karanth et al. (2014) and unpublished data; [b]Harihar et al. (2011); [c]Rayan and Linkie (2015)
[d]Abundance estimated by summing pixel densities over park boundary/administrative area
[e]Abundance estimated as the "number of individuals exposed to the sampling"
[f]No details provided on how abundance was estimated and what area it pertains to

available in an ad hoc manner to sample disjunct areas, unthinkingly sacrificing analytical gains from a more careful study design. Overall, we believe that it is a better strategy to obtain one set of reliable estimates of tiger population parameters through data integration and borrowing of information across space and time, as opposed to scattering resources across multiple areas to conduct surveys with low inferential strength.

In many open population capture–recapture studies, the probability of observing a particular tiger capture history is a function of two key sets of parameters: the probability of capture and the probability of that individual surviving between

Table 10.4 Results from spatial capture-recapture analyses of the same data-set from a camera trap survey in Nagarahole – North Wayanad, under different modeling approaches

SCR modeling approach used	Borchers and Efford (2008)	Dorazio (2013)	Royle et al. (2014)	Royle et al. (2009)
Software used	R package secr	R, code provided by RM Dorazio	JAGS, code from Royle et al. (2014)	R package SPACECAP
State space S (km^2)	1125			
Administrative area (km^2)	858			
Trap array area (km^2)	504			
No. of trap locations	161			
Mean trap spacing (m)	1534			
Duration of trapping (days)	30			
Effort (trap days)	4760			
No. of individuals photo-captured n	82			
No. of capture events	305	302	305	302
Encounter process model	Poisson	Binomial	Poisson	Binomial
Detection function	Half normal	Half normal	Half normal	Half normal
Baseline capture probability $\widehat{g(0)}$ $\left(\widehat{\mathrm{SE}}\right)$	0.026	0.025 (0.002)	0.026 (0.002)[a]	0.025 (0.002)[a]
Scale parameter $\widehat{\sigma}\left(\widehat{SE}\right)$ (m)	1843 (37)	1871 (70)	1832 (64)	1877 (70)
Density $\widehat{D}\left(\widehat{SE}\right)$ (/ 100 km^2)	11.1 (1.2)	11.0 (1.2)	10.9	11.0 (0.7)
$\widehat{D}$ 95% Confidence Interval	8.9–13.8	8.7–13.5[b]	8.9–13.5[b]	9.6–12.4[b]
Zero-inflation parameter $\widehat{\psi}\left(\widehat{SE}\right)$	NA	NA	NA	0.26
Effectively sampled area (km^2)	929	931	928	931
Abundance within administrative area $\widehat{N}_{admin}$	103	102	101	102
Abundance within state space $\widehat{N}_{super}$	NA	NA	NA	124

[a]Baseline encounter rate λ_0; [b]95% Credible Interval

primary sampling occasions and remaining within the sampled area. This survival is referred to as "apparent survival" (φ), and its complement (1- φ) does not distinguish between losses due to death and permanent emigration from the sampled area. As with most closed model estimation, the data for open model analyses are the capture histories, and maximum likelihood estimates are those parameter values that maximize the probability of having obtained the observed set of capture histories (Chap. 9).

The original Cormack–Jolly–Seber (CJS; Cormack 1964, Jolly 1965, Seber 1965) model permits only estimation of apparent survival and capture probabilities, whereas the Jolly–Seber (JS) model that includes an additional assumption of equal capture probabilities for tigers that have and have not been previously caught can estimate abundance and recruitment as well. A major problem with these earlier models for analyses of tiger capture data is that they produce biased estimates of abundance in the presence of individual heterogeneity or trap response behavior, which are sometimes observed. However, trap response does not induce bias in survival estimators, and heterogeneity produces bias in survival that is relatively small (see summary in Williams et al. 2002).

In Pollock's robust design (Pollock 1982, Pollock et al. 1990, Williams et al. 2002), this problem is overcome by integrating the sampling at two temporal scales: the primary sampling periods separated by years and between which the population is assumed to be "open" to gains and losses and within each of these several secondary sampling occasions between which the population is assumed to be "closed" to gains and losses. Under Pollock's (1982) original two-step approach, robust design analyses estimate survival across the years using CJS-type estimators and abundance each year basically using closed models (Nichols 2005), including those that account for heterogeneity. Thereafter, recruitment into the tiger population can also be estimated by combining estimates of survival and time-specific abundance. Kendall et al. (1995) proposed an improved likelihood-based robust design approach that simultaneously combines data from primary and secondary samples. Such joint modeling enables borrowing information across years, potentially reducing the number of model parameters and increasing their precision (Table 10.5).

As a result of all these advances, even temporary emigration (Kendall and Nichols 1995, Kendall et al. 1997) and transience (Pradel et al. 1997) can be estimated and dealt with. We note that "transience probability" can be viewed as the expected proportion (among all captures of new animals) of tigers that have a near-zero probability of being recaptured, because of movement out of the study area. Temporary emigration is likely to be a function of study area size: in open study areas without hard edges, some individuals may be absent during some of the primary occasions simply because they are occupying parts of their home ranges that lie outside the trapped area. Furthermore, improvements in likelihood-based estimators that can incorporate individual heterogeneity in capture probabilities (e.g., Pledger 2000, Pledger et al. 2003) have been very important for tiger studies (see Karanth et al. 2006).

Table 10.5 Assessing tiger population dynamics from camera trap surveys under robust design

Site	Year t	Time interval Δt	Abundance $\widehat{N}_t\left(\widehat{SE}\right)$	Annual survival $\widehat{\varphi}_t\left(\widehat{SE}\right)$	Interval survival $\widehat{\varphi}_t^{\Delta t}\left(\widehat{SE}\right)$	Recruitment $\widehat{B}_t\left(\widehat{SE}\right)$	Mean annual growth rate $\widehat{\lambda}\left(\widehat{SE}\right)$
Nagarahole, India	1991	0.667	9 (0.0)	0.77 (0.051)	0.85 (0.040)	– [a]	1.03 (0.02)[b]
	1992	1.333	7 (2.6)	0.77 (0.051)	0.72 (0.061)	– [a]	
	1993	0.75	11 (5.5)	0.77 (0.051)	0.83 (0.043)	– [a]	
	1994	0.917	21 (3.2)	0.77 (0.051)	0.80 (0.048)	– [a]	
	1995	1.25	12 (0.0)	0.77 (0.051)	0.73 (0.059)	– [a]	
	1996	1.167	27 (1.4)	0.77 (0.051)	0.75 (0.056)	3 (3.2)	
	1997	0.583	20 (3.2)	0.77 (0.051)	0.87 (0.037)	0 (3.0)	
	1998	1.25	17 (1.7)	0.77 (0.051)	0.73 (0.059)	11 (2.8)	
	1999	1.083	23 (1.7)	0.77 (0.051)	0.77 (0.051)	14 (2.9)	
	2000	–	30 (2.1)	–	–	–	
Huai Kha Khaeng, Thailand	2005	1.408	51 (1.85)	0.80 (0.08)	0.73 (0.08)	–[a]	0.99 (0.23)[c]
	2006	1.0438	58 (2.06)	0.80 (0.08)	0.79 (0.08)	0 (5.03)	
	2007	1.0109	41 (1.32)	0.96 (0.06)	0.95 (0.06)	5 (2.86)	
	2008	0.9151	43 (1.11)	0.87 (0.05)	0.88 (0.05)	9 (2.67)	
	2009	1.0822	47 (1.27)	0.59 (0.08)	0.57 (0.08)	17 (4.22)	
	2010	0.974	44 (1.37)	0.86 (0.05)	0.86 (0.05)	0 (2.70)	
	2011	1.0192	35 (1.00)	0.90 (0.09)	0.90 (0.09)	24 (3.40)	
	2012	–	56 (1.37)	–	–	–	

[a]Not estimated due to changes in study area; [b]estimated for the period 1996–2000; [c]estimated for the period 2006–2012

To understand tiger population dynamics fully, several parameters in addition to abundance are biologically relevant and need to be estimated. These include survival, losses, recruitment, temporary emigration (probability of a tiger not being present during some of the primary sampling periods), and transience (probability of a dispersing individual just passing through). Of course, the key "nuisance parameter" of detection probability must necessarily be estimated. Program MARK offers a large suite of flexible models, including most of those mentioned above, and they can be used under a robust design analysis of secondary samples derived from closed CR models.

Bayesian models for spatial capture–recapture analyses of data from demographically open populations have been developed recently (Gardner et al. 2010, Chandler and Clark 2014, Ergon and Gardner 2014, Raabe et al. 2014, Schaub and Royle 2014, Whittington and Sawaya 2015) and are undergoing refinements. These will, in the future, enable analyses under a fully Bayesian approach to explore tiger population dynamics.

10.4.5 Understanding and Using Capture–Recapture Analyses

Analyses of tiger capture–recapture data under various closed and open CR models covered here and in Chap. 9 rely on a variety of fairly complex models implemented in different software options available. They generate analytical outputs and results that are also fairly complex and use a variety of notations. These analytical options are not always well understood by practitioners, who often simply choose default options offered in the software. In the Appendix, we provide brief definitions and descriptions of notation used in the Dorazio (2013) modeling approach.

10.5 Conclusion

We note that in spite of all the advanced tools for surveying tiger populations that are available and massive investments by conservation agencies, there have been only two rigorous capture–recapture studies of long-term population dynamics of tiger populations. In India, Karanth et al. (2006, 2011) combined conventional closed model and open models analyses in a likelihood-based, robust design analysis. Another study in Thailand combined Bayesian SCR closed-population analyses with a likelihood-based open model analysis of primary samples (Duangchantrasiri et al. 2016). Table 10.5 summarizes results of these two studies, which clearly demonstrate the power and utility of rigorous capture–recapture modeling approaches. We hope these examples will encourage tiger conservation practitioners to more widely adopt rigorous SCR and Robust Design methods presented here for monitoring tiger populations. Regardless of the type of analysis employed, we emphasize that careful thought and planning are critical to reliable assessments of tiger population dynamics.

Appendix

Example: Spatial Capture-Recapture Analysis Based on the Approach Developed by Dorazio (2013)

A. Notation Used

Parameters in the output:

n_0: Number of undetected individuals in the state space
σ: Scale parameter (m)
α_1: Logarithm of baseline encounter probability
β_1: Logarithm of density

Real parameters to be reported, with calculation:

$N = n_0 + n$: Total number of individuals within the state space
$p_0 = e^{\alpha_1}$: Maximum probability of capture (when trap and activity center coincide)
$\lambda = e^{\beta_1}$: Density (/100km^2)

B. Input Files

- GIS shapefiles for tiger habitat within state space (shown mapped in a GIS along with trap array below)

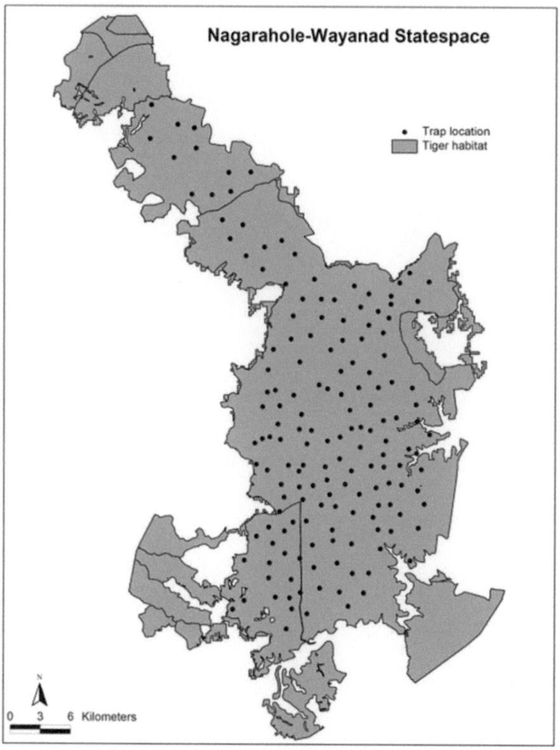

Captures File

The matrix is of dimension $n \times j$ where n is the number of captured individuals and j is the number of trap locations. Each element of the matrix specifies the number of captures of each individual in each trap location.

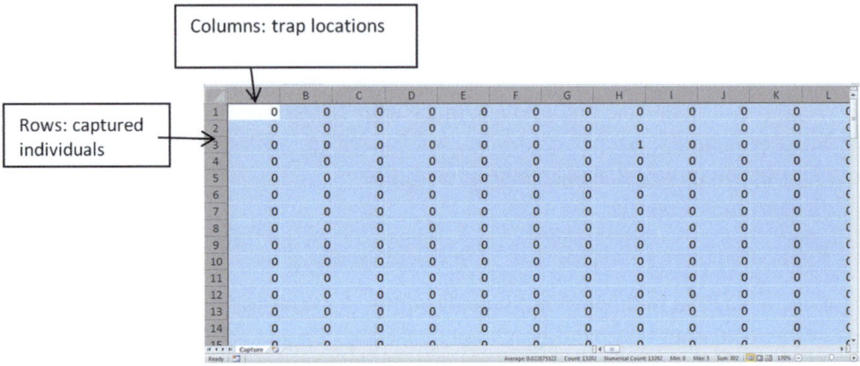

	A	B	C	D	E	F	G	H	I	J	K
1	0	0	0	0	0	0	0	0	0	0	0
2	0	0	0	0	0	0	0	0	0	0	0
3	0	0	0	0	0	0	0	1	0	0	0
4	0	0	0	0	0	0	0	0	0	0	0
5	0	0	0	0	0	0	0	0	0	0	0
6	0	0	0	0	0	0	0	0	0	0	0
7	0	0	0	0	0	0	0	0	0	0	0
8	0	0	0	0	0	0	0	0	0	0	0
9	0	0	0	0	0	0	0	0	0	0	0
10	0	0	0	0	0	0	0	0	0	0	0
11	0	0	0	0	0	0	0	0	0	0	0
12	0	0	0	0	0	0	0	0	0	0	0
13	0	0	0	0	0	0	0	0	0	0	0
14	0	0	0	0	0	0	0	0	0	0	0

– Trap file, with deployment matrix

The first three columns are the trap location ID and the X and Y coordinates in UTM projection. The remaining columns indicate whether a trap location was active (1) or not (0) in each sampling occasion.

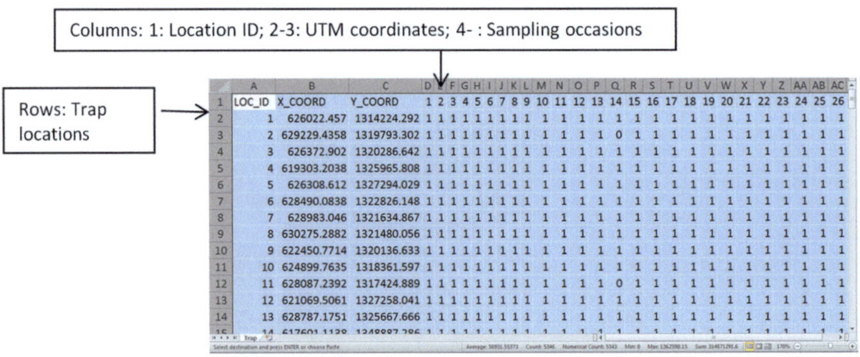

	A	B	C	D	E	F	G	H	I	J	K	L	M	N	O	P	Q	R	S	T	U	V	W	X	Y	Z	AA	AB	AC
1	LOC_ID	X_COORD	Y_COORD	1	2	3	4	5	6	7	8	9	10	11	12	13	14	15	16	17	18	19	20	21	22	23	24	25	26
2	1	626022.457	1314224.292	1	1	1	1	1	1	1	1	1	1	1	1	1	1	1	1	1	1	1	1	1	1	1	1	1	1
3	2	629229.4358	1319793.302	1	1	1	1	1	1	1	1	1	1	1	1	1	0	1	1	1	1	1	1	1	1	1	1	1	1
4	3	626372.902	1320286.642	1	1	1	1	1	1	1	1	1	1	1	1	1	1	1	1	1	1	1	1	1	1	1	1	1	1
5	4	619303.2038	1325965.808	1	1	1	1	1	1	1	1	1	1	1	1	1	1	1	1	1	1	1	1	1	1	1	1	1	1
6	5	626308.612	1327294.029	1	1	1	1	1	1	1	1	1	1	1	1	1	1	1	1	1	1	1	1	1	1	1	1	1	1
7	6	628490.0838	1322826.148	1	1	1	1	1	1	1	1	1	1	1	1	1	1	1	1	1	1	1	1	1	1	1	1	1	1
8	7	628983.046	1321634.867	1	1	1	1	1	1	1	1	1	1	1	1	1	1	1	1	1	1	1	1	1	1	1	1	1	1
9	8	630275.2882	1321480.056	1	1	1	1	1	1	1	1	1	1	1	1	1	1	1	1	1	1	1	1	1	1	1	1	1	1
10	9	622450.7714	1320136.633	1	1	1	1	1	1	1	1	1	1	1	1	1	1	1	1	1	1	1	1	1	1	1	1	1	1
11	10	624899.7635	1318361.597	1	1	1	1	1	1	1	1	1	1	1	1	1	1	1	1	1	1	1	1	1	1	1	1	1	1
12	11	628087.2392	1317424.889	1	1	1	1	1	1	1	1	1	1	1	1	1	0	1	1	1	1	1	1	1	1	1	1	1	1
13	12	621069.5061	1327258.041	1	1	1	1	1	1	1	1	1	1	1	1	1	1	1	1	1	1	1	1	1	1	1	1	1	1
14	13	628787.1751	1325667.666	1	1	1	1	1	1	1	1	1	1	1	1	1	1	1	1	1	1	1	1	1	1	1	1	1	1

– R script file, showing a part of the MCMC algorithm

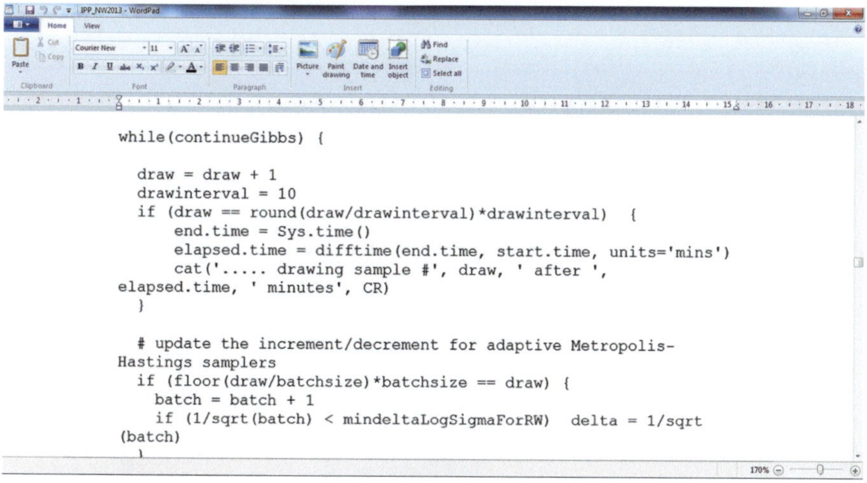

```
while(continueGibbs) {

  draw = draw + 1
  drawinterval = 10
  if (draw == round(draw/drawinterval)*drawinterval)  {
      end.time = Sys.time()
      elapsed.time = difftime(end.time, start.time, units='mins')
      cat('..... drawing sample #', draw, ' after ',
elapsed.time, ' minutes', CR)
  }

  # update the increment/decrement for adaptive Metropolis-
Hastings samplers
  if (floor(draw/batchsize)*batchsize == draw) {
    batch = batch + 1
    if (1/sqrt(batch) < mindeltaLogSigmaForRW)  delta = 1/sqrt
(batch)
  }
```

C. Output Files

– Density and abundance maps

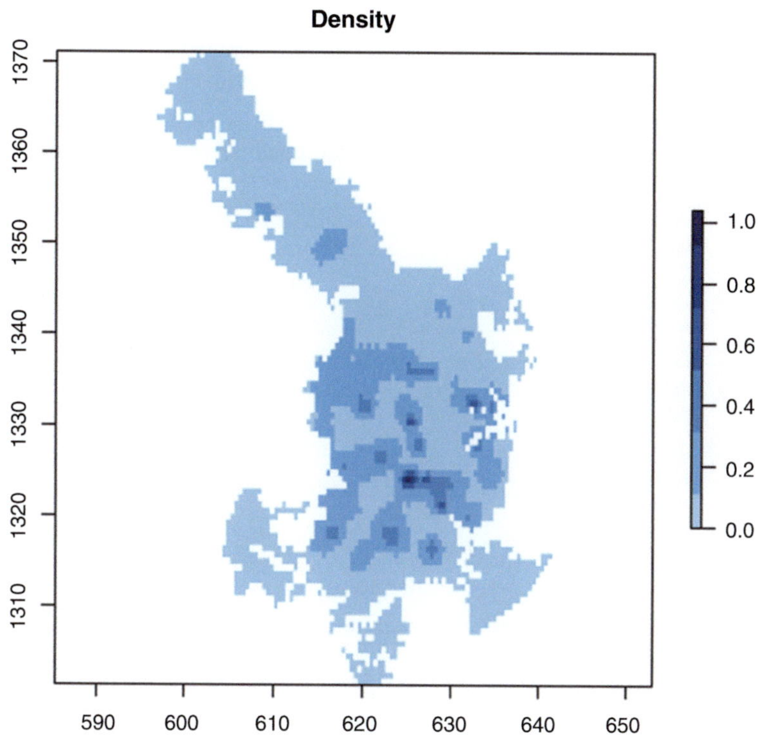

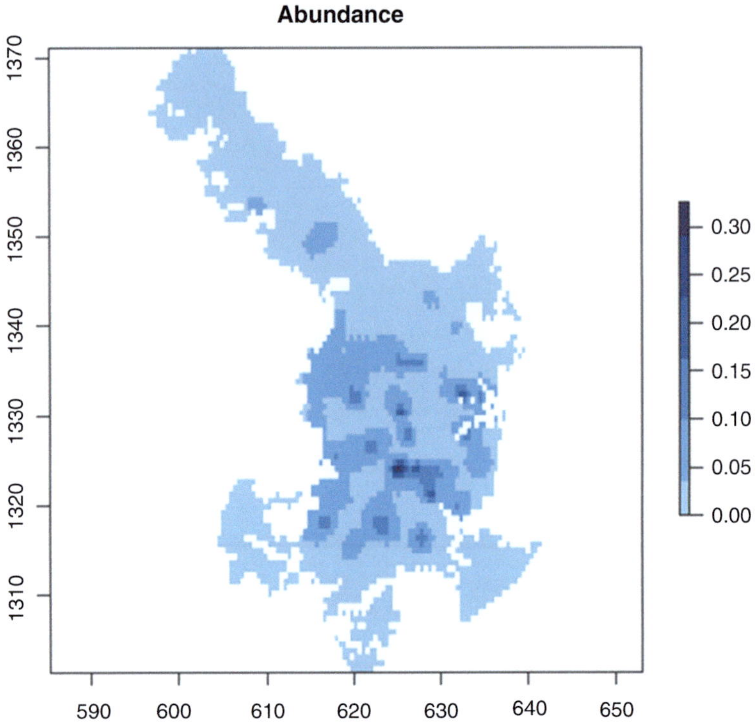

– Markov chain (mc.csv), with iterations in rows and parameters in columns

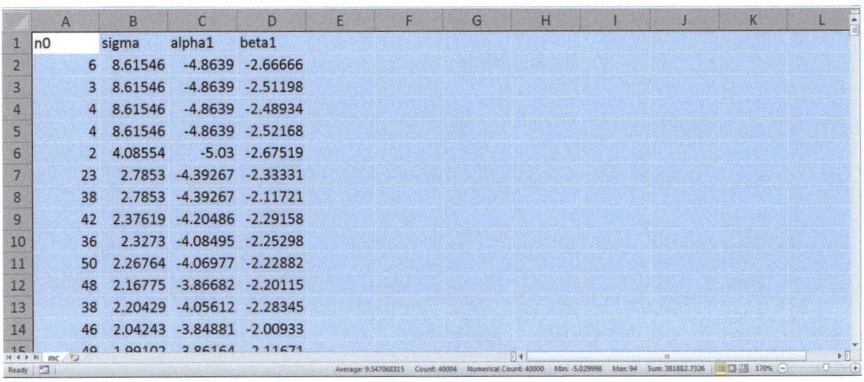

	A	B	C	D
1	n0	sigma	alpha1	beta1
2	6	8.61546	-4.8639	-2.66666
3	3	8.61546	-4.8639	-2.51198
4	4	8.61546	-4.8639	-2.48934
5	4	8.61546	-4.8639	-2.52168
6	2	4.08554	-5.03	-2.67519
7	23	2.7853	-4.39267	-2.33331
8	38	2.7853	-4.39267	-2.11721
9	42	2.37619	-4.20486	-2.29158
10	36	2.3273	-4.08495	-2.25298
11	50	2.26764	-4.06977	-2.22882
12	48	2.16775	-3.86682	-2.20115
13	38	2.20429	-4.05612	-2.28345
14	46	2.04243	-3.84881	-2.00933
15	[illegible]	[illegible]	[illegible]	[illegible]

– Summary files, showing posterior mean, median, and 95% credible intervals for parameters of interest

	Mean	50%	2.50%	97.50%
n0	42.236	42	27	60
sigma	1.871	1.869	1.741	2.016
alpha1	−3.688	−3.686	−3.869	−3.512
beta1	−2.216	−2.215	−2.439	−2.005

– Summary of Markov chain standard error (MCSE); this indicates how many significant digits can be considered in summaries of the posterior distributions

	Mean	50%	2.50%	97.50%
n0	0.1265	0.14	0.1804	0.3089
sigma	0.0013	0.0014	0.0018	0.0028
alpha1	0.0014	0.0017	0.0023	0.0025
beta1	0.0016	0.0018	0.0036	0.0028

– Plotting posterior distribution of model parameters: descriptions of parameters given above

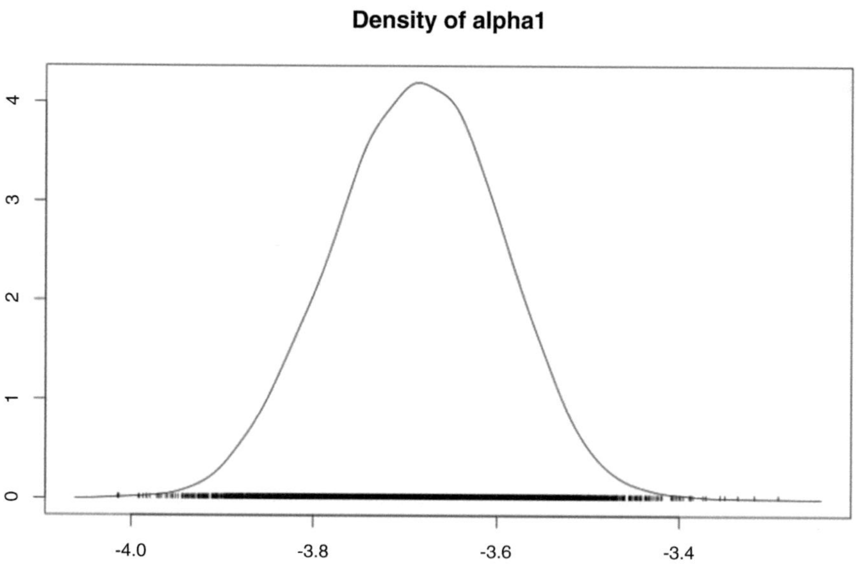

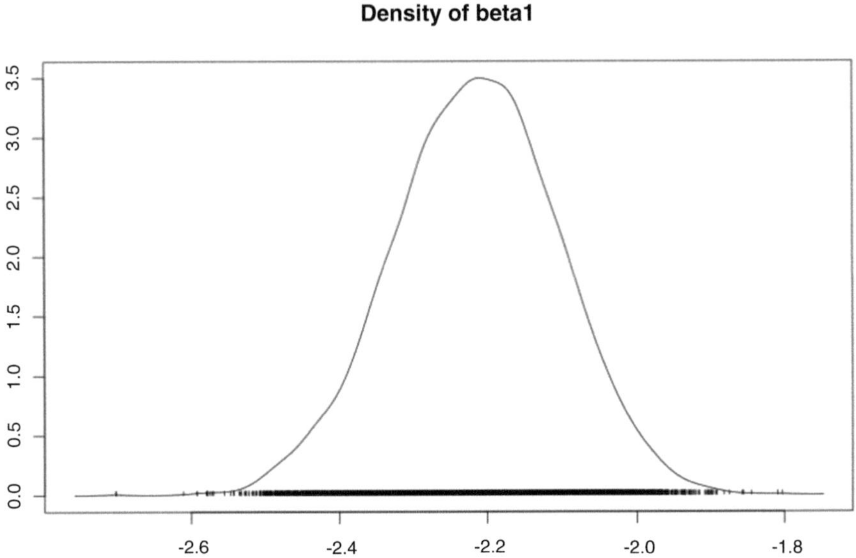
Density of beta1
0.0
0.5
1.0
1.5
2.0
2.5
3.0
3.5
-2.6
-2.4
-2.2
-2.0
-1.8

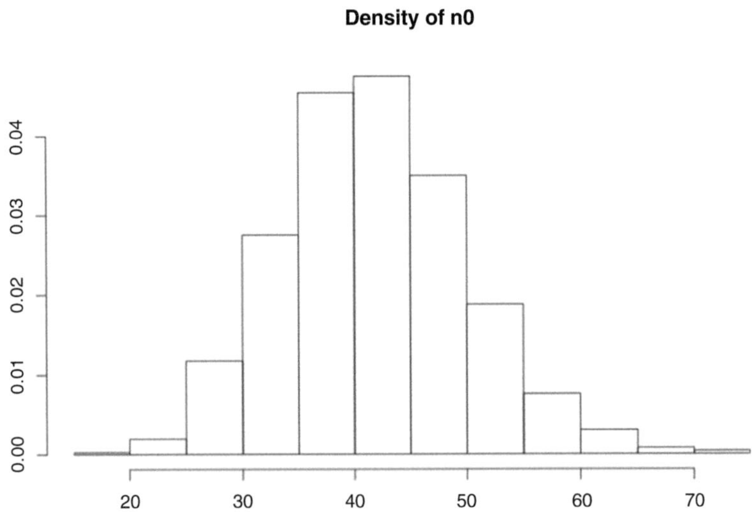
Density of n0
0.00
0.01
0.02
0.03
0.04
20
30
40
50
60
70

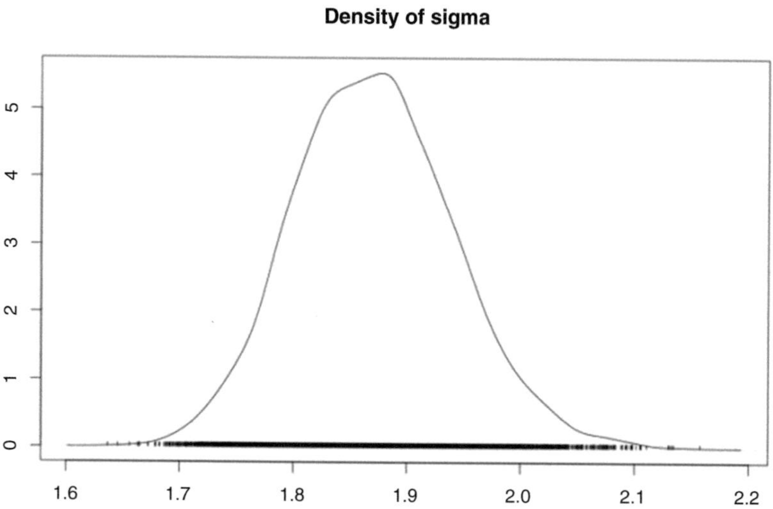

References

Amstrup SC, McDonald TL, Manly BFJ (2005) Handbook of capture–recapture analysis. Princeton University Press, Princeton

Augustine B, Royle JA, Kelly M, Satter C, Alonso, R, Boydston E, Crooks K (2016) Spatial capture–recapture with partial identity: an application to camera traps. bioRxiv 056804 https://doi.org/10.1101/056804

Bhagavatula J, Singh L (2006) Genotyping fecal samples of Bengal tiger *Panthera tigris tigris* for population estimation: a pilot study. BMC Genet 7:48

Bolger DT, Vance B, Morrison TA, Farid H (2011) Wild-ID user guide: pattern extraction and matching software for computer-assisted photographic mark recapture analysis. Dartmouth College, Hanover

Borchers DL, Efford MG (2008) Spatially explicit maximum likelihood methods for capture–recapture studies. Biometrics 64:377–385

Borchers DL, Distiller G, Foster R, Harmsen B, Milazzo L (2014) Continuous time spatially explicit capture–recapture models, with an application to a jaguar camera-trap survey. Methods Ecol Evol 5:656–665

Bubnicki JW, Churski M, Kuijper DP (2016) Trapper: an open source web-based application to manage camera trapping projects. Methods Ecol Evol 7:1209–1216

Chandler RB, Clark JD (2014) Spatially explicit integrated population models. Methods Ecol Evol 5:1351–1360

Cormack RM (1964) Estimates of survival from the sighting of marked animals. Biometrika 51:429–438

Crall JP, Stewart CV, Berger-Wolf TY, Rubenstein DI, Sundaresan SR (2013) Hotspotter: patterned species instance recognition. In Applications of Computer Vision (WACV), 2013 IEEE Workshop, pp. 230–237

Development Core Team R (2012) R: a language and environment for statistical computing. R Foundation for Statistical Computing, Vienna

Dorazio RM (2013) Bayes and empirical Bayes estimators of abundance and density from spatial capture–recapture data. PLoS One 8:e84017

Dorazio RM, Karanth KU (2017) A hierarchical model for estimating the spatial distribution and abundance of animals detected by continuous-time recorders. PloS One 12:e0176966
Duangchantrasiri S, Umponjan M, Pattanavibool A, Chaiwattana S, Maneerat S, Kumar NS, Jathanna D, Srivathsa A, Karanth KU (2016) Dynamics of a low-density source population of tigers in Southeast Asia in the context of improved law enforcement. Conserv Biol 30:639–648
Efford MG (2004) Density estimation in live-trapping studies. Oikos 106:598–610
Efford MG (2011) secr: spatially explicit capture–recapture models. R package version 2.3.1
Efford MG, Dawson DK, Robbins CS (2004) DENSITY: software for analysing capture-recapture data from passive detector arrays. Anim Biodivers Conserv 27:217–228
Efford MG, Borchers DL, Byrom AE (2009) Density estimation by spatially explicit capture–recapture: likelihood-based methods. In: Thompson DL, Cooch EG, Conroy MJ (eds) Modeling demographic processes in marked populations. Springer, New York, pp 255–269
Ergon T, Gardner B (2014) Separating mortality and emigration: modelling space use, dispersal, and survival with robust design spatial capture–recapture data. Methods Ecol Evol 5(12): 1327–1336
ESRI (2011) ArcGIS desktop: release 10. Environmental Systems Research Institute, Redlands
Fegraus EH, Lin K, Ahumada JA, Baru C, Chandra S, Youn C (2011) Data acquisition and management software for camera trap data: A case study from the TEAM Network. Ecol. Informatics 6:345–353
Gardner B, Reppucci J, Lucherini M, Royle JA (2010) Spatially explicit inference for open populations: estimating demographic parameters from camera-trap studies. Ecology 91: 3376–3383
Gilks WR, Thomas A, Spiegelhalter DJ (1994) A language and program for complex bayesian modelling. Statistician 43:169–178
Goodrich JM, Miquelle DG (2010) Tiger telemetry. In: Tilson R, Nyphus PJ (eds) Tigers of the world: the science, politics and conservation of *Panthera tigris*, 2nd edn. Elsevier, New York, pp 263–276
Goodrich JM, Miquelle DG, Smirnov EN et al (2010) Spatial structure of Amur (siberian) tigers (*Panthera tigris altaica*) on sikhote-alin biosphere Zapovednik, Russia. J Mammal 91:737–748
Gopalaswamy AM, Royle JA, Delampady M et al (2012a) Density estimation in tiger populations: combining information for strong inference. Ecology 93:1741–1751
Gopalaswamy AM, Royle JA, Hines JE et al (2012b) Program SPACECAP: software for estimating animal density using spatially explicit capture–recapture models. Methods Ecol Evol 3:1067–1072
Harihar A, Pandav B, Goyal SP (2011) Responses of leopard Panthera pardus to the recovery of a tiger *Panthera tigris* population. J Appl Ecol 48:806–814
Hernandez-Blanco JA, Naidenko SV, Chistopolova MD et al (2015) Social structure and space use of Amur tigers (*Panthera tigris altaica*) in southern Russian far east based on GPS telemetry data. Integr Zool 10:365–375
Hiby L, Lovell P (1990) Computer aided matching of natural markings: a prototype system for grey seals. Rep Int Whaling Commission 12:57–61
Hiby L, Lovell P, Patil N et al (2009) A tiger cannot change its stripes: using a three-dimensional model to match images of living tigers and tiger skins. Biol Lett 5:383–386
Ivan JS, Newkirk ES (2015) CPW photo warehouse: a custom database to facilitate archiving, identifying, summarizing and managing photo data collected from camera traps. Methods Ecol Evol 7:499–504
Jędrzejewski W, Puerto MF, Goldberg JF et al (2017) Density and population structure of the jaguar (*Panthera onca*) in a protected area of Los Llanos, Venezuela, from 1 year of camera trap monitoring. Mammal Res 62:9–19
Jhala YV, Qureshi Q, Gopal R (eds) (2015) The status of tigers, copredators & prey in India 2014. National Tiger Conservation Authority, New Delhi & Wildlife Institute of India, Dehradun. TR2015/021
Jolly GM (1965) Explicit estimates from capture–recapture data with both death and immigration-stochastic model. Biometrika 52:225–247

Karanth KU, Nichols JD (1998) Estimation of tiger densities in India using photographic captures and recaptures. Ecology 79:2852

Karanth KU, Nichols JD (2002) Monitoring tigers and their prey: a manual for wildlife researchers, managers and conservationists in tropical Asia. Centre for Wildlife Studies, Bangalore

Karanth KU, Nichols JD (2010) Non-invasive survey methods for assessing tiger populations. In: Tilson R, Nyhus PJ (eds) Tigers of the world: the science, politics and conservation of *Panthera tigris*, Second edn. Academic Press, London/Burlington/San Diego, pp 241–262

Karanth KU, Stith BM (1999) Prey depletion as a critical determinant of tiger population viability. In: Seidensticker J, Christie S, Jackson P (eds) Riding the tiger: tiger conservation in human-dominated landscapes. Cambridge University Press, Cambridge, pp 100–113

Karanth KU, Sunquist ME (2000) Behavioural correlates of predation by tiger (*Panthera tigris*), leopard (*Panthera pardus*) and dhole (*Cuon alpinus*) in Nagarahole, India. J Zool 250:255–265

Karanth KU, Nichols JD, Kumar NS, Hines JE (2006) Assessing tiger population dynamics using photographic capture–recapture sampling. Ecology 87:2925–2937

Karanth KU, Nichols JD, Kumar NS (2011) Estimating tiger abundance from camera trap data: field surveys and analytical issues. In: O'Connell AF, Nichols JD, Karanth KU (eds) Camera traps in animal ecology: methods and analyses. Springer, Tokyo, pp 97–117

Kendall WL, Nichols JD (1995) On the use of secondary capture–recapture samples to estimate temporary emigration and breeding proportions. J Appl Stat 22:751–762

Kendall WL, Pollock KH, Brownie C (1995) A likelihood-based approach to capture–recapture estimation of demographic parameters under the robust design. Biometrics 51:293–308

Kendall WL, Nichols JD, Hines JE (1997) Estimating temporary emigration using capture–recapture data with Pollock's robust design. Ecology 78:563–578

Kerley L, Salkina G (2007) Using scent-matching dogs to identify individual Amur tigers from scats. J Wildl Manag 71:1349–1356

Kerley LL, Goodrich JM, Miquelle DG, Smirnov EN, Nikolaev IG, Quigley HB, Hornocker MG (2003) Reproductive parameters of wild female Amur (Siberian) tigers (*Panthera tigris altaica*). J Mammal 84:288–298

Krishnappa YS, Turner WC (2014) Software for minimalistic data management in large camera trap studies. Ecol. Informatics 24:11–16.

Link WA, Barker RJ (2010) Bayesian inference: with ecological applications. Academic Press, San Diego

Mondol S, Karanth KU, Kumar NS et al (2009) Evaluation of non-invasive genetic sampling methods for estimating tiger population size. Biol Conserv 142:2350–2360

Nichols JD (2005) Modern open-population capture–recapture models. In: Amstrup SC, McDonald TL, Manly BFJ (eds) Handbook of capture–recapture analysis. Princeton University Press, Princeton, pp 88–122

Niedballa J, Sollmann R, Courtiol A, Wilting A (2016) camtrapR: an R package for efficient camera trap data management. Methods Ecol Evol 7:1457–1462

NTCA (2012) A protocol on phase IV monitoring. In: Technical document No.1/2011. National Tiger Conservation Authority. Available via http://projecttiger.nic.in/WriteReadData/CMS/Protocol_Phase_IV_Monitoring_r.pdf. Accessed 4 Jan 2017

O'Connell AF, Nichols JD, Karanth KU (2011) Camera traps in animal ecology: methods and analyses. Springer, Tokyo

Otis DL, Burnham KP, White GC, Anderson DR (1978) Statistical-inference from capture data on closed animal populations. Wildl Monogr 62:7–135

Pledger S (2000) Unified maximum likelihood estimates for closed capture–recapture models using mixtures. Biometrics 56:434–442

Pledger S, Pollock KH, Norris JL (2003) Open capture–recapture models with heterogeneity: I. Cormack-Jolly-Seber model. Biometrics 59:786–794

Plummer M (2003) JAGS: a program for analysis of Bayesian Graphical Models using Gibbs Sampling. In: Hornik K, Leisch F, Zeileis A (eds). Proceedings of the 3rd international workshop on distributed statistical computing; Vienna, Austria. ISSN 1609-395X, URL http://www.ci.tuwien.ac.at/Conferences/DSC-2003/Proceedings/

Pollock KH (1982) A capture–recapture design robust to unequal probability of capture. J Wildl Manag 46:752–757

Pollock KH, Nichols JD, Brownie C, Hines JE (1990) Statistical inference for capture–recapture experiments. Wildl Monogr 107:3–97

Pradel R, Hines JE, Lebreton JD, Nichols JD (1997) Capture–recapture survival models taking account of transients. Biometrics 53:60–72

Raabe JK, Gardner B, Hightower JE (2014) A spatial capture–recapture model to estimate fish survival and migration patterns from linear telemetry arrays. Can J Fish Aquat Sci 71(1): 120–130

Rayan DM, Linkie M (2015) Conserving tigers in Malaysia: a science-driven approach for eliciting conservation policy change. Biol Conserv 184:18–26

Rexstad E, Burnham KP (1991) User's guide for interactive program CAPTURE. Abundance estimation of closed animal populations. Colorado State University, Fort Collins

Royle JA, Dorazio RM (2008) Hierarchical modeling and inference in ecology: the analysis of data from populations, metapopulations, and communities. Academic Press, San Diego

Royle JA, Karanth KU, Gopalaswamy AM, Kumar N (2009) Bayesian inference in camera trapping studies for a class of spatial capture–recapture models. Ecology 90(11):3233–3244

Royle JA, Chandler RB, Sollmann R, Gardner B (2014) Spatial capture–recapture. Academic Press, San Diego

Schaub M, Royle JA (2014) Estimating true instead of apparent survival using spatial Cormack–Jolly–Seber models. Methods Ecol Evol 5:1316–1326

Seber GAF (1965) A note on multiple-recapture census. Biometrika 52:249–259

Smith JLD (1993) The role of dispersal in structuring the Chitwan tiger population. Behaviour 124:165–195

Stanley TR, Burnham KP (1999) A closure test for time-specific capture–recapture data. Environ Ecol Stat 6:197–209

Sunquist ME (1981) The social organization of tigers (Panthera tigris) in Royal Chitwan National Park, Nepal. Smithson Contrib Zool 336

Sunquist ME, Karanth KU, Sunquist F (1999) Ecology, behaviour and resilience of the tiger and its conservation needs. In: Seidensticker J, Christie S, Jackson P (eds) Riding the tiger: tiger conservation in human-dominated landscapes. Cambridge University Press, Cambridge, pp 5–18

Sutherland, C, Royle JA, Linden DW (2016) oSCR: Multi-session sex-structured spatial capture–recapture models. R package version 0.30.1. https://github.com/jaroyle/oSCR

Swanson A, Kosmala M, Lintott C, Simpson R, Smith A, Packer C (2015) Snapshot Serengeti: high-frequency annotated camera trap images of 40 mammalian species in an African savanna. Sci. Data 2

Tobler MW (2013) Camera base 1.6. http://www.atrium-biodiversity.org/tools/camerabase/

Wegge P, Pokheral CP, Jnawali SR (2004) Effects of trapping effort and trap shyness on estimates of tiger abundance from camera trap studies. Anim Conserv 7:251–256

White GC, Burnham KP (1999) Program MARK: survival estimation from populations of marked animals. Bird Study 46(Supplem):120–138

White GC, Anderson DR, Burnham KP, Otis DL (1982) Capture–recapture and removal methods for sampling closed populations. Los Alamos National Laboratory Publication LA-8787-NERP. Los Alamos, New Mexico, USA

Whittington J, Sawaya MA (2015) A comparison of grizzly bear demographic parameters estimated from non-spatial and spatial open population capture–recapture models. PLoS One 10:e0134446

Williams BK, Nichols JD, Conroy MJ (2002) Analysis and management of animal populations. Academic Press, San Diego

Concepts and Practices: Assessing Tiger Population Dynamics Using Genetic Captures

11

Samrat Mondol, Uma Ramakrishnan, Olutolani Smith, and Devcharan Jathanna

11.1 Introduction

Capture–recapture (CR) models (Chap. 9) represent a powerful suite of approaches for estimating animal population parameters such as abundance, density, survival, population growth rate, among others. What makes CR particularly useful is the ability to define "captures" broadly. Capture–recapture sampling may entail physical capture and tagging of animals; "photo-capturing" animals with individually identifiable markings using camera trap surveys; searching a defined area of interest with handheld cameras to photo-capture individually identifiable animals; obtaining "captures" of individuals from deoxyribonucleic acid (DNA) samples collected using fixed hair snares; or obtaining DNA from noninvasive samples such as scats encountered by searching along forest trails and roads. Noninvasive genetic CR surveys are particularly useful for monitoring tiger populations in situations and sites where camera trap surveys (Chap. 10) are not feasible due to lack of access,

S. Mondol (✉)
Wildlife Institute of India, Dehradun, India
e-mail: samratmondol@gmail.com; samrat@wii.gov.in

U. Ramakrishnan
National Centre for Biological Sciences, TIFR, Bengaluru, India
e-mail: uramakri@ncbs.res.in; uramakri@gmail.com

O. Smith
Tiger Program, Panthera, New York, NY, USA

Department of Genetics, Evolution & Environment, University College London, London, UK
e-mail: osmith@panthera.org

D. Jathanna
Wildlife Conservation Society, India Program, Bengaluru, India

Centre for Wildlife Studies, Bengaluru, India
e-mail: devcharan@gmail.com

K.U. Karanth, J.D. Nichols (eds.), *Methods For Monitoring Tiger And Prey Populations*, DOI 10.1007/978-981-10-5436-5_11

equipment, or risk of damage or theft of camera traps. Potential disadvantages of DNA-based CR include the requirement for a specialized, well-equipped laboratory and skilled geneticists for DNA extraction, amplification, genotyping, and identification of individuals, as well as the requirement for capture–recapture models that account for genetic misidentification, which typically have lower precision due to uncertainties induced by such misidentification. While noninvasive DNA-based individual identification has previously been used to derive the minimum number alive in wild tiger populations (Wasser et al. 1997; Bhagavatula and Singh 2006), when used in conjunction with capture–recapture modeling, it can be used to reliably estimate population parameters, as shown using conventional CR (e.g., Mondol et al. 2009) or SCR (e.g. Gopalaswamy et al. 2012a) approaches.

However, errors in DNA-based individual identification can cause substantial bias in estimates of abundance, density, and other population parameters (Lukacs and Burnham 2005a) and need to be addressed either through laboratory protocols to screen samples that may be subject to erroneous identification or by incorporating the misidentification process—and the consequent "ghost" individuals created—in subsequent statistical modeling. It is also important to carefully consider key factors affecting the observation process (i.e., survey design, field survey protocols) that generate the genetic capture data (see Chaps. 9 and 10).

Estimation based on genetic captures requires multilocus (*locus*, the position of a gene on a chromosome) genotype data from genetic markers such as microsatellite or single nucleotide polymorphism (SNP) panels for unambiguous identification of individual animals (Taberlet and Luikart 1999, Broquet et al. 2007). This chapter deals with the details of genetic-based individual identification in the context of capture–recapture modeling and estimation of tiger populations.

11.2 Field Collection and Storage of Fecal DNA

In this section, we focus on field and laboratory protocols necessary to provide capture data amenable to reliable tiger population estimation using the methods described in Chap. 9.

11.2.1 Sources of Noninvasive DNA in the Field

For any monitoring program based on genetic samples, one would ideally prefer to use biological samples such as blood or tissue, which yield sufficient, high-quality molecular data. However, for endangered and elusive species such as tigers, it is not possible (or desirable) to capture and draw blood/tissue samples from dozens of individuals at the large spatial scales necessary for meaningful estimation of population parameters. Noninvasive sources of DNA are clearly required and, in conjunction with appropriate survey designs, field protocols, and analytical frameworks, can permit long-term studies of tiger population dynamics. Here we examine advantages and disadvantages of different noninvasive sources of DNA for monitoring wild tiger populations:

1. Feces: Tiger feces (scats) are the most widely used biological material for DNA extraction across tiger range. The use of fecal DNA in capture–recapture-based monitoring of other wildlife populations has been a standard practice for over two decades, including a number of studies of tigers (e.g., Mondol et al. 2009, Gopalaswamy et al. 2012a, Sugimoto et al. 2012). Fecal samples, however, contain a mixture of DNA from the focal animal (tiger), prey species consumed, and gut microbes, as well as other bacterial, fungal, and environmental DNA. The host DNA comes from intestinal gut cells that are sloughed off while the scat is passing through the digestive system and is generally concentrated on the surface of the scat, particularly at the tip, and in areas of the scat covered with mucus. The advantages of using fecal DNA include relative ease of sample collection and potentially large sample size and extensive area coverage in a short time. Disadvantages include poor quality and quantity of DNA in scats (particularly in old samples) and problems of safe collection, storage, and shipping in tropical, humid environments. Despite these shortcomings, scat samples are often the most suitable DNA source for noninvasive tiger population monitoring.
2. Urine and scent marks: Urine is known to be a good source of DNA from wild animals (Valiere and Taberlet 2000, Inoue et al. 2007). Tigers additionally spray scent from the anal gland, mixed with urine, on boles of trees and leaves of bushes along their trails as a means of olfactory communication (see Chap. 2). These scent marks can be detected easily due to their unique smell. Although only validated with captive tigers, Caragiulo et al. (2015) show that DNA derived from tiger scent is as good as DNA derived from scats. However, there has been no study of the potential of urine or scent-derived DNA for genetic studies of wild tiger populations. Population estimation may also prove to be difficult based on such samples, which would only be obtained opportunistically and infrequently.
3. Hair: Shed or extracted body hair is another common source of noninvasively obtained DNA in wildlife genetics studies. DNA is obtained only from the hair roots and has been extracted from plucked hairs obtained using hair snares (e.g., Gardner et al. 2010) or naturally shed hair (e.g., Morin and Woodruff 1992) in mammal species. While hair samples have been used for individual identification of captive tigers and species identification in forensic cases, no study of tigers has been conducted to date using DNA derived from hair samples collected in the field. It may also be possible to collect tiger hair from scratch trees, kills of prey, and resting sites. However, as with scent and urine samples, such opportunistic samples may not meet data requirements for rigorous population estimation. Potential problems with hair samples include poor quantity and quality of DNA and contamination (e.g., if multiple individual tigers use the same scratch trees).

Table 11.1 Example field data form for fecal DNA sample collection in the field

Scat collection field data form	
Sample ID:	Date:
Collection time:	Geographical location:
Strata:	Condition:
Scat diameter:	Remarks:
Collector:	GPS location:

11.2.2 Field Collection and Storage of Samples

Appropriate DNA sample collection media and protocols are areas of active research in noninvasive wildlife genetics. As discussed earlier, DNA from fecal samples is present mostly on the top layer of the scat, originating from the outer layer of the intestine's inner walls. As soon as these sloughed-off cells leave the host's body, degradation of DNA begins, resulting in poor quality and quantity of host DNA in old fecal samples. This degradation process is accelerated by exposure to sunlight, humidity, and rain. These environmental degradation processes make appropriate sampling and storage in the field a critical consideration.

A number of different approaches have been used for fecal sample collection including simple drying, use of silica gel, ethanol, TNE buffer, RNAlater (for DNA stability), and a number of commercially available preservatives (DNA/RNA Shield, Norgen Biotek stool sampling tubes). Appendix 1 outlines protocols for the commonly used simple drying and ethanol collection protocols for fecal DNA samples. A sample data collection form is given in Table 11.1.

Sample Storage Appropriate storage of field DNA samples is critical to retain adequate quantity and quality of DNA prior to transportation to the laboratory but may need resources which may not always be readily available in field camps. Prolonged exposure to sun, humidity, or high temperatures increases DNA degradation rate. Irrespective of the field sample collection approach, scat samples should ideally be refrigerated, which requires a continuous electricity source. If continuous refrigeration is not available, the next best option is to air-dry the samples in dark, dry conditions, followed by refrigeration in the laboratory. Samples collected in ethanol or TNE can be stored at room temperature until they are transferred to the laboratory refrigerator.

11.3 Laboratory Protocols for Noninvasive Genetic Samples

We describe each of the steps outlined in Fig. 11.1 in detail below:

DNA Extraction In the laboratory, DNA extraction protocols will vary depending on the sample collection method and medium (e.g., samples collected dry or in a liquid). For samples collected in any liquid medium (ethanol, TNE buffer, or

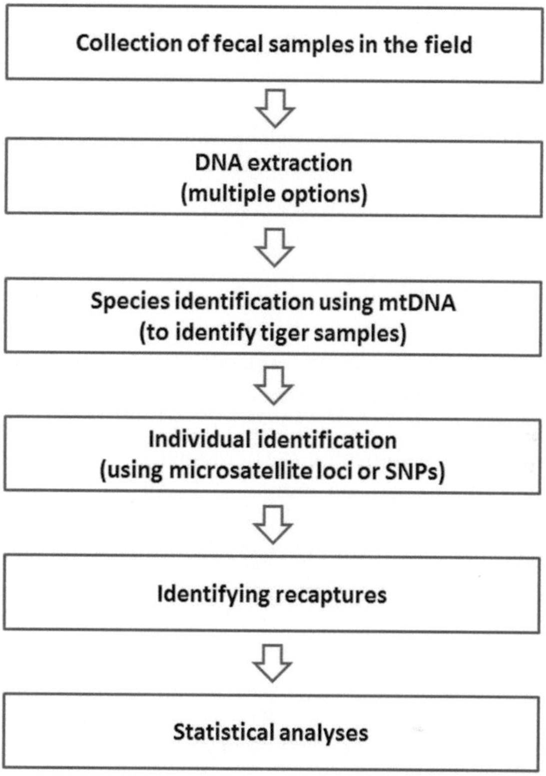

Plate 11.1 Standard work flow in the laboratory for tiger population estimation from fecal samples

RNAlater), the top layer should be scraped with a spatula into a tube for DNA extraction. With dry samples, the top layer can either be swabbed or scraped for subsequent processing. DNA extraction can be done with commercially available "Stool DNA Extraction Kits" or by using modified and validated protocols with standard tissue DNA kits. Processing a negative control along with samples is important in detecting potential contamination. Detailed protocols for different extraction methods can be found in Appendix 2.

Species Identification Distinguishing carnivore species based on signs such as tracks or on fecal morphology (size, shape, scent, and dietary contents) poses significant challenges (Zuercher et al. 2003; Prugh and Ritland 2005) and can be error-prone in areas with multiple similar-sized, co-occurring species (Davison et al. 2002; Fernandes et al. 2008). More accurate and reproducible DNA-based species identification has become a preferred alternative (Beja-Pereira et al. 2009; Goossens and Bruford 2009) and is being increasingly used to study rare, threatened, and cryptic populations. Species identification is an important first step to exclude samples from nontarget species prior to individual identification (Paetkau 2003) in genetic (S)CR studies, or it can be used to confirm species identity for occupancy modeling

(Chaps. 4 and 5). For tigers, a number of different mitochondrial DNA (mtDNA)-based molecular approaches have been used for species identification (Mukherjee et al. 2007, Mondol et al. 2015). Broadly, species identification is carried out by amplifying specific regions of the mitochondrial DNA using a polymerase chain reaction (PCR), which produces multiple copies of the target regions, selected based on their utility for distinguishing between species. Alternatively, regions of mtDNA can be sequenced and species identity assigned based on comparisons with known sequences saved in publicly available databases (e.g., NCBI).

Individual Identification Individual identification from DNA requires a set of genetic markers that, taken together, differ for every individual of a species. For example, in photographic identification of tigers or thumb imprint-based identification of humans, the patterns differ in every individual. Similarly, in genetic approaches we use a set of biological markers that can provide unambiguous individual identification. "Microsatellites" (Fig. 11.2), the biomarkers most widely used for mammalian individual identification, are small repetitive fragments of nuclear DNA found across the mammalian genome, with alleles characterized by the number of repeats. They are neutral (not subject to natural selection), and high mutation rates make them highly polymorphic, which is particularly important for individual identification. Microsatellites have additional advantages, including the facts that they are codominant (both alleles are detectable), abundant (found in every mammal examined to date), distributed across the genome, and that the same loci may be present in closely related species (allowing markers identified for one species to be used for other related species). Among disadvantages, laboratory protocols may generate erroneous genotypes (see *Sources of error and types of genetic misidentification* below), scoring loci involves some degree of subjectivity, and laboratory protocols are extremely time-consuming and expensive. In this section we will try to provide a detailed understanding of the use of microsatellite markers in individual identification of tigers (see Bhagavatula and Singh 2006, Mondol et al. 2009, 2015 for further information).

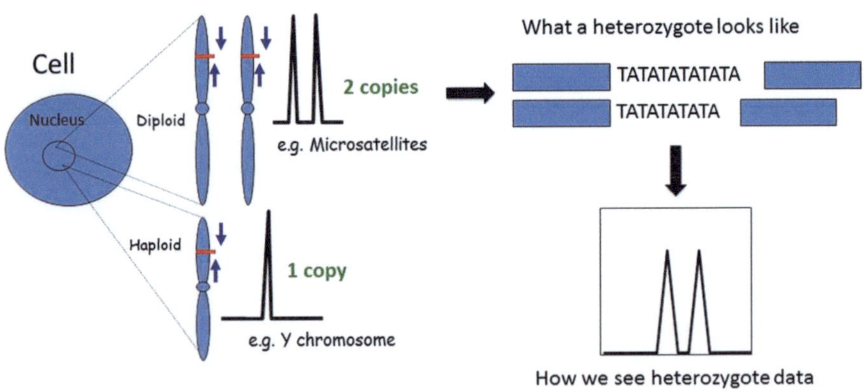

Plate 11.2 Basic illustration of microsatellite markers

1. *Selection, Testing, and Validation of Microsatellites*

The best practices in selecting a set of polymorphic microsatellite markers are described by Mondol et al. (2009). Some important points include:

- Standardization and validation of microsatellites requires good quality samples such as blood or tissue, which may be obtainable from captive tigers.
- A large number of microsatellites should be assessed at the outset, from which the geneticist selects the best panel, based on polymorphic information content and expected heterozygosity (i.e., low P_{ID}; see below). As different microsatellite markers are polymorphic in different parts of the tiger's range, selection of the optimal panel should be based on tissue samples from captive individuals that are genetically similar to the target population of interest.

Selection of an optimal panel of loci for microsatellite-based individual identification (to be used in subsequent (S)CR analyses) requires a trade-off between (a) selecting a larger number of loci with higher power to distinguish individuals (which means that amplification may fail at one or more loci for more samples, which need to be discarded) and (b) selecting a smaller number of loci with lower power to distinguish individuals (P_{ID}; see *Sources of error and types of genetic misidentification below*) but which are likely to be amplified in a larger number of samples. While statistical power to distinguish different individuals cannot be compromised, a large number of captures (and recaptures, including spatial recaptures) are nonetheless required for reliably fitting statistical models to the capture data. Similarly, ghost identities (described below) are more likely to occur in panels with larger numbers of loci. Empirical work shows that optimal panels consist of a small set (~8–12) of highly polymorphic microsatellites, where all alleles are well characterized.

- After setting the final panel of polymorphic markers and standardizing it with captive tiger samples, the panel should be tested using field-collected wild tiger scat samples.
- Identifying individual tigers from fecal sample extracts is done by amplifying the selected microsatellites using primers during a PCR. Gel electrophoresis is then carried out on the PCR product, where the amplified DNA fragments separate based on length, forming distinct bands on the gel. A molecular weight size marker, containing DNA fragments of known size and run as one of the samples on the gel, helps determine the DNA lengths corresponding to each band. It is important that no bands are observed for the negative control included in each gel, to ensure no contamination of samples. A genotyping machine with greater power to resolve small differences in fragment lengths is used to accurately determine fragment lengths (Fig. 11.3), which are then "scored" by a geneticist and recorded as the observed genotype.
- In most noninvasive studies, the variability in DNA quality and quantity is very large across samples, producing genotype information of varying quality.

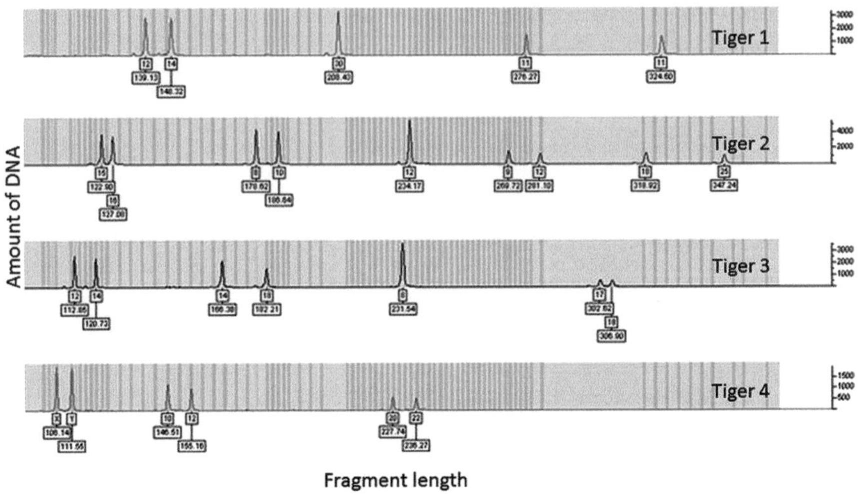

Plate 11.3 Example microsatellite panel and individual identification using microsatellite markers

To eliminate genotyping errors, it is necessary to amplify and genotype each locus multiple times from each sample and only use samples (in subsequent statistical analyses) in which some specified minimum consensus is obtained across multiple instances of genotyping. This procedure, known as the multiple tube approach (Navidi et al. 1992, Tablerlet et al. 1996), has been widely used in noninvasive wildlife research. For example, Mondol et al. (2009, 2013, 2015) showed that four independent amplifications of each locus from each sample, followed by screening for at least three identical genotypes from these four repetitions, result in reliable individual identification. Another approach is to use a modeling framework that incorporates the possibility of genotyping error. The added cost here is reduced precision of estimates.

– During all testing steps, estimates of genotyping error need to be calculated. Genotyping errors mainly occur during PCR amplification from poor quality and quantity DNA sources, such as scat, hair, urine, or scent samples. Any locus with high genotyping error rate should be removed from the panel. Unfortunately, genotyping errors are sometimes difficult to compute without some notion of "truth," often unavailable in field studies.

2. *Identifying Recaptures*

Once the panel of microsatellite loci is amplified from all field-collected scat samples, the genotypes derived from the samples are compared to identify samples with identical genotypes at the selected loci. This can be performed by software program CERVUS (Marshall et al. 1998). Samples are considered to belong to the same individuals if they have identical genotypes at the selected loci. Because samples are referenced in space and time, these captures can then be used to create input data for conventional or spatial capture–recapture analyses, in the appropriate format (see Chaps. 9 and 10).

11.4 Novel Approaches to Identifying Individuals

The last decade has witnessed unprecedented changes in DNA sequencing technologies. These technological changes have allowed us to sequence genomes faster and cheaper than ever before (Schuster 2008). The ability to sequence genomes also allows us to better understand individuals, populations, and even species. Genome-wide data have tremendous application in conservation (e.g., Steiner et al. 2013, Allendorf et al. 2010), including investigating connectivity, identifying individuals, quantifying inbreeding, and disease susceptibility (Steiner et al. 2013) in populations of endangered species. The tiger genome was sequenced a few years ago (Cho et al. 2013), allowing us to understand more about tiger genetic variability but also providing the initial framework to develop tools for tiger conservation.

Careful analyses of genome-level data reveal that the most common polymorphic DNA markers (more polymorphic markers improve statistical power to distinguish individuals) are single nucleotide polymorphisms (SNPs). These are single base mutations, which makes scoring them easy and the data on individuals more reliable. SNPs can also be used for individual identification. Typing of 10,000 SNPs in several Indian tigers (Natesh et al., in revision) reveals that some combinations of SNPs can be more powerful than microsatellites at identifying individuals (Fig. 11.4).

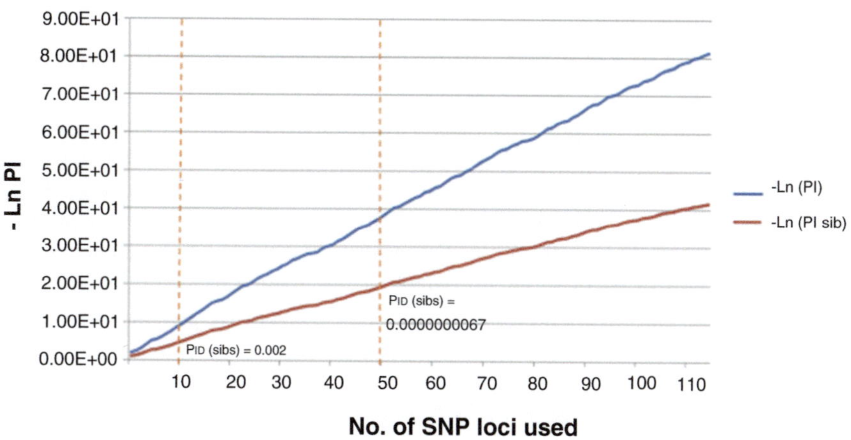

Plate 11.4 Power to distinguish individuals as a function of number of SNP loci used

Based on comparisons to Mondol et al. (2009), 20 SNPs have the same power to discriminate individuals as 30 microsatellites, and 96 SNPs make an identification error once in 10^{16} individuals, even if they were all siblings! Future methods for conservation genetics of tigers are sure to be based on SNP data. This is particularly true if we wish to develop a common, range-wide set of genetic markers for tigers. Tigers have a very large geographic range, and different microsatellite markers are polymorphic in different locations. This would necessitate a large set of microsatellite markers if we wish to develop a pan-range common individual identification protocol. Unfortunately, there is a trade-off between the number of loci and error (see Mondol et al. 2009). This would make it difficult to use a large set of microsatellite markers reliably. SNP markers are less prone to genotyping error, and so in the future, it may be possible to develop a large panel of common SNP markers that could be used to individually identify tigers across their range.

11.5 Identification of Sex from DNA and Use in (S)CR Modeling

In addition to allowing identification of species and individuals, DNA (including from noninvasive sources) also allows determination of sex. Incorporating gender in (S)CR modeling helps account for heterogeneity in detection parameters between genders (e.g., Sollmann et al. 2011), investigate intersex differences in vital rates (e.g., Lebreton et al. 1992, Horak and Lebreton 1998), and estimate gender ratios in addition to population state variables such as abundance or density (e.g., Broekhuis and Gopalaswamy 2016, Elliot and Gopalaswamy 2016). When sex is identified from fecal DNA for carnivores such as tigers, care is needed to ensure that sex identification primers do not amplify prey DNA (Waits 2004).

11.6 Sources of Error and Types of Genetic Misidentification

There is a substantial amount of literature on sources of error in genetic identification of individuals, particularly based on noninvasive samples. Taberlet et al. (1999), Paetkau (2003), Pompanon et al. (2005), Lukacs and Burnham (2005a), Waits (2005), Waits and Paetkau (2005), and Broquet et al. (2007) review a number of laboratory procedures and protocols, discuss sources and consequences of error, and make recommendations for detecting and avoiding such errors. Broadly, errors in genetic identification of individuals from noninvasive samples may occur due to:

1. *Amplification failure*, where the PCR fails to produce copies of the required part of the genome. As (S)CR models explicitly account for imperfect capture probability, this is fully accounted for in analyses using these models but may reduce the number of captures or individuals included in the analyses.

2. *Allelic dropout*, where one allele at a locus fails to amplify, causing a sample that is heterozygous at a particular locus to appear homozygous. One way of addressing this is by repeating the genotyping multiple times and obtaining consensus in the observed genotypes (e.g., obtaining the same results at least three out of four times). An alternative is to incorporate this probability of misclassification/error directly into the modeling.
3. *False alleles* could be caused by mutation early in the PCR that gives rise to spurious alleles. These arise relatively rarely and could also be addressed by obtaining consensus in multiple genotyping of the same PCR sample.
4. The *shadow effect* is when different individuals share the same alleles over a panel of loci. This problem can be overcome by using a set of microsatellite loci with high power to resolve individuals, which entails selection of polymorphic loci and increasing the number of loci considered, thereby decreasing the probability that two different individuals share the same alleles at those loci (also known as probability of identity, P_{ID}), often assessed conservatively assuming that the two individuals are siblings ($P_{ID(sibs)}$).

Ultimately, these different types of genotyping error lead to either loss of sample(s) from analysis (i above), creation of "ghost" individuals (ii and iii above), or incorrectly inferring that two different individuals are the same due to the shadow effect (iv above). If undetected ghost individuals persist into statistical analyses, the number of individuals captured M is inflated. Additionally, ghost individuals are captured only once, leading to false capture histories containing only a single capture each time a misidentification occurs, while the capture histories of the true individuals have additional zeroes introduced each time a misidentification occurs. This leads to underestimation of overall detection probability p^*. Using the canonical estimator (Chap. 3), which forms the basis for all population estimation models covered in this volume, abundance is estimated as $\hat{N} = M/p^*$, where $\widehat{N}$ is estimated abundance, M is the total number of individuals captured, and p^* is the probability that an animal is captured at least once. When M is inflated and p^* is underestimated, the ratio M/p^* is severely overestimated (Lukacs and Burnham 2005a). Therefore, it is critical to ensure that genotyping errors do not vitiate subsequent analyses, either by screening samples and including only those determined to be error-free (e.g., through consensus in the multiple tubes approach) or by adopting a modeling framework that explicitly accounts for genetic misidentification.

11.7 Capture–Recapture Models Dealing with Genetic Misidentification

Lukacs and Burnham (2005b) developed a model that estimates the probability that a genotype is identified correctly when observed for the first time, based on the disproportionate number of individuals captured only once. Link et al. (2010) later developed a modeling framework that dispenses with some of the restrictive assumptions in the Lukacs and Burnham (2005b) formulation, and the treatment by Yoshizaki et al. (2011) better dealt with "ghost" individuals created by genotyping errors, and reduced bias in estimates, particularly as genotyping error rates increase. All these methods assume that ghost individuals are never recaptured, since the probability of observing the same DNA sequence by error is very low (Link et al. 2010). Miller et al. (2005) developed a model that utilizes multiple recaptures of an individual's DNA within sampling occasions, but their approach assumes no errors in genetic identification.

There has been no attempt to address genetic misidentification within an SCR framework. As pointed out by Royle et al. (2014), SCR models explicitly include the locations of individual animals and captures, and this should help substantially resolve uncertainties due to misidentification. Augustine et al. (2016) developed an SCR model that addresses the partial identity problem (e.g., when we obtain only left flank or only right flank photo-captures for some individuals in a camera trap study and are unable to determine if and which of these single flank captures belong to the same individual) based on locations of captures, and a similar perspective should help in the context of genetic misidentification.

11.8 Types of Genetic Surveys and Corresponding Analytical Frameworks

Surveys for fecal or other types of DNA samples could be conducted using various survey designs and field protocols, and the modeling approach used for estimation should be carefully selected based on how well it describes the particular observation processes.

11.8.1 Nonspatial CR

An example of nonspatial CR estimation of tiger abundance based on fecal DNA-based identification is by Mondol et al. (2009), who carried out a study in which tiger scats were sampled in Bandipur Tiger Reserve, Karnataka, India. The study area contained 18 scat search routes designed to cover the entire 671 km^2 study area with no "holes" where a tiger could move around and not be exposed to the sampling. The field teams searched the routes sequentially, covering the entire study area over a period of 7 days, and this was repeated over six such weekly sampling occasions. These surveys yielded data for standard nonspatial CR analyses, with weekly sampling occasions (scats collected on days 1–7 were assigned to sampling occasion 1, scats collected on days 8–14 to sampling occasion 2, and so on). This design ensured full spatial coverage of the study area in each sampling occasion and a survey duration (42 days) that was sufficiently short to ensure population closure. The data were analyzed using the M_h Jackknife estimator in program CAPTURE and yielded abundance estimates that were very similar to estimates derived from camera trap surveys conducted 15 weeks earlier in the same study area.

11.8.2 Spatial CR

As described in Chap. 9, SCR approaches incorporate the spatial locations of captures and the spatio-temporal schedule of effort (e.g., where and when camera traps were deployed) in the analysis, in addition to information on which individual was captured on which sampling occasion. In surveys for fecal DNA, scat samples are typically found along search routes, unlike camera trap surveys in which detections of animals can only occur at fixed camera trap locations. Several options are available to incorporate these elements of the observation process in the modeling, as described below:

1. *Areal searches, polygon detector models:* When searches (e.g., for scat samples) are conducted with uniform sampling intensity and encounter probability across a defined polygon of interest, the approaches developed by Royle and Young (2008) and Efford (2011) may be appropriate. The formulation by Royle and Young (2008) is implemented in a Bayesian framework and can be analyzed using software program WinBUGS (Gilks et al. 1994) by modifying the code provided in the paper. Efford's (2011) approach uses maximum likelihood estimation and is implemented in the R package secr (Efford 2016).
2. *Detections at fixed, discrete sampling stations:* When "captures" of DNA samples occur at an array of point locations, such as hair snares (hair snags), standard SCR analyses used for camera trap data can be applied. SCR models (see Chap. 9 for a fuller description) include (i) an observation component describing the capture data conditional on the activity centers (including an encounter process model that specifies the type of data obtained at each trap and a detection function describing the decline in capture probability with increasing distance between trap and individual activity center) and (ii) the process component describing the number and distribution of activity centers within a defined state space (see Royle et al. 2014, Chap. 9). The Bernoulli encounter process model should be used for hair snare surveys (individual encountered or not at a trap during a sampling occasion, rather than 0,1,2,3 . . . encounters), as it is not possible to separately identify multiple visits to a trap within a sampling occasion (Gardner et al. 2010, Royle et al. 2014). Such analyses can be implemented using R packages SPACECAP (Gopalaswamy et al. 2012b), scrBayes (https://github.com/jaroyle/SCRbayes), or secr (Efford 2016) and require careful preparation of a captures file (specifying which individual was captured on which sampling occasion at which location), a traps file (with locations of each trap and information on which traps were active on which sampling occasions), and a state-space file (defining a large state space of interest encompassing the trap array, with non-habitat areas clipped out; see Chap. 10 and its Appendix for details).
3. *Detections from search paths, hazard to encounter:* When searches are carried out along predetermined search routes, and "captures" of DNA samples occur at different distances from this route, the observation component of the SCR model should additionally describe the locations of individuals (or scats) conditional on their activity centers, in addition to components that model the data conditional

on the locations and the process component describing the number and locations of activity centers (Royle et al. 2011, 2014). Such an approach may be useful if searches are carried out along fixed routes using scat detection dogs which can detect scats deposited some distance away from these routes. Analysis using this approach may be carried out by modifying the R and WinBUGS code provided by Royle et al. (2011, 2014).

4. *Detections on search paths, segments as detectors:* When searches for DNA samples such as scat are carried out along routes, and "captures" occur on the route, one may break the route up into a series of short segments and treat the midpoint of each segment as a fixed detector. The segments should be short enough that one retains adequate resolution in spatial locations of captures of individuals. An example of this approach can be seen in Gopalaswamy et al. (2012a). Analyses using this approximation can be implemented using standard SCR software such as R packages SPACECAP (Gopalaswamy et al. 2012b), scrBayes (https://github.com/jaroyle/SCRbayes), and secr (Efford 2016); the latter contains functions to discretize search paths into detectors.
5. *Detections on search paths, grid cells as detectors:* Similar to the approach described above, searches carried out along routes can be overlaid on a raster layer of pixels (grid cells) covering the area of interest, with the center point of each pixel treated as a "trap." Because captures are more likely in pixels that receive greater search effort, the length of the search route within each pixel (on the log scale) can be used as a covariate for baseline encounter probability, which is set = 0 for pixels that are not searched. Similarly, captures of DNA samples are referenced with respect to the pixels in which they are encountered. Pixel size needs to be small enough to retain spatial resolution in (re)capture data but large enough that detections do not occur across pixels (Royle et al. 2014) and could potentially be the same grid system used to define the state space. This approach has been used for unstructured spatial surveys, with "captures" obtained using biopsy darts from mountain lions treed by dogs (Russell et al. 2012), from fisher scats detected by dogs (Thompson et al. 2012), or photographic captures of cheetahs (Broekhuis and Gopalaswamy 2016) and lions (Elliot and Gopalaswamy 2016) by field survey teams searching along routes within the area of interest. The analyses can be carried out using R packages scrBayes (https://github.com/jaroyle/SCRbayes) or secr (Efford 2016).

Of the different designs—and corresponding analytical approaches—described above (see Fig. 11.5), polygon searches (approach 1 above) are unlikely to be useful for fecal DNA surveys of tigers and other large carnivores, particularly since it is difficult to apply uniform search effort at the spatial scales required to sample tiger populations. To our knowledge, hair snare surveys (2 above) have not been attempted for tigers anywhere across their distributional range. The hazard to encounter approach (3 above) may be used if searches are carried out along routes and scats or other DNA samples are detected at different distances from the search path. Alternatively, in such a situation, one may rasterize the area of interest and

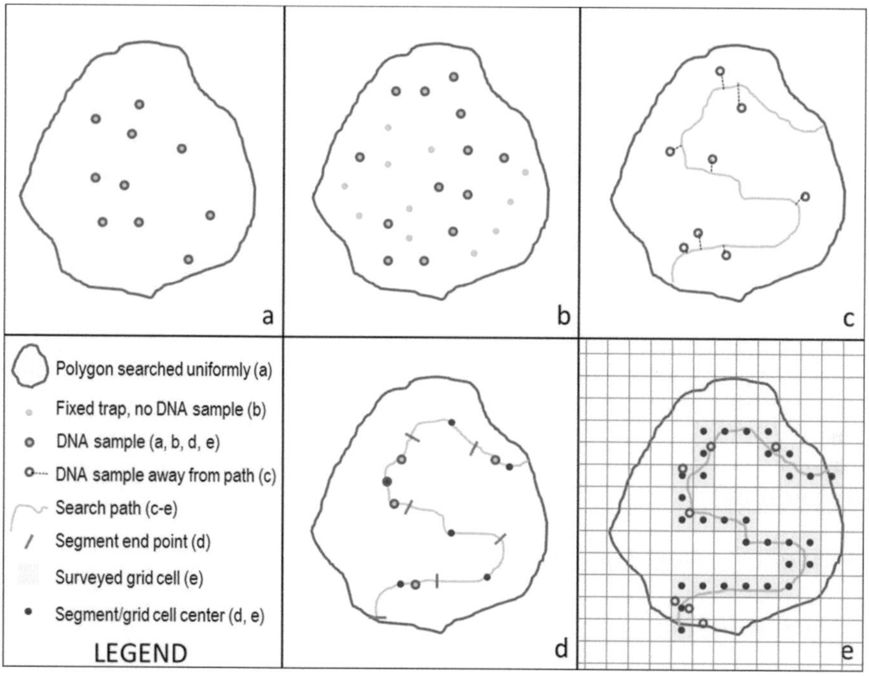

Plate 11.5 Schematic representation of different DNA-based SCR survey designs

treat centers of grid cells as "traps" (5 above). For the most common situation where surveys (e.g., for scat samples) are carried out by searching forest trails and detections occur on (not away from) these routes, midpoints of segments along these routes may be treated as traps (approach 4 above).

11.9 Summary and Conclusion

Noninvasive fecal DNA surveys, in conjunction with CR (particularly SCR) modeling, can be an important tool in monitoring and studying tiger populations, particularly where camera trap surveys cannot be used. Genetic CR surveys, however, require substantial resources including laboratory facilities, reagents, and trained geneticists. Protocols for field sample collection, storage, DNA extraction, species identification, and individual identification have by now been well established and need to be followed very strictly—both in the field and in the laboratory—to minimize errors from contamination, DNA degradation, amplification failure, allelic dropout, false alleles, or shadow effects. However, such errors cannot be eliminated completely and can seriously bias estimates of population parameters such as abundance, density, or survival. Genetic misidentification needs to be addressed through strict screening of samples or through the use of CR models that incorporate the various ways genetic misidentification may have affected the dataset in hand. The modeling framework used should also reasonably describe the sampling and ecological realities, particularly in the case of SCR, where captures are not collapsed across space. Recent developments in the use of SNPs may allow biologists to identify a large panel of common SNP markers that could be used to individually identify tigers across their range.

Overall, the approaches based on identifying individual tigers offer an alternative, sometimes highly relevant, approach to understanding tiger dynamics using capture–recapture methods outlined in Chap. 9. This is particularly so in many contexts where deployment of cameras is challenging or additional data is needed on sex, reproductive status, relatedness among individuals, or to answer biogeographic and evolutionary questions.

Appendices

Appendix 1: Field Collection Protocols

(a) Simple Drying

Requirements:

1. GPS
2. Butter paper
3. Plastic gloves
4. Ziploc bags (small and large)
5. Writing pad
6. Permanent marker
7. Pen

1. Collect fresh/relatively fresh scats in the field. If large numbers of samples are available, then it is better to collect samples that are less than 3–4 days old. However, in low-density areas, collect all available samples.
2. Use a fresh pair of gloves for each sample and place the sample in butter paper.
3. Put the sample in a large Ziploc bag.
4. Record the following details in a preformatted datasheet:
   - (i) Date of collection
   - (ii) Putative species
   - (iii) Sample age (old/fresh)
   - (iv) Latitude, longitude (from GPS unit)
   - (v) Location, beat, range, PA (all if possible)
   - (vi) Collector name and contact details
5. Place the datasheet inside another large Ziploc along with the Ziploc containing the scat sample. Label both Ziploc bags with sample ID (in standard format) using a permanent marker and record sample ID and other details in a notebook.

(b) Ethanol Collection

Requirements:

1. GPS
2. Scat collection vial
3. Sterile plastic spoon
4. Ethanol (100%)
5. Plastic gloves
6. Ziploc bags (small and large)
7. Writing pad
8. Permanent marker
9. Pen
10. Sticker labels

1. Wear fresh gloves before collecting each scat.
2. Open the ethanol-filled sample collection vial.
3. Use the spoon to scoop the sample into the collection vial. If the scat is fresh and it is easy to distinguish the top layer, please collect from there. If not collect from both the upper and lower parts of the scat. Where possible, collect the tip that emerges last during defecation.
4. Please collect as much of the scat as possible in a single vial. For fresh scats try to collect as much of the top layer as possible. If needed, use more than one vial.
5. After scat collection, record sample number, date, collection area, and your name on the vial.
6. Tightly close the vial, ensuring that the alcohol does not leak. Place the vial in a Ziploc bag and seal it. Write the sample number and the GPS location on this zip lock bag with the marker or pencil (pencil marks stay even if the alcohol leaks).
7. Please fill the collection sheet, including sample number, age of the scat (fresh, old), diameter, GPS location, collector name and contact information, and any other relevant remarks.
8. Place the collection sheet inside an outer Ziploc bag.
9. Place both bags in a third, bigger Ziploc bag, to ensure that even if leakage occurs, the alcohol does not erase the sample information. Write just the sample number on this bag.
10. Discard the spoon and the gloves in a bag or container.
11. Always use a new collection kit for every new sample, even if you think two scats are from the same individual, to eliminate cross-sample contamination.

Appendix 2: DNA Extraction Protocols for Scat Samples

(a) Dry Samples: Swabbing Extraction

1. Label two 2 ml microfuge tubes for each field sample with lab ID. Include one negative control for every 11 tubes.
2. Swab each scat with a sterile cotton swab applicator, after soaking in 1X PBS solution for ~10 seconds. Try to swab the entire surface of the scat, using all

sides of the swab and targeting all areas that seem to have mucus. Insert swab into labeled tube and break the stick. Repeat the swabbing process. Close the tube carefully.
3. Add 300 μl ATL buffer in all tubes.
4. Add 30 μl Proteinase K to all tubes.
5. Vortex all tubes for 1 minute for proper mixing of the contents.
6. Put all tubes in the 56 °C incubator overnight. During incubation, the ATL buffer lyses cells and the Pro-K digests proteins. During incubation vortex the samples several times.
7. Centrifuge tubes for 1 minute at the end of the incubation.
8. Carefully remove swabs with clean forceps, ensuring no cross-contamination among tubes.
9. Add 330 μl AL buffer in each tube.
10. Add 330 μl 100% EtOH to each sample.
11. Vortex the tubes for ~20 seconds each tube.
12. Spin down tubes briefly to get liquid off lids (~1 minute).
13. Set up one Qiagen spin column in a collection tube (lidless tube for collecting waste) for each sample and three more 2 ml collection tubes in the same column for each sample. Also set up one 1.5 ml microfuge tube for each sample for the final elution.
14. Add 500 μl of each sample to its corresponding spin column, and centrifuge lysate through column into catch tube, ~30 sec/spin. Do this twice or until all of the lysate has been loaded on to column.
15. Add 500 μl of wash buffer AWI to the column in the third catch tube. Centrifuge for 30 seconds and move column to the fourth (last) catch tube.
16. Add 500 μl of wash buffer AW2 to the column and centrifuge for 3 minutes. Move the column to its corresponding final tube for elution.
17. Add 100 μl Qiagen AE buffer to the column. Incubate samples for 10 minutes, and then centrifuge at top speed for 2 minutes.
18. Add another 100 μl Qiagen AE buffer to the column. Incubate samples for 10 minutes, and then centrifuge at top speed for 2 minutes. Store the final elution tubes in freezer for long-term storage of DNA.

(b) Dry/Wet Samples: Scraping Extraction

1. Label two 2 ml microfuge tubes for each field sample with lab ID. Include one negative control for every 11 tubes.
2. Scrape top layer of each scat with sterile blade/spatula. Try to scrape the entire surface of the scat. Place scrape pieces into both labeled tubes. Close the tube carefully. Use approximately 200 mg of scrape material in each tube.
3. Add 300 μl ATL buffer in all tubes.
4. Add 30 μl Proteinase K to all tubes.
5. Vortex all tubes for 1 minute for proper mixing of the contents.

6. Put all tubes in the 56 °C incubator overnight. During this incubation, the ATL buffer lyses cells and the Pro-K digests proteins. During incubation, vortex the samples several times.
7. Centrifuge tubes for 1 minute at the end of the incubation.
8. Carefully remove swabs with clean forceps, ensuring no cross-contamination among tubes.
9. Add 330 μl AL buffer in each tube.
10. Add 330 μl 100% EtOH to each sample.
11. Vortex each tube for ~20 seconds.
12. Spin down tubes briefly to get liquid off lids (~1 min).
13. Set up one Qiagen spin column in a collection tube (lidless tube for collecting waste) for each sample and three more 2 ml collection tubes in the same column for each sample. Also set up one 1.5 ml microfuge tube for each sample for the final elution.
14. Add 500 μl of each sample to its corresponding spin column, and centrifuge lysate through column into catch tube, ~30 seconds per spin. Do this twice (until all of lysate has been loaded on to column).
15. Add 500 μl of wash buffer AWI to the column in the third catch tube. Centrifuge for 30 sec and move column to the fourth (last) catch tube.
16. Add 500 μl of wash buffer AW2 to the column. Centrifuge for 3 minutes. Move the column to its corresponding final tube for elution.
17. Add 100 μl Qiagen AE buffer to the column. Incubate samples for 10 minutes, and then centrifuge at top speed for 2 minutes.
18. Add another 100 ul Qiagen AE buffer to the column. Incubate samples for 10 minutes, and then centrifuge at top speed for 2 minutes. Store the final elution tubes in freezer for long-term storage of DNA.

(c) Dry/Wet Samples: Swabbing/Scraping Extraction with Commercial Kits

1. Process the scat samples similar to the above two methods into tubes, and then follow the protocol specified in the kit. The Qiagen Stool DNA kit is one of the best available commercial kits for extraction from fecal samples.

References

Allendorf FW, Hohenlohe PA, Luikart G (2010) Genomics and the future of conservation genetics. Nat Rev Genet 11(10):697–709

Augustine B, Royle JA, Kelly M, Satter C et al (2016) Spatial capture-recapture with partial identity: an application to camera traps. bioRxiv 056804; doi: https://doi.org/10.1101/056804

Beja-Pereira AL, Oliveira R, Alves PC, Schwartz MK, Luikart G (2009) Advancing ecological understandings through technological transformations in noninvasive genetics. Mol Ecol Resour 9(5):1279–1301

Bhagavatula J, Singh L (2006) Genotyping fecal samples of bengal tiger *Panthera tigris tigris* for population estimation: a pilot study. BMC Genet 7(1):48

Broekhuis F, Gopalaswamy AM (2016) Counting cats: spatially explicit population estimates of cheetah (*Acinonyx jubatus*) using unstructured sampling data. PLoS One 11(5):e0153875

Broquet T, Menard N, Petit E (2007) Noninvasive population genetics: a review of sample source, diet, fragment length and microsatellite motif effects on amplification success and genotyping error rates. Conserv Genet 8:249–260

Caragiulo A, Pickles RSA, Smith JA, Smith O, Goodrich J, Amato G (2015) Tiger (*Panthera tigris*) scent DNA: a valuable conservation tool for individual identification and population monitoring. Conserv Genet Resour 7:681–683

Cho YS, Hu L, Hou H, Lee H, Xu J et al (2013) The tiger genome and comparative analysis with lion and snow leopard genomes. Nat Commun 4:2433

Davison A, Birks JDS, Brookes RC, Braithwaithe TC, Messenger JE (2002) On the origin of faeces: morphological versus molecular methods for surveying rare carnivores from their scats. J Zool 257:141–143

Efford MG (2011) Estimation of population density by spatially explicit capture–recapture analysis of data from area searches. Ecology 92(12):2202–2207

Efford MG (2016) secr: Spatially explicit capture-recapture models. R package version 2.10.3. http://cran.r-project.org/package=secr

Elliot NB, Gopalaswamy AM (2016) Towards accurate and precise estimates of lion density. Conserv Biol 31:934–943

Fernandes CA, Ginja C, Pereira I, Tenreiro R, Bruford MW, Santos-Reis M (2008) Species-specific mitochondrial DNA markers for identification of non-invasive samples from sympatric carnivores in the iberian peninsula. Conserv Genet 9(3):681–690

Gardner B, Royle JA, Wegan MT et al (2010) Estimating black bear density using DNA data from hair snares. J Wildl Manag 74(2):318–325

Gilks WR, Thomas A, Spiegelhalter DJ (1994) A language and program for complex bayesian modelling. Statistician 43:169–178

Goossens B, Bruford MW (2009) Non-invasive genetic analysis in conservation. In: Bertorelle G, Bruford MW, Hauffe HC, Rizzoli A, Vernesi C (eds) Population genetics for animal conservation. Cambridge University Press, Cambridge, pp 167–201

Gopalaswamy AM, Royle JA, Delampady M et al (2012a) Density estimation in tiger populations: combining information for strong inference. Ecology 93(7):1741–1751

Gopalaswamy AM, Royle JA, Hines JE, Singh P, Jathanna D et al (2012b) Program SPACECAP: software for estimating animal density using spatially explicit capture–recapture models. Methods Ecol Evol 3(6):1067–1072

Horak P, Lebreton JD (1998) Survival of adult great tits Parus major in relation to sex and habitat; a comparison of urban and rural populations. Ibis 140(2):205–209

Inoue E, Inoue-Murayama M, Takenaka O et al (2007) Wild chimpanzee infant urine and saliva sampled noninvasively usable for DNA analyses. Primates 48(2):156–159

Lebreton JD, Burnham KP, Clobert J, Anderson DR (1992) Modeling survival and testing biological hypotheses using marked animals: a unified approach with case studies. Ecol Monogr 62(1):67–118

Link WA, Yoshizaki J, Bailey LL, Pollock KH (2010) Uncovering a latent multinomial: analysis of mark–recapture data with misidentification. Biometrics 66(1):178–185

Lukacs PM, Burnham KP (2005a) Review of capture–recapture methods applicable to noninvasive genetic sampling. Mol Ecol 14(13):3909–3919

Lukacs PM, Burnham KP (2005b) Estimating population size from DNA-based closed capture–recapture data incorporating genotyping error. J Wildl Manag 69(1):396–403

Marshall TC, Slate J, Kruuk LEB, Pemberton JM (1998) Statistical confidence for likelihood-based paternity inference in natural populations. Mol Ecol 7:639–655

Miller CR, Joyce P, Waits LP (2005) A new method for estimating the size of small populations from genetic mark–recapture data. Mol Ecol 14(7):1991–2005

Mondol S, Karanth KU, Kumar NS et al (2009) Evaluation of non-invasive genetic sampling methods for estimating tiger population size. Biol Conserv 142(10):2350–2360

Mondol S, Bruford MW, Ramakrishnan U (2013) Demographic loss, genetic structure and the conservation implications for indian tigers. Proc R Soc Lond [Biol] 280(1762):20130496

Mondol S, Kumar NS, Gopalaswamy A, Sunagar K, Karanth KU, Ramakrishnan U (2015) Identifying species, sex and individual tigers and leopards in the Malenad-Mysore Tiger Landscape, Western Ghats, India. Conserv Genet Resour 7(2):353–361

Morin PA, Woodruff DS (1992) Paternity exclusion using multiple hypervariable microsatellite loci amplified from nuclear DNA of hair cells. In: Martin RD, Dixson AF, Wickings EJ (eds) Paternity in primates: Genetic tests and theories. Karger, Basel, pp 63–81

Mukherjee N, Mondol S, Andheria A, Ramakrishnan U (2007) Rapid multiplex PCR based species identification of wild tigers using non-invasive samples. Conserv Genet 8:1465–1470

Navidi W, Arnheim N, Waterman MS (1992) A multiple-tubes approach for accurate genotyping of very small DNA samples by using PCR: statistical considerations. Am J Hum Genet 50(2):347–359

Paetkau D (2003) An empirical exploration of data quality in DNA-based population inventories. Mol Ecol 12(6):1375–1387

Pompanon F, Bonin A, Bellemain E, Taberlet P (2005) Genotyping errors: causes, consequences and solutions. Nat Rev Genet 6(11):847–846

Prugh LR, Ritland CE (2005) Molecular testing of observer identification of carnivore feces in the field. Wildl Soc Bull 33:189–194

Royle JA, Young KV (2008) A hierarchical model for spatial capture–recapture data. Ecology 89(8):2281–2289

Royle JA, Kery M, Guelat J (2011) Spatial capture-recapture models for search-encounter data. Methods Ecol Evol 2(6):602–611

Royle JA, Chandler RB, Sollmann R, Gardner B (2014) Spatial capture-recapture. Academic Press, Waltham

Russell RE, Royle JA, Desimone R et al (2012) Estimating abundance of mountain lions from unstructured spatial sampling. J Wildl Manag 76(8):1551–1561

Schuster SC (2008) Next-generation sequencing transforms today's biology. Nat Methods 5(1):16

Sollmann R, Furtado MM, Gardner B, Hofer H, Jácomo AT, Tôrres NM, Silveira L (2011) Improving density estimates for elusive carnivores: accounting for sex-specific detection and movements using spatial capture–recapture models for jaguars in central Brazil. Biol Conserv 144(3):1017–1024

Steiner CC, Putnam AS, Hoeck PE, Ryder OA (2013) Conservation genomics of threatened animal species. Annu Rev Anim Biosci 1(1):261–281

Sugimoto T, Nagata J, Aramilev VV, McCullough DR (2012) Population size estimation of Amur tigers in russian far east using noninvasive genetic samples. J Mammal 93(1):93–101

Taberlet P, Luikart G (1999) Non-invasive genetic sampling and individual identification. Biol J Linn Soc 68:41–55

Taberlet P, Griffin S, Goossens B, Questiau S, Manceau V, Escaravage N, Waits LP, Bouvet J (1996) Reliable genotyping of samples with very low DNA quantities using PCR. Nucleic Acids Res 24(16):3189–3194

Taberlet P, Waits LP, Luikart G (1999) Noninvasive genetic sampling: look before you leap. Trends Ecol Evol 14(8):323–327

Thompson CM, Royle JA, Garner JD (2012) A framework for inference about carnivore density from unstructured spatial sampling of scat using detector dogs. J Wildl Manag 76(4):863–871

Valiere N, Taberlet P (2000) Urine collected in the field as a source of DNA for species and individual identification. Mol Ecol 9(12):2150–2152

Waits LP (2005) Using noninvasive genetic sampling to detect and estimate abundance of rare wildlife species. In: Thompson WL (ed) Sampling rare or elusive species. Island Press, Washington, DC, pp 211–228

Waits LP, Paetkau D (2005) Noninvasive genetic sampling tools for wildlife biologists: a review of applications and recommendations for accurate data collection. J Wildl Manag 69(4):1419–1433

Wasser SK, Houston CS, Koehler GM, Cadd GG, Fain SR (1997) Techniques for application of fecal DNA methods to field studies of ursids. Mol Ecol 6:1091–1097

Yoshizaki J, Brownie C, Pollock KH, Link WA (2011) Modeling misidentification errors that result from use of genetic tags in capture–recapture studies. Environ Ecol Stat 18(1):27–55

Zuercher G, Gipson P, Stewart G (2003) Identification of carnivore feces by local peoples and molecular analyses. Wildl Soc Bull 31(4):961–970

Concepts: Integrating Population Survey Data from Different Spatial Scales, Sampling Methods, and Species

12

Robert M. Dorazio, Mohan Delampady, Soumen Dey, and Arjun M. Gopalaswamy

12.1 Introduction

Conservationists and managers are continually under pressure from the public, the media, and political policy makers to provide "tiger numbers," not just for protected reserves, but also for large spatial scales, including landscapes, regions, states, nations, and even globally. Estimating the abundance of tigers within relatively small areas (e.g., protected reserves) is becoming increasingly tractable (see Chaps. 9 and 10), but doing so for larger spatial scales still presents a formidable challenge. Those who seek "tiger numbers" are often not satisfied by estimates of tiger occupancy alone, regardless of the reliability of the estimates (see Chaps. 4 and 5). As a result, wherever tiger conservation efforts are underway, either substantially or nominally, scientists and managers are frequently asked to provide putative large-scale tiger numbers based either on a total count or on an extrapolation of some sort (see Chaps. 1 and 2).

The sheer size of a tiger population's physical habitat and the clustering of individuals within that habitat present formidable challenges to biologists and managers charged with estimating the number and spatial distribution of tigers. For

R.M. Dorazio (✉)
Wetland and Aquatic Research Center – United States Geological Survey (USGS), Gainesville, FL, USA
e-mail: bdorazio@usgs.gov

M. Delampady • S. Dey
Statistics and Mathematics Unit, Indian Statistical Institute, Bengaluru, India
e-mail: mohan@isibang.ac.in; soumenstat89@gmail.com

A.M. Gopalaswamy
Statistics and Mathematics Unit, Indian Statistical Institute, Bengaluru, India

Department of Zoology, University of Oxford, Oxford, UK
e-mail: arjungswamy@gmail.com; https://www.isibang.ac.in; https://www.zoo.ox.ac.uk/

K.U. Karanth, J.D. Nichols (eds.), *Methods For Monitoring Tiger And Prey Populations*, DOI 10.1007/978-981-10-5436-5_12

example, during the past two centuries, the geographic range of tigers has declined by approximately 93% owing to fragmentation of physical habitat, depletion of prey, and direct hunting (Karanth et al. 2004; Walston et al. 2010). Remaining populations of tigers are composed of interconnected clusters of individuals distributed within a large landscape of several thousand square kilometers (Karanth et al. 2011). These clusters, often located within small areas or protected reserves, typically span a few hundred square kilometers.

The abundance of tigers can be estimated accurately for relatively small portions of a population using intensive surveys of uniquely identified individuals (via distinct stripe patterns or fecal DNA) (Royle et al. 2009; Gopalaswamy et al. 2012b); however, these surveys are impractical to apply to entire populations owing to logistical constraints and financial limitations. Other kinds of surveys based on the detection of tiger signs (tracks, scats) are less costly, easier to implement, and therefore more practical for larger areas (Karanth et al. 2011).

Surveys of tiger signs have primarily been used to estimate probabilities of occurrence of tigers. Statistical approaches for estimating spatial variation in abundance from sign surveys do exist (Royle and Dorazio 2008, chapter 4), but to our knowledge these approaches have rarely been used in the analysis of tiger signs. Instead, some researchers have attempted to establish a correlation between observed encounter rates of tiger signs and rigorous estimates of tiger abundance with the intention of predicting abundance at locations where only tiger signs are observed (Kindberg et al. 2009; Jhala et al. 2011). The proclaimed "success" of these so-called index-calibration experiments has been questioned on theoretical grounds (Gopalaswamy et al. 2015). As an alternative, Dey et al. (2017) recently proposed a model-based approach for relating sign-based estimates of tiger abundance to estimates of tiger abundance based on camera-trap surveys.

The approach of Dey et al. (2017) provides a step in the right direction, particularly in circumstances where the underlying data are unavailable. What is needed, however, is a common statistical framework for the *joint analysis* of data collected during intensive surveys of relatively small areas (e.g., camera-trap surveys) and data collected during surveys of tiger signs over much larger areas that overlap the intensively surveyed areas. Models for the joint analysis (or integration) of different types of survey data have been developed previously (e.g., see Buckland et al. 2007; Conroy et al. 2008; Schaub and Abadi 2011), but none of these models includes a spatially explicit description of the population, which is needed when surveys use different spatial scales – that may or may not overlap – and different sampling methods (Chandler and Clark 2014).

Before elaborating on this idea, we note that the challenges associated with a joint analysis of data from intensive and less-intensive surveys of tigers also apply to other species. For example, line-transect surveys are often used to obtain accurate estimates of abundance of tiger prey species (Karanth et al. 2004, also see Chapters 6–8). These surveys, being labor-intensive, are generally conducted within relatively small areas. However, prey species also are sampled over much larger areas using occupancy-based surveys that require only the presence or absence of individuals to be observed (Kawanishi and Sunquist 2004; Killivalavan 2010; Gopalaswamy et al. 2012a; Linkie et al. 2013; Vongkhamheng et al. 2013). As with the analysis of tiger signs, occupancy-based surveys of prey have been used to estimate probabilities of occurrence of prey over the area surveyed. A combined analysis of data observed in the different types of surveys (line-transect and occupancy) has not been attempted.

In this chapter we describe a conceptual framework for the joint analysis of survey data collected using different spatial scales, different methods of sampling, and multiple species. In a community of interacting species, such as a predator-prey network, it is reasonable to expect the abundances of different species to be correlated. The strength of this correlation will depend, of course, on the relative effects of factors that can influence abundance, including habitat quality or availability. However, knowing the abundance of one species is likely to help predict the abundance of another species when interactions exist between individuals of each species (Dorazio et al. 2015).

We intentionally limit our description of the statistical framework for the joint analysis of different types of survey data to its conceptual underpinnings. Implementation of this framework is an ambitious undertaking and extends beyond the scope of this book. Our goal here is to describe the framework in sufficient detail that it may inspire future research in the development of methods for analyzing survey data obtained using different spatial scales, sampling methods, and species. We also summarize the potential benefits of this approach for the conservation and management of tigers and their prey.

12.2 A Hierarchical Approach for the Analysis of Different Types of Spatially Referenced Data

To develop a statistical framework for the joint analysis of different types of survey data, we adopt a hierarchical modeling approach similar to that used in the analysis of spatially explicit, capture-recapture data (Royle et al. 2014, also see Chapter 9). One component of the hierarchy is used to specify the number and spatial distribution of latent (unobserved) individual activity centers for each species and the movements of individuals about those centers. A second component is used to specify how the locations and methods selected for sampling induce various kinds of observations that depend on the locations and activity centers of individuals. This approach was used to analyze the detections of a single species, the Louisiana black bear (*Ursus americanus luteolus*) (Chandler and Clark 2014). In the following sections, we extend this approach to multiple species and to different methods of sampling.

12.2.1 A Spatial Point Process for the Locations of Multiple Species

Models of spatially explicit, capture-recapture data have been formulated using relatively simple spatial point processes (binomial or Poisson) (Royle et al. 2014). These processes are specified using a first-order intensity function $\lambda(\mathbf{s})$ to denote the limiting expected density of individual activity centers at location $\mathbf{s}$. The parameters of the first-order intensity function specify the effects of spatially varying covariates, such as habitat, on the expected density of activity centers.

To model the number and spatial distribution of two or more interacting species, we require a *multivariate*, spatial point process – specifically, a Markov point process formulated to specify pairwise interactions among individuals of each species (Cressie 1993; Högmander and Särkkä 1999; Diggle 2014). In this process the activity center of an individual is determined, at least in part, by its position relative to the activity centers of all other individuals.

Without going into too much technical detail, we describe the underpinnings of a Markov point process for two interacting species, say A and B. (Extensions of this process to three or more species are straightforward.) Let $\mathbf{s}_A = (\mathbf{s}_{A,1}, \ldots, \mathbf{s}_{A,N_A})'$ denote the activity centers of N_A individuals of species A living within some finite region of interest $S \subset \mathbb{R}^2$ (i.e., $\mathbf{s}_{A,i} \in S$). Similarly, let $\mathbf{s}_B = (\mathbf{s}_{B,1}, \ldots, \mathbf{s}_{B,N_B})'$ denote the activity centers of N_B individuals of species B also living in region S. If the abundances and activity centers of these two species follow a Markov point process, their probability density is proportional to

$$f(N_A, \mathbf{s}_A, N_B, \mathbf{s}_B) \propto \frac{1}{N_A!\,N_B!} \times f_A \times f_B \times f_{AB}$$

where

$$f_A = \prod_i \lambda_A(\mathbf{s}_{A,i}) \times \prod_i \prod_{j>i} h_A(\mathbf{s}_{A,i}, \mathbf{s}_{A,j})$$

$$f_B = \prod_k \lambda_B(\mathbf{s}_{B,k}) \times \prod_k \prod_{l>k} h_B(\mathbf{s}_{B,k}, \mathbf{s}_{B,l})$$

$$f_{AB} = \prod_i \prod_k h_{AB}(\mathbf{s}_{A,i}, \mathbf{s}_{B,k})$$

The intensity functions λ_A and λ_B specify the effects of spatially varying covariates on the expected densities of species A and B, respectively. The *pair-potential functions* h_A and h_B specify the effects of a pairwise interaction between two individuals of species A and between two individuals of species B, respectively (i.e., a pairwise interaction between conspecifics). In contrast, the pair-potential function h_{AB} specifies the effect of a pairwise interaction between an individual of species A and an individual of species B. Considerable flexibility exists for choosing the functional forms of h_A, h_B, and h_{AB}, but these interactions are usually formulated as functions of the *distance* between locations of individuals. Interactions between individuals can be positive and induce spatial clustering, or the interactions can be negative and induce spatial regularity or repulsion. This versatility is useful because it allows the effects of territoriality, predation, and other kinds of ecological interactions to be formulated with the same type of model specification. Pair-potential functions generally contain one or more parameters that determine the spatial extent and type of interaction between individuals. The values of these parameters are estimated in conjunction with other parameters of the model.

While the Markov point process allows us to model the activity centers of tigers and prey species, it does not specify movements of individuals during the period of sampling. We assume that each individual moves randomly about its activity center during the period of sampling. In some cases, as with detections of tigers in camera-trap surveys, the locations where individuals can be detected are fixed. In these cases the model of an individual's movements is implicit because trap locations are selected so that the same individual can be detected at multiple traps as a consequence of its movements. In other cases an explicit model may be used to specify an individual's locations during repeated surveys of the same area or sample unit. For example, if individuals are detected in line-transect surveys or in searches of discrete areas, we might use a bivariate Normal($\mathbf{s}, \sigma^2\mathbf{I}$) distribution to model the locations of an individual with activity center $\mathbf{s}$.

12.2.2 Modeling Survey Data Obtained Using Different Methods and Spatial Scales

Thus far, we have described a fairly general model for the abundance and spatial distribution of individuals, though neither abundance nor the activity centers of individuals is directly observable. In this section we describe an approach for modeling different kinds of observations obtained using different kinds of surveys. The key idea in developing such models is to specify the distribution of an observable quantity, such as a detection or a count, *conditional on the latent locations or activity centers of individuals*. By adopting this approach, each set of survey data contributes information about the population of individuals in a consistent and coherent fashion. Furthermore, because the data are modeled conditionally, different sources of survey data need not be independent, a requirement that may be difficult to satisfy when surveys using different methods overlap in space and time.

As described earlier (see Introduction), several kinds of surveys are often used to sample tiger populations and their prey. Considerable differences exist in the spatial scales, sampling methods, and types of observations used in these surveys. Despite these differences, observations may be classified into one of two groups. One group corresponds to information observed at the level of an individual animal. Detections of an individual in a camera-trap survey or in a line-transect survey are examples. For this group observations are modeled as a function of the distance between the individual and the detector (person or device). For example, the number of detections Y of an individual during J days of exposure to a camera trap might be assumed to have a binomial distribution as follows:

$$Y \mid \mathbf{s} \sim \text{Binomial}(J, p(\mathbf{s}, \mathbf{x}))$$

where the individual's daily detection probability $p(\mathbf{s}, \mathbf{x})$ is specified as a function of the distance between the individual's activity center $\mathbf{s}$ and the location of the camera trap $\mathbf{x}$. Similarly, the detection of an individual in a line-transect survey might be assumed to have a Bernoulli distribution, where the detection probability $p(\mathbf{u}, \mathbf{x})$ is specified as a function of the perpendicular distance between the individual's location $\mathbf{u}$ and the transect line $\mathbf{x}$. The model of line-transect data also could be expanded by assuming $\mathbf{u} \sim \text{Normal}(\mathbf{s}, \sigma^2\mathbf{I})$. The key point in both of these surveys, however, is that each observation is modeled conditional on an individual's activity center $\mathbf{s}$ or location $\mathbf{u}$.

A second group of observations in animal surveys occurs when information pertains to an unknown number of individuals. For example, these observations are common when the region of interest S is partitioned into a finite number of disjoint (nonoverlapping) sample units and surveys are conducted within a subset of those units. Common examples include occupancy surveys – which can include surveys of animal signs (tracks, scats) – and count-based surveys of animals based on single- or double-observer protocols (see Chap. 8). In these surveys each observation is modeled conditional on the latent abundance of individuals in a sample unit. For example, suppose each of K sample units (denoted by $C_1, \ldots, C_K$) is surveyed on J occasions using a particular sampling protocol. The number of individuals present and available to be detected during the jth survey of unit C_k is

$$N_{kj} = \sum_{i=1}^{N} I(\mathbf{u}_{ij} \in C_k)$$

where N is the total number of individuals living in region S and where $\mathbf{u}_{ij} \sim \text{Normal}(\mathbf{s}_i, \sigma^2\mathbf{I})$. Note that $I(z)$ is an indicator function whose value equals one if expression z is true (or zero if z is false). In a single-observer, point-count survey, the number of individuals detected during the jth survey of unit k would be modeled as a binomial outcome as follows:

$$n_{kj} \mid N_{kj} \sim \text{Binomial}(N_{kj}, p_k)$$

where p_k is the probability of detection per individual during the survey. Similarly, in an occupancy survey, a binary indicator of the detection of at least one individual would be modeled as a Bernoulli outcome as follows:

$$Y_{kj} \mid N_{kj} \sim \text{Bernoulli}(q_k\, I(N_{kj} > 0))$$

where q_k is the probability of detecting at least one individual during the survey. We emphasize that the key feature common to both surveys is that each observation is modeled conditional on the latent abundance of individuals in the sample unit and that this abundance, being spatially referenced, depends on the locations of individuals during sampling.

12.3 Discussion

In this chapter we briefly described a general framework for the joint analysis of survey data collected using different spatial scales, different methods of sampling, and multiple species. Our framework is intended primarily for the analysis of surveys of tigers and their prey. In India and in other parts of Asia, vast amounts of financial resources and human effort have been expended to establish rigorous monitoring programs for these species. Some of these monitoring programs have been in place for decades, and it seems prudent now to conduct a retrospective analysis of the data acquired in these programs.

Our model-based framework provides a reasonable starting point for this task. At the very least, it should be possible to estimate the abundance and spatial distribution of an entire population of tigers by conducting a joint analysis of data observed in camera-trap surveys and occupancy-based surveys. Similarly, the distribution and abundance of individual prey species should be estimable using a joint analysis of data observed in line-transect surveys and occupancy-based surveys. More challenging, perhaps, is to complete a combined analysis of data obtained in surveys of tigers, other large predators (leopards, dholes), and their prey. However, the results of this type of analysis could reveal patterns of spatial overlap between tigers, leopards, and individual prey species, as well as quantitative relationships between the densities of prey and their habitat. Both results are important to decisions regarding the conservation and management of these species. It seems possible also to develop point-process models that specify changes in the distribution of predator and prey species over time as a result of the biological interactions between individuals. This would allow hypotheses about the nature of these interactions to be specified mechanistically in terms of the model's parameters.

References

Buckland ST, Newman KB, Fernández C, Thomas L, Harwood J (2007) Embedding population dynamics models in inference. Stat Sci 22:44–58

Chandler RB, Clark JD (2014) Spatially explicit integrated population models. Methods Ecol Evol 5:1351–1360

Conroy MJ, Runge JP, Barker RJ, Schofield MR, Fonnesbeck CJ (2008) Efficient estimation of abundance for patchily distributed populations via two-phase, adaptive sampling. Ecology 89:3362–3370

Cressie NAC (1993) Statistics for spatial data. John Wiley & Sons, New York

Dey S, Delampady M, Parameshwaran R, Kumar NS, Srivathsa A, Karanth KU (2017) Bayesian methods for estimating animal abundance at large spatial scales using data from multiple sources. J Agric Biol Environ Stat 22:111–139

Diggle PJ (2014) Statistical analysis of spatial and spatio-temporal point patterns, third edition. CRC Press, Boca Raton

Dorazio RM, Connor EF, Askins RA (2015) Estimating the effects of habitat and biological interactions in an avian community. PLoS One 10:e0135987

Gopalaswamy AM, Karanth KU, Kumar NS, Macdonald DW (2012a) Estimating tropical forest ungulate densities from sign surveys using abundance models of occupancy. Anim Conserv 15:669–679

Gopalaswamy AM, Royle JA, Delampady M, Nichols JD, Karanth KU, MacDonald DW (2012b) Density estimation in tiger populations: combining information for strong inference. Ecology 93:1741–1751

Gopalaswamy AM, Delampady M, Karanth KU, Kumar NS, Macdonald DW (2015) An examination of index-calibration experiments: counting tigers at macroecological scales. Methods Ecol Evol 6:1055–1068

Högmander H, Särkkä A (1999) Multitype spatial point patterns with hierarchical interactions. Biometrics 55:1051–1058

Jhala YV, Qureshi Q, Gopal R (2011) Can the abundance of tigers be assessed from their signs? J Appl Ecol 48:14–24

Karanth KU, Nichols JD, Kumar NS, Link WA, Hines JE (2004) Tigers and their prey: predicting carnivore densities from prey abundance. Proc Natl Acad Sci 101:4854–4858

Karanth KU, Gopalaswamy AM, Kumar NS, Vaidyanathan S, Nichols JD, MacKenzie DI (2011) Monitoring carnivore populations at the landscape scale: occupancy modelling of tigers from sign surveys. J Appl Ecol 48:1048–1056

Kawanishi K, Sunquist ME (2004) Conservation status of tigers in the primary rainforest of Peninsular Malaysia. Biol Conserv 120:329–344

Killivalavan R (2010) Assessing potential tiger habitats in Cauvery Wildlife Sanctuary, Karnataka using occupancy modeling approaches. Master's thesis, National Centre for Biological Sciences, Bangalore

Kindberg J, Ericsson G, Swenson JE (2009) Monitoring rare or elusive large mammals using effort-corrected voluntary observers. Biol Conserv 142:159–165

Linkie M, et al (2013) Cryptic mammals caught on camera: assessing the utility of range wide camera trap data for conserving the endangered Asian tapir. Biol Conserv 162:107–115

Royle JA, Dorazio RM (2008) Hierarchical modeling and inference in ecology. Academic Press, Amsterdam

Royle JA, Nichols JD, Karanth KU, Gopalaswamy AM (2009) A hierarchical model for estimating density in camera-trap studies. J Appl Ecol 46:118–127

Royle JA, Chandler RB, Sollmann R, Gardner B (2014) Spatial capture-recapture. Academic Press, Amsterdam

Schaub M, Abadi F (2011) Integrated population models: a novel analysis framework for deeper insights into population dynamics. J Ornithol 152:S227–S237

Vongkhamheng C, Johnson A, Sunquist ME (2013) A baseline survey of ungulate abundance and distribution in northern Lao: implications for conservation. Oryx 47:544–552

Walston J, Robinson JG, Bennett EL, Breitenmoser GAB da Fonseca, Goodrich J, Gumat M, Hunter L, Johnson A, Karanth KU, Leader-Williams N, MacKinnon K, Miquelle D, Pattanavibool A, Poole C, Rabinowitz A, Smith JLD, Stokes EJ, Stuart SN, Vongkhamheng C, Wibisono H (2010) Bringing the tiger back from the brink – the six percent solution. PLoS Biol 8:e1000485

Assessing Landscape Connectivity for Tigers and Prey Species: Concepts and Practice

13

Divya Vasudev, James D. Nichols, Uma Ramakrishnan, Krishnamurthy Ramesh, and Srinivas Vaidyanathan

13.1 Introduction

Tiger conservation efforts are shifting from an exclusive focus on single reserves to include large heterogeneous landscapes. Consequently, monitoring now encompasses both between-habitat ecological processes and within-habitat demographic processes, the most important among these being connectivity. In this chapter, we provide an overview of current theory and conceptual frameworks of connectivity modelling, examining their applicability to the conservation of tiger population connectivity. We visualize large landscapes where tiger habitat fragments are

D. Vasudev (✉)
Wildlife Conservation Society, India Program, Bengaluru, India

Centre for Wildlife Studies, Bengaluru, India
e-mail: vasudev.divya@gmail.com

J.D. Nichols
Crofton, MD, USA
e-mail: jamesdnichols2@gmail.com

U. Ramakrishnan
National Centre for Biological Sciences, TIFR, Bengaluru, India
e-mail: uramakri@ncbs.res.in; uramakri@gmail.com

K. Ramesh
Wildlife Institute of India, Dehradun, India

Department of Forest and Conservation Sciences, University of British Columbia, Vancouver, BC, Canada
e-mail: ramesh@wii.gov.in

S. Vaidyanathan
Foundation for Ecological Research, Advocacy and Learning (FERAL), Auroville, Tamil Nadu, India
e-mail: srinivasv@feralindia.org

K.U. Karanth, J.D. Nichols (eds.), *Methods For Monitoring Tiger And Prey Populations*, DOI 10.1007/978-981-10-5436-5_13

described in terms of their size and quality, while linkages among them are quantified by either actual transition (movement) rates or the probability of dispersal between patches. We briefly review long-term studies of tiger dispersal biology from radiotelemetry, photographic capture-recapture and landscape genetic approaches. We describe methods for drawing inferences about tiger dispersal biology, and about current and potential connectivity patterns. As connectivity is still a growing field of study, we do not stress prescriptive rules but suggest an array of methods that can shed light on tiger connectivity. These include fine-scale tracking of individual dispersal events through radiotelemetry; estimation of population-level connectivity through multistate capture-recapture or dynamic occupancy models and landscape genetics; modelling of potential connectivity, through resistance-based models, individual-based models and stochastic dynamic optimization; and spatial conservation prioritization. We also describe incorporation of remotely sensed data into landscape-scale analyses.

We discuss the current state of tiger connectivity conservation, noting that there seem to be few isolated populations, with evidence of long-distance dispersal across multiple-use and densely populated landscapes, and conclude the chapter with a prospectus for future conservation of connectivity, emphasizing the utility of a thorough understanding of tiger dispersal behaviour for managing a heterogeneous matrix for landscape-scale tiger conservation.

13.2 A Landscape Perspective on Tiger Conservation

13.2.1 From Protected Areas to Landscapes

The last couple of decades have seen a global shift of the scope of animal conservation from single isolated protected areas (PAs) or reserves to large landscapes comprised of habitat patches interspersed within a heterogeneous human-dominated matrix (or lands outside of species' optimal habitat; see Table 13.1 for a definition of select terms used in this chapter). Habitat fragmentation is a global phenomenon, and threatened species are increasingly pocketed into small, isolated habitat fragments. Simultaneously, there has been recognition of the importance of ecological processes that occur in the matrix, primary among these being connectivity among habitat fragments (Taylor et al. 1993; Doerr et al. 2011). While the relative importance of fragmentation and the establishment of landscape connectivity, in comparison with maintaining habitat integrity, has been much debated, there is no doubt that processes such as dispersal that occur within the matrix can have a serious impact on species persistence (Doerr et al. 2011).

Connectivity enhances species persistence through demographic rescue effects (Brown and Kodric-Brown 1977), colonization of empty habitat fragments (Hanski 1998), inbreeding avoidance (Frankham 2005; Hostetler et al. 2013) and metapopulation dynamics (Hanski 1998); it can shape communities through influencing extinction-colonization dynamics (MacArthur and Wilson 1967) and inter-species interactions (e.g. competition, Yackulic et al. 2014, and disease dynamics,

Table 13.1 Operational definitions of terms used in this chapter that are relevant to tiger connectivity

Term	Definition
Landscape	An area that is spatially heterogeneous in at least one factor of interest; here we specifically refer to a heterogeneous area comprised of habitat fragments interspersed in a mosaic of matrix types
Habitat fragment (or patch)	A well-defined area with local conditions suited to support a population of the focal species
Matrix	All nonoptimal habitat areas of the landscape
Connectivity	Functional linkages between populations or habitat fragments
Structural connectivity	Physical linkages formed of the species' habitat between fragments
Functional connectivity	Linkages between habitat fragments accounting for the interaction between species traits and landscape characteristics. Taylor et al. (1993) define this as 'the degree to which a landscape facilitates or impedes animal movement'
Population connectivity	Movement of individuals or genes between two populations. It can be quantified as transition rates or emigration and immigration rates
Potential connectivity	The probability of successful dispersal, given a decision taken by an individual to emigrate
Actual or realized dispersal	Measurements or estimates of actual movement of animals among fragments
Dispersal	Movement of genes or animals across space. Here, we specifically refer to movement across heterogeneous landscapes, comprised of three stages, namely, emigration, search for new habitat and immigration
Matrix resistance	The inverse of permeability of a given matrix type to animal movement
Network	A graph-theoretic representation of a collection of interconnected habitat fragments, where the fragments are represented as nodes, while linkages between fragments are represented as edges
Connectivity conservation	All research and action taken to conserve linkages among populations or habitat patches. To date, the predominant strategy for connectivity conservation has been the demarcation of corridors
Corridors	Linear or narrow sections of habitat that may facilitate dispersal between habitat fragments

Source: Vasudev et al. (2015), Taylor et al. (1993), Lindenmayer and Fischer (2007), Clobert et al. (2012), Turner and Gardner (2001)

McCallum and Dobson 2002). Connectivity is also increasingly being perceived as a tool to ameliorate impacts of climate change and land-use dynamics (Doerr et al. 2011). Consequently, the last few decades have seen a surge of landscape ecology studies, resulting in an improved understanding of animal movement and dispersal (Nathan et al. 2008; Vasudev et al. 2015), new methods for observing and analysing animal movement routes (e.g. McRae et al. 2008; Langrock et al. 2012; Ramesh et al. 2016) and novel applications of spatial conservation planning (Minor and Urban 2007; Moilanen et al. 2009; Fletcher et al. 2013), providing methods for improved landscape-scale conservation.

There are multiple aspects of tiger biology that are pertinent to their landscape-scale conservation. First, tigers, being apex predators, live at low population densities (<10–15 individuals/100 km^2; Karanth et al. 2004, Chap. 2), thus requiring large tracts of habitat for long-term demographic viability. Such large contiguous tracts of natural habitat are decreasing (Wikramanayake et al. 1998; Sanderson et al. 2010). Therefore, sustaining demographic and genetic linkages among tiger populations is crucial (Frankham 2005; Kenney et al. 2014). Second, landscape connectivity can potentially lead to recovery of low-density tiger populations as well as colonization of newly protected habitats, both key requirements for increasing overall tiger numbers (Walston et al. 2010; Joshi et al. 2013). Third, the demographic stage of transience or dispersal, involving search for new territories, presents high risk for individual tigers (Smith 1993). Through long-term photographic capture-recapture studies (see Chaps. 9 and 10), Karanth et al. (2006) estimated that approximately 20% of individual tigers >1 year of age are lost from even healthy tiger populations annually through emigration and mortality, suggesting naturally high dispersal rates in this territorial species. Such high emigration rates may reduce population persistence as tiger 'source sites' (Walston et al. 2010) become surrounded by a hostile matrix, increasing dispersal-related mortality rates. Fourth, conservation programmes need to encompass the issue of connectivity, keeping in view other conservation needs, such as human-tiger conflict mitigation (Treves and Karanth 2003; Malviya and Ramesh 2015; Goswami and Vasudev 2017). Last, human impacts originating from the surrounding matrix will impact tiger populations within any protected area, as the species is known to be sensitive to human presence and activities (Chundawat et al. 2016). Taken together, it is clear that effectively addressing landscape connectivity and matrix heterogeneity issues will enhance the efficacy of tiger conservation programmes.

13.2.2 Landscape-Scale Conservation of Tigers

Over time, tiger conservation focus has shifted from large Tiger Conservation Units, consisting of a single habitat, putatively insular fragments (Wikramanayake et al. 1998), to a more realistic scenario of Tiger Conservation Landscapes (TCLs; Sanderson et al. 2010). Sanderson et al. (2010) describe 76 TCLs situated across the species' range, of varying degrees of conservation priority and viability. However, as initially described, even these TCLs did not sufficiently emphasize the heterogeneity and conservation utility of the matrix, which is central to the more recent field of landscape-scale conservation. Furthermore, there is a need for TCLs to be demarcated based on 'critical scales', relevant to tiger conservation (see Fletcher et al. (2013) for an assessment of critical scales for endangered species). Genetic studies (Sharma et al. 2013a; Ramesh et al. 2016), as well as observation of dispersal events from camera trap data (K. U. Karanth, unpublished data; Jhala et al. (2015)), demonstrate higher levels of connectivity among populations than originally visualized, indicating that some TCLs may need to be combined into larger conservation landscapes. Finally, while the need for connectivity has been emphasized (Sanderson et al. 2010), future conservation plans also need to balance landscape connectivity conservation with the protection of source populations (Walston et al. 2010).

With novel methods available to model, assess and conserve functional connectivity, in addition to the earlier purely structural approaches (such as the focus on forested corridors in Sharma et al. 2013a and in the practice of tiger connectivity conservation), it is feasible to incorporate species' dispersal biology into connectivity conservation (Vasudev et al. 2015). Tiger connectivity conservation practices largely rely on structural connectivity, stressing contiguity of habitat or the preservation of forested corridors. However, recently movement ecology-based assessments of tiger dispersal among populations have emerged (e.g. Kanagaraj et al. 2013; Ramesh et al. 2016).

In this chapter, we focus on newer methods that can potentially achieve more informed connectivity conservation for tigers. We first contextualize tiger connectivity conservation, within broader theoretical developments. We review knowledge on tiger movement behaviour and describe quantitative methods for making inferences on connectivity based on this knowledge. We conclude with a discussion on current tiger monitoring practices related to connectivity issues and suggest future directions.

13.2.3 Tiger Connectivity: Concepts and Theory

The term *connectivity* has been used variously in the past (Lindenmayer and Fischer 2007). There has particularly been much confusion on how metrics of landscape connectivity patterns translate to ecological processes (Lindenmayer and Fischer 2007; Zeller et al. 2012). Early definitions of connectivity largely stemmed from descriptions of habitat contiguity in the landscape (*structural connectivity*; Turner and Gardner 2001). Taylor et al. (1993) defined landscape connectivity as the 'degree to which a landscape facilitates or obstructs animal movement', implicitly acknowledging that connectivity is realized through an interaction of individual dispersers with landscape elements, relating to *functional connectivity*. Even this definition remains a description of the landscape, rather than of a species-specific ecological process(es). *Population connectivity* is the movement of individuals or genes between resource patches and, as such, is a species-focused definition of connectivity. Throughout this chapter, we refer to either the actual movement of animals between patches or the potential for such movement when using the term connectivity, unless otherwise specified.

Connectivity for tigers is realized through dispersal, which is a complex of processes that includes permanent emigration, a transient phase involving search for a new habitat patch, and immigration. Gene flow (*genetic connectivity*) is then effected by dispersal followed by the successful settlement and reproduction of immigrants into a population (sometimes termed as effective dispersal). While

Table 13.2 Commonly used metrics and their relation to patterns and processes of connectivity

Metric	Definition	Relates to
Neighbourhood metrics	Euclidean distance between patches or number of neighbourhood patches within a threshold distance	Structural connectivity
Cost- or resistance-weighted distance	Euclidean distances between patches weighted by the permeability of different matrix types	Functional connectivity, potential connectivity
Transition, emigration or immigration rates	Rates of movement of individuals between populations	Population connectivity, actual connectivity
Genetic distance measures such as Fst or Dsp, or genetic clustering	Measures of historic or contemporary genetic distance or differentiation between populations, or detection of spatial genetic structure	Genetic connectivity, actual connectivity

gene flow and population connectivity have often been viewed interchangeably in practice, various factors may cause differences between the rate of movement and the extent of gene flow among populations. In Table 13.2, we relate frequently used metrics to each of these above-mentioned formulations of connectivity.

There has been a shift away from structural connectivity and a move towards incorporating dispersal behaviour into connectivity conservation (Vasudev et al. 2015). New movement ecology paradigms show movement routes as determined by an interaction between physiological capability, navigability and motivation of the moving individual, with the external environment (Nathan et al. 2008). Vasudev et al. (2015) point out a number of factors that can limit successful dispersal of individuals, categorized as (a) spatial factors, or those that limit dispersal by virtue of their spatial location in the landscape; (b) environmental factors, including biotic (e.g. food resources) and abiotic limits (e.g. terrain); and (c) intrinsic species-, population-, group- or individual-specific factors that impede animal movement. There is now evidence from natural populations that various factors ranging from road-induced mortality (Kramer-Schadt et al. 2004) to configuration of the matrix (Revilla and Wiegand 2008; Ramesh et al. 2016) to species' social structure (Riley et al. 2006) impact landscape connectivity of large carnivores. At the same time, there is a rising concern that expert opinion, often used in connectivity models and conservation, may not adequately reflect how animals traverse heterogeneous dynamic landscapes (Zeller et al. 2012). Similarly, remotely sensed covariates (Sawyer et al. 2011) or indices of habitat suitability (Vasudev and Fletcher 2015) used as proxies for matrix permeability also may not accurately reflect ways in which animals traverse landscapes. It is therefore imperative for conservation to be firmly based on scientific knowledge of tiger movements in the matrix and explicitly incorporate empirical monitoring data relevant to such movement.

Both metapopulation theory and source-sink dynamics have been applied in the past to issues of large carnivore landscape conservation. While a metapopulation in its simplest form is a set of interrelated populations, many of the theoretical and methodological advancements over the last few decades have focused on turnover dynamics (Hanski 1998). Further, Hanski (1998) distinguishes a metapopulation approach as one that simplifies the landscape into a binary habitat/non-habitat mosaic, whereas landscape ecology delves more into matrix heterogeneity. In more recent years, network theory, which focuses on the study of discrete, linked objects (here, habitat patches or populations), has been applied to the study of species connectivity (Minor and Urban 2007); many of the metrics used in both network and metapopulation theories are based on the same fundamental principles and relate to the same ecological processes. Network theory in itself continues to focus on a binary habitat/non-habitat dichotomy, but the linkages between populations can be configured to incorporate matrix heterogeneity using models that simulate animal movement (e.g. circuit models, McRae et al. 2008; individual-based models, Revilla and Wiegand 2008). Network theory can also incorporate characteristics of habitat patches, as well as movement metrics between patches. In doing so, it provides a sound basis for spatial conservation planning in large, heterogeneous landscapes.

Defining linkages, or transition matrices, between habitats in a network addresses the fundamental question of how one quantifies connectivity. Connectivity can be quantified as movement rates between populations or habitat patches, typically estimated in measurements of actual connectivity. This quantity can then be decomposed into the probability of an individual making a decision to emigrate based on causal factors intrinsic to populations (e.g. density-dependent dispersal), and a conditional probability of successful dispersal, impacted almost wholly by factors emerging from the matrix (Vasudev et al. 2015). Depending on the model/metric, potential dispersal may relate to either the former or the latter. Here again, we highlight that 'successful dispersal' in the demographic context, or movement of an individual into a fragment, is different from a genetic context, implying movement followed by successful reproduction, and the two contexts cannot be used interchangeably. Taken together, we envisage tiger conservation landscapes described by population-specific characteristics (nodes in a network), as well as a transition matrix parameterized by the probability of successful dispersal between population pairs (linkages).

13.3 Tiger Dispersal Biology: What We Know and Knowledge Gaps

13.3.1 Tiger Dispersal Kernels

A fundamental piece of information on dispersal is the distance that a species naturally moves while searching for new territories, either as point measures or dispersal kernels, typically characterized via negative exponential, half-normal or Weibull distributions.

Radio tracking provides direct information on tiger movement routes, which however may be biased due to very small sample sizes and the inability to sample all possible destinations at increasing distances from origins. Smith (1993) reported a mean dispersal distance of 33 km for adult males (N = 10 individuals) and 10 km for females (N = 4 individuals) in Chitwan, Nepal, which may be an underestimate because of limitations of VHF telemetry employed at the time. Multiple genetic studies have demonstrated lack of genetic structure in fragmented tiger populations, indicating (at least) occasional past and recent movement (and post-dispersal reproduction) of tigers across large distances (e.g. Reddy et al. 2012; Joshi et al. 2013; Sharma et al. 2013b). Intensive and long-term photographic capture of tigers in the Western Ghats has provided evidence of tigers dispersing 200–300 km (Karanth KU, unpublished data). A radio-tracking study in Central India of wild-caught tigers released into a new landscape recorded dispersal events from 20 km to 375 km (Ramesh et al. 2016). Joshi et al. (2013), through genetic assignment tests, suggest that tigers may move as far as 650 km, but definitely half that distance, i.e. 375 km, across human-modified lands in Central India. In general, available evidence suggests that tigers currently can occasionally move approximately 375 km through human-modified lands.

13.3.2 Movement Through the Matrix

Despite being a species much studied, knowledge on how tigers respond to landscape features, how they navigate the matrix or how they choose and settle into new territories is limited. Joshi et al. (2013) show movement from high-density to low-density populations, reinforcing that a search for vacant territories drives dispersal in the species. Many connectivity studies have used either expert opinion or tiger habitat suitability as a proxy for matrix permeability to movement (e.g. Carroll and Miquelle 2006; Kanagaraj et al. 2013), an assumption that may not always be valid (see Vasudev and Fletcher 2015). Using data on movement paths of radio-collared tigers, Ramesh et al. (2016) found that open agricultural matrix types were avoided by dispersing tigers compared to forested areas with complex terrain.

Evidence from landscape genetic studies suggests that the maintenance of vegetation cover, or the presence of tenuous habitat corridors, facilitates tiger movement (Joshi et al. 2013; Sharma et al. 2013a), either through the presence of wild prey or cover from human persecution. Karanth et al. (2004, 2011) show that tiger presence and abundance are predominantly driven by abundance of ungulate prey (Chap. 2). Further, that tigers perceive risk in human presence is supported by their typical avoidance of settlements and villages while traversing the landscape (Carroll and Miquelle 2006; Joshi et al. 2013; Ramesh et al. 2016). Joshi et al. (2013) also inferred that roads were detrimental to landscape connectivity through landscape genetic approaches. It is unknown what the mortality rates for tigers are in the matrix, though evidence from some landscapes suggests that these rates may be very high (Goodrich et al. 2008).

While the maintenance of forested corridors remains the leading strategy to connect tiger populations, simulation models (Kanagaraj et al. 2013) suggest that narrow corridors may inhibit tigers from locating movement routes (similar to 'target effects' that lead to decreased immigration into small islands; MacArthur and Wilson 1967). Kanagaraj et al. (2013) also demonstrate that connectivity is sensitive to animal movement parameters, showing (a) that quantifying parameters such as dispersal distance and matrix permeability is vital to modelling and predicting tiger connectivity and (b) that small changes in land-cover or behavioural adaptation of tigers to fragmented landscapes can have major impacts on connectivity.

13.4 Methods for Monitoring Landscape Connectivity

13.4.1 What Are We Quantifying?

Connectivity emerges from a complex of processes, initiating with animals making decisions to disperse from source populations and ending at either settlement in destination habitats or successful reproduction, depending on the context of the study. Actual events of dispersal can be detected through genetic analyses, tracking marked animals across populations, or colonization events. These events typically record *actual* or *realized connectivity.* On the other hand, there also exist models for *potential connectivity*, which only deals with the probability of an animal successfully completing a dispersal event (Table 13.1).

Realized connectivity depends on two sets of factors: one, factors arising from within a population/habitat that incite a decision to disperse, often population-specific characteristics such as population density, and second, those that arise from characteristics *outside* of the source fragment. Measures of actual connectivity typically integrate the decision of animals to disperse with the probability of successful dispersal. Metapopulation models, for instance, take into account population characteristics such as patch area in measures of connectivity. Similarly occupancy models incorporate the state of 'occupancy' of a fragment while assessing connectivity. Cost- or resistance-based potential connectivity models, on the other hand, typically differ from this treatment in focussing on characteristics of the matrix, and thus modelling the probability of successful dispersal, conditional on a decision taken to disperse.

Dispersal in itself is a complex of processes, typically divided into emigration, a transient search phase and immigration. In general, models can either focus on predominantly the emigration/immigration phases, i.e. on identifying the source and destination fragments. Information on the source and destination phases can then be used to indirectly infer on the transient search phase. Sampling for these models is focussed on animal populations or habitat fragments. Alternatively, models can focus on the transient phase of dispersal, using empirical data to infer on how animals make movement decisions while in the matrix. Often, the former source/destination models estimate actual connectivity, while the latter deal with potential connectivity.

While ideally one would want information on all three phases of dispersal, with the power to make inferences on factors impacting different stages of dispersal, as well as on the impacts of dispersal (or the lack thereof) on species conservation status, there is probably no large mammal species worldwide for which such data exist. Being a growing field, there are a number of methods for estimating various parameters of interest to connectivity, with little agreement on an ideal approach. We present multiple methods below that are all applicable for the study of tiger connectivity in fragmented landscapes. First, we describe a method, radiotelemetry, to record entire dispersal events of tigers at fine temporal and spatial scales; while this method provides a whole suite of information on connectivity, it suffers from small sample sizes and typically provides inferences at the level of individuals, rather than populations. We then describe source/destination models for assessing actual connectivity, including extensions of the capture-recapture models described in Chap. 5, occupancy models described in Chap. 4 and landscape genetic approaches. We note here that occupancy models and landscape genetics contrast in that the former deal with demographic connectivity, while the latter are based on genetic connectivity. Following this, we describe a suite of approaches for modelling potential connectivity, including a discussion on parameterizing potential connectivity models with movement-relevant data. We extensively discuss the appropriate use of remotely sensed data, widely used for studies conducted at the scale of large landscapes. We conclude with a discussion on spatial conservation prioritization approaches. Each of these approaches provides a different flavour of data, all relevant to complete the picture of tiger connectivity.

13.4.2 Observing Dispersal Events Through Radiotelemetry

Tigers disperse from natal areas mostly around the age of 2 years, and dispersal characteristics vary between sexes and different age classes depending on the demographic structure of the population and configuration of the landscape. Empirical data on individual responses to habitat features at multiple scales are useful to draw robust inferences on dispersal patterns and processes. Direct observations on tiger dispersal are near impossible due to their elusive nature and the large spatial scale of movement. Radiotelemetry serves as a useful tool for fine-scale and systematic observation of tiger movement and provides insights into tiger movement behaviour required for modelling dispersal and connectivity.

While telemetry provides valuable fine-scale movement data, it has some limitations. (a) Traditional ground tracking involving Very High-Frequency (VHF) telemetry requires close proximity to radio-collared animals for accurate data collection. Animal positions are then obtained by acquiring VHF transmissions from radio-collars using a hand-held antenna from three or more sampling points to triangulate the location of the animal. This method is very prone to position errors as tigers can rarely be approached at close proximities. New technologies in the form of Geographic Positioning System (GPS)-based collars and Global System for Mobile Communication (GSM) networks provide much more power for tracking animal movement in inaccessible areas and across large spatial scales. With these technologies, data on the location of the collared animal can be obtained at regular predetermined intervals through satellites with reasonably high accuracy. These collars also have automated drop-off options such that the collar can be retrieved without recapturing the focal animal, after a predetermined period of time. (b) Radiotelemetry studies invariably suffer from small sample size, as telemetry data typically come from few individuals (Patterson et al. 2008). (c) Further, the fine-scale movement data obtained from radiotelemetry (and other methods of recording animal movement) are subject to nonindependence as the location of animals at any time step is dependent on its location at previous time steps, though these are addressable through the use of Markov models (Langrock et al. 2012).

Due to small sample sizes typically associated with telemetry data, it is recommended that the following issues be carefully taken into consideration. First, it is important to carefully select the individual tiger to be collared, based on study objectives. If the objective of the study is to observe dispersal movements of transient tigers in search of new territories, it is ideal to radio-collar an individual of 12–18 months of age, within its natal area, that is likely to disperse. Second, the time interval between fixes of the animal's position should be determined; this is based on an assessment of animal activity and storage capacity of the device. Third, collar-retrieving strategies should be decided beforehand. For tigers, as they are shy of human presence, drop-off collars are recommended (unless the objective is to replace with a new collar); the time frame over which the animal is to be tracked is determined beforehand but can be modified subsequent to collaring. Lastly, it is important to regularly monitor the animal using GPS and/or VHF signals. It is crucial to have field teams to be able to go to the last recorded location at short notice in case satellite signals are lost, and immediately locate the animal using VHF signals; the reasons for loss of signals—collar damage or malfunctioning—can then be determined and corrected.

Radio tracking provides direct information on animal dispersal routes. In combination with data on potential covariates that may influence tiger movement, it is possible to obtain insights into tiger responses to landscape elements (Ramesh et al. 2016). Dispersal routes are typically analysed by breaking up observed paths into 'steps' at specified temporal resolution. Path analysis includes quantifying step lengths and turning angles of the dispersing animal and how these parameters change with varying matrix characteristics (Morales et al. 2004). Conditional logistic models have been used to assess why tigers (and other animals) choose certain movement paths, in comparison to a set of available (randomly chosen) alternatives (e.g. Ramesh et al. 2016). More recently, analytical models have been developed that can handle the spatial autocorrelation that is inherent in movement data using Markov processes (Langrock et al. 2012). These models can additionally take the 'state' of the animal into account, differentiating between temporal phases showing different animal movement behaviours, such as foraging/resting/directed movement phases.

The scale at which animals make movement decisions depends on multiple factors, such as perceptual window of the species and the spatial scale, or distance, at which it responds to perceived threats or attractors in the landscape. Ramesh et al. (2016) show that tigers tend to respond to different variables at different spatial scales; they responded at much finer scales to human presence and local topology and at coarser scales to the presence of (open) forests. This highlights two important points: (a) spatial scale is an important consideration while recording, modelling and predicting connectivity, and (b) connectivity conservation, apart from being spatially informed, needs to incorporate information on the scale at which dispersal is impacted by threats.

Radiotelemetry also provides crucial individual-level information that can provide insights into variation among dispersers (Ramesh et al. 2016); such variation is expected to have significant impacts on connectivity patterns across landscapes (Vasudev et al. 2015). These advantages notwithstanding, it must be noted that radiotelemetry typically provides individual-level information at relatively small sample sizes, to the scale of a few individuals. It is only for a few large cat species that a relatively larger number of individuals have been radio-collared for inference at population or landscape scales (e.g. Florida panther, Maehr et al. 2002, $n = 27$; Iberian lynx, Revilla and Wiegand 2008, $n = 30$).

13.4.3 Marked Animals: Photographic Capture of Dispersal Events

Non-invasive captures and recaptures of tigers (and other animals) from photographs (Chaps. 9 and 10) or DNA (Chap. 11) offer the advantages of obtaining larger sample sizes of individuals and conducting population-level studies, in contrast to telemetry studies that provide more fine-grained data on a few individuals. Capture-recapture studies conducted simultaneously at multiple locations provide information about animal movement. If an identified animal is 'caught' at time t in one location and then again at time $t + 1$ at another location, we know that the animal survived the interval and that it moved. However, the issue of non-detection will cause us to miss (not detect) some movements, complicating inferences about rates and probabilities of movement (Williams et al. 2002:335–336). These complications led to the development of multistate capture-recapture models, which are extensions of the open-population capture-recapture models (Chap. 9).

Multistate capture-recapture models were developed for open populations by Arnason (1972) but saw very little use. They were 'rediscovered' in the late 1980s (Hestbeck et al. 1991), were developed for general use (Schwarz et al. 1993) and have become an important tool in population ecology (Lebreton et al. 2009). Initial uses of multistate models (Arnason 1972; Hestbeck et al. 1991) were for the problems of interest in this chapter, movements of individuals among different locations. It was quickly recognized that the concept of 'state' of an animal at capture was much more general than just spatial location (Nichols et al. 1992), but we restrict multistate models in this chapter to inferences about movement from one location to another.

The sampling situation for which multistate models were developed entails sampling at multiple locations for each of K sampling occasions. Data can still be summarized as capture histories (or detection histories), but multistate models require additional information. Rather than each history being a vector of 1s (indicating detection) and 0s (indicating non-detection), detections must now provide the extra information that the animal was not only detected but was detected in a specific location. For a two-location study, for example, we might find the following capture history:

$$1\ 0\ 2\ 2$$

This history indicates that the animal was initially captured at location 1. It was not captured at occasion 2, but it was captured at location 2 at occasions 3 and 4.

Modelling of these more complicated capture histories requires some new notation, and we denote location with superscripts. We can simply extend the open CJS models of Chap. 5 by adding location superscripts to the capture and survival parameters:

p_t^r = probability that an animal alive in location r and occasion t is captured/detected then;

ϕ_t^{rs} = probability that an animal alive in location r at occasion t is alive in location s at occasion $t + 1$.

With these parameters, we can model multistate encounter history data.

Consider the probability associated with history 1 0 2 for a study of 3 sampling periods and 2 locations:

$$\Pr(1\ 0\ 2 \mid \text{release at location 1 in occasion 1}) = \left[\phi_1^{11}\left(1 - p_2^1\right)\phi_2^{12} + \phi_1^{12}\left(1 - p_2^2\right)\phi_2^{22}\right]p_3^2$$

The uncertainty associated with the location of the animal at occasion 2 led to the additive terms in brackets. One possible sequence of events, denoted by the first additive term, is that the animal stayed in location 1 at occasion 2, was not detected at that time and then moved to location 2 between occasions 2 and 3. The second possibility (the second additive term) is that the animal moved to location 2 between occasions 1 and 2, was not detected there at occasion 2, and then stayed in location 2 at occasion 3. There is no uncertainty associated with the location of the animal at time 3; hence the detection parameter p_3^2.

The parameters φ_t^{rs} incorporate both survival and movement, and in some cases it is possible and useful to separate these two processes. If survival for an interval is determined primarily by the location of an animal at the beginning of the interval, then φ_t^{rs} can be rewritten as (Hestbeck et al. 1991):

$$\phi_t^{rs} = S_t^r \psi_t^{rs},$$

where

S_t^r = probability that an animal alive at location r in occasion t survives and remains in study system at occasion $t+1$.

ψ_t^{rs} = probability that an animal is in location s at occasion $t+1$, given that it was in location r at occasion t and survived until occasion $t+1$.

The above decomposition of ϕ_t^{rs} into survival and movement components is much more commonly used than the ϕ_t^{rs} parameterization. Both parameterizations incorporate movement between locations, provide estimable quantifications of the linkages between populations as described in Sect. 2.3 of this chapter and are thus relevant to connectivity.

Multistate models have not seen much use with data from tigers or prey species, although such work has begun. For example, Karanth and others have planned a multistate analysis to draw inferences about movement of tigers between Nagarahole and Bandipur National Parks. These models have been used for other species to test hypotheses about the relationship between movement probabilities and distance between two locations (e.g. Skvarla et al. 2004), intervening matrix characteristics (Coffman et al. 2001; Skvarla et al. 2004) and corridors (Coffman et al. 2001). All of these questions are relevant to the concept of connectivity and should prove useful in future studies of tigers and perhaps prey.

The two movement parameters defined above, ϕ_t^{rs} and ψ_t^{rs}, refer to the probability of an animal in one location at sampling occasion t moving and being at a different location at sampling occasion $t+1$. It is also possible to estimate other parameters associated with connectivity using both single state and multistate data. For example, it is sometimes of interest to be able to decompose recruitment of new adults in a sampled population at some occasion $t+1$ into two components, surviving adults and young that were members of the sampled population at occasion t and immigrants that moved into the sampled population from some other location. This is possible using single state models for open populations (Nichols and Pollock 1990).

Additional questions related to connectivity include the following: what is the probability that a randomly selected individual at location r at occasion $t+1$ was in location s at occasion t. The parameter defined in this manner, γ_{t+1}^{rs}, can be estimated by reversing the temporal sequence of multistate capture histories, in which case the estimates of ϕ_t^{rs} are now interpreted as γ_{t+1}^{rs} (Nichols 2016). For example, consider reverse-time modelling of the capture history shown above, 1 0 2:

$$\Pr(1\ 0\ 2 \mid \text{last capture at period 3 in location 2}) = \left[\gamma_3^{22}\left(1-p_2^2\right)\gamma_2^{21} + \gamma_3^{21}\left(1-p_2^1\right)\gamma_2^{11}\right]p_1^1$$

So we condition on (use as a starting point) the capture at occasion 3 in location 2. The summed terms indicate the uncertainty about whether the animal was in location 1 or 2 during occasion 2 when it was not caught.

This reverse-time parameter, γ_{t+1}^{rs}, is especially relevant to the concepts of source and sink and, more generally, to the contributions of metapopulation components to each other and to the entire metapopulation (Runge et al. 2006). In fact, γ_{t+1}^{rs} directly estimates the contribution of subpopulation s to the population growth rate of subpopulation r, defined as $\lambda_t^r = N_{t+1}^r / N_t^r$, where λ_t^r is growth rate of subpopulation r and N_t^r is abundance of subpopulation r at occasion t. We believe that direct inferences about contributions of different locations to an overall system of locations have the potential to be very useful in a conservation setting (Runge et al. 2006).

13.4.4 Occupancy Dynamics and Movement

Surveys of animal presence and absence at multiple locations have long been used to draw inferences about animal distribution patterns. Because animals that are present at a location are not always detected via surveys, occupancy models were developed to incorporate the thinking of Chap. 3, and associated detection issues, into inferences about species distribution (Chap. 4, also MacKenzie et al. 2006). As described in Chap. 4, dynamic (multi-season) occupancy models include two vital rate parameters, colonization and extinction. Colonization γ_t reflects the probability that a site not occupied by a focal species at sampling occasion t is occupied at sampling occasion $t + 1$. This colonization parameter thus describes movement into a site from somewhere outside of that site. However, unlike the capture-recapture approaches described above, the specific origin of colonizing individuals is not specified in dynamic occupancy models. Extinction ε_t is the probability that a site that is occupied at occasion t is not occupied at occasion $t + 1$. The complement of extinction, $1 - \varepsilon_t$, may thus incorporate a so-called rescue effect (Brown and Kodric-Brown 1977) in which a site may go locally extinct and then become recolonized before the next sampling occasion.

Inferences about movement based on dynamic occupancy modelling are often based on site-specific covariates that affect rates of colonization. Dynamic occupancy models have not seen use for tigers and prey, to our knowledge, although we expect this to change, as the second occupancy survey of the Malenad Tiger Landscape in Karnataka, India has just been completed (see Karanth et al. (2011) for description of the first survey of this landscape). However, examples from other species and systems provide an indication of what sorts of inferences are possible with these models. For example, we frequently expect that probabilities of a site being colonized will be influenced by such site-specific covariates as distance to a potential source population and nature of the intervening habitat matrix.

Specialized autologistic occupancy models include not only site-specific covariates that may be related to movement but also potential effects of occupancy of neighbouring sites (Yackulic et al. 2012; Eaton et al. 2014). For example, it would be reasonable to expect a higher probability of colonization, and a lower probability of extinction (rescue effect), for a focal site surrounded by occupied neighbouring sites than one surrounded by unoccupied neighbours. Although no marked individual animals are involved, the expectation is based on the likelihood of movement into the focal site from neighbouring sites. This hypothesis can be tested by defining a 'neighbourhood' (sites adjacent to, or within some radius of, the focal site) and then assessing occupancy at these sites. Focal site occupancy (for single season models) or rates of extinction and/or colonization (for multi-season models) can then be written as a function of neighbourhood occupancy, for example, as

$$\text{logit}\,(\theta_{i,t}) = \beta_0 + \beta_1 X + \beta_2 \overline{\psi}_{i,t}^{N_i}$$

where:

$\theta_{i,t}$ is the modelled parameter (occupancy, extinction, colonization) of focal site i at occasion t.
β_j are model parameters describing relationships.
X is an environmental covariate.
$\overline{\psi}_{i,t}^{N_i}$ is the average occupancy of neighbours of focal site i.

Such autologistic models have been used with a priori specified neighbourhoods (e.g. Yackulic et al. 2012). Another approach is to specify neighbourhoods of differing sizes (different distances from focal site) and allow model selection to help determine the appropriate neighbourhood size (Eaton et al. 2014). These models are especially useful for systems experiencing non-equilibrium occupancy dynamics (Yackulic et al. 2012) and can be useful for determining the scale of movement in landscapes.

In summary, despite the absence of marked individual animals and documented movements, occupancy modelling can be used to draw indirect inferences about movement. These inferences are directly related to the concept of connectivity and should be useful for surveys that cover large spatial scales.

13.4.5 Landscape Genetics

Tigers have a very broad geographic distribution (Chaps. 1 and 2). It is pertinent to ask whether tigers in different protected areas and landscapes are genetically different. Population genetics is the theoretical framework that allows us to quantify these differences. The faecal DNA approach allows us to identify individual tigers non-invasively (see Chap. 11), and also to identify how gene frequencies change across tiger populations.

The most important population genetic process in relation to connectivity is gene flow, or movement of genes between populations. Population genetics can help us quantify whether two populations of tigers are similar or different. When gene flow is high, populations will be similar, but when gene flow is low and they have been separated for some time, they are likely to be more different. This similarity or

Plate 13.1 (**a**) As connectivity decreases, genetic dissimilarity between patches increases. (**b**) This may be affected by distance between patches, or the intervening landscape

difference is reflected in a statistic called F_{st} (Fig. 13.1a; see Slatkin (1987) for a review on population structure and F_{st}).

Landscape genetics integrates these concepts of *population genetics* with those in *landscape ecology*. As discussed earlier, population genetics can help us quantify the effects of movement on gene pools, as shown in Fig. 13.1a. But a closer look at this illustration shows that we are ignoring what lies between the two populations.

Imagine if the two populations were equidistant to one another, but what lies between these populations is different (Fig. 13.1b). In this case, if you collected genetic data from two sets of populations that were the same distance apart, but one had more genetic connectivity than the other, you could infer that the first landscape was conducive to movement, while the second was not. In the illustrative example shown in Fig. 13.1b, forest cover enhances connectivity while urban settlements hamper it. This suggests that populations are not simply isolated by distance (as in Slatkin 1987), but in addition by *resistance* offered by the intervening matrix (detailed further in Sect. 4.6 of this chapter). The use of a graph-theoretic approach to model movement across a landscape of varying *resistance* is detailed in McRae et al. (2008), and these approaches are implemented in software programs such as Circuitscape. In these models, genetic samples are collected from different locations in the landscape (for details on collection and processing of genetic samples of tigers, see Chap. 11), and connectivity among these different locations is modelled in a manner akin to the flow of current along circuits (more details provided in Sect. 4.6), the underlying assumption being that more gene flow would lead to more similar genetic composition among samples (e.g. Joshi et al. 2013).

While the concept of resistance to movement (as opposed to linear distance) is easy to understand, parameterizing such resistance poses a larger challenge. Landscape genetics provides a valuable framework to parameterize resistance surfaces, using information on genetic dissimilarity across space. The first step to estimating a resistance surface in a landscape would be to list a set of environmental variables that may resist or impede movement. These could include land-use types, elevation or climatic variables. Testing which of these environmental variables drive genetic dissimilarity across sites then provides inputs on their contribution to resistance. Essentially, the genetic data can be used to test the validity of potential models of resistance surfaces; note that this would provide information specifically on genetic connectivity. The protocols and toolkits for developing landscape genetic approaches are outlined in several reviews including Zeller et al. (2012) and Hall and Beissinger (2014). Several other papers review landscape genetics in general (e.g. Manel and Holderegger 2013). These estimated resistances can then be used to predict potential dispersal across landscapes through various methods detailed in Sect. 4.6.

Current models that assess the actual (realized) rates of movement or connectivity among populations using genetics are fundamentally based on two concepts, both emerging from the genetic (dis)similarity among collected samples. One, they may be based on partitioning samples (or individuals) into clusters showing genetic similarity. These models are able to further identify the most appropriate number of clusters in a landscape (Pritchard et al. 2000) and thus highlight isolated populations that have had no recent immigration. Clustering of genetic samples of tigers from across India leads us to believe that the westernmost population in Ranthambhore forms a separate cluster, or is isolated and genetically differentiated from populations in the rest of the country (U Ramakrishnan, unpublished data). Second, one may assign genetic samples from specific individuals to reference populations, which are defined based on the entire set of samples obtained in a landscape.

Dispersers, or their progeny, can then be identified when the population of origin, the population to which the sample is assigned, differs from the population of sampling, the population from which the sample was physically collected (Piry et al. 2004). Using assignment tests, Joshi et al. (2013) identified dispersers in a Central Indian landscape and an example of a long-distance dispersal (approximately 650 km).

13.4.6 Modelling Potential Population Connectivity

Modelling potential connectivity is useful as (a) inferences about actual connectivity are based on infrequent events that are difficult to detect and, thus, may be subject to substantial uncertainty; (b) modelling potential connectivity allows us to isolate matrix effects on connectivity from those arising within populations, whereas this distinction is less tangible in measures of actual or realized connectivity (though this distinction can also be made through modelling population-related effects; e.g. Nichols and Kendall 1995); and (c) predicting potential connectivity within a scenario-testing framework can be useful for prioritizing conservation interventions situated in the matrix.

Animals make decisions on where to move in the landscape based on various factors. The permeability of different land-cover types to animal movement has been quantified through either the *cost* of movement (Adriaensen et al. 2003) or the *resistance* the matrix offers for animal movement (Mcrae et al. 2008). Cost-distance modelling is based on the cumulative cost of traversing through a particular movement route (Adriaensen et al. 2003). These analyses are typically conducted within a Geographic Information Systems (GIS) platform, whereby each pixel (or location) in an image is associated with a 'cost' to movement. There are many possible movement routes between two habitat fragments. The total cost of moving along each of these routes (or the dispersal cost) is simply the added cost of each pixel located along the route. The movement route between a source and destination habitat patch that minimizes the cumulative cost of dispersal is then chosen as the 'least-cost path'. A threshold dispersal cost is sometimes used to define isolation, whereby it is assumed that an animal cannot move along a movement path that imposes a dispersal cost greater than this threshold.

Least-cost modelling has been very popular as an approach that (a) takes into account matrix heterogeneity, (b) is spatially explicit, and (c) is based on the cumulative nature of dispersal costs. However, least-cost modelling makes a number of assumptions, which are not likely to hold for tigers and most endangered species. (1) The model assumes that the disperser has complete knowledge of the landscape. In reality, most animals perceive their environment in their immediate neighbourhood, determined by the animals' navigation capabilities, or perceptual range, as well as the environment, and this shapes movement routes (Nathan et al. 2008). (2) Least-cost modelling assumes that the disperser has chosen its destination, immediately upon emigration, an assumption that is unlikely to be true. Animals, in fact, use a number of cues, including the presence of conspecifics, to determine suitable destination fragments to immigrate into. (3) It assumes that animals choose movement paths optimizing the cost of movement, while there is evidence that this need not be true. (4) Least-cost modelling focuses on a single optimal path, whereas path redundancy can increase overall landscape connectivity and resilience to land-cover change. While path redundancy has been partly accounted for through selection of multiple paths with low cumulative dispersal costs, it is still not explicitly quantified in cost-distance modelling, that is, two habitat patches joined by two movement paths are not quantified to be more connected than when joined by one movement path, through the cost-distance approach. Despite these issues, cost-distance modelling continues to be used for connectivity modelling of tigers (e.g. Carroll and Miquelle 2006; Areendran et al. 2012; Rathore et al. 2012) and other endangered species (e.g. Cushman et al. 2016).

Some of the restrictive assumptions mentioned above are relaxed in another set of models, derived from electrical circuit theory (McRae et al. 2008). These models simulate animal movement as current flowing along a circuit map of 'nodes', while the resistance between the two nodes provides information on local matrix permeability. The current flowing between nodes depends on the resistance, which mirrors the assumption that animals are more likely to move along paths of lower resistance. Modelled in such a manner, the simulated dispersal is more akin to a random walk and hence more formally based in movement ecology. These models then quantify cumulative 'resistance' along all potential dispersal routes between source and destination habitat patches. Unlike cost-distance models, path redundancy—shown to be important for connectivity—*is* accounted for in circuit models such that connectivity between two patches is quantified to be greater when there are multiple movement paths between them. Results of the model include 'resistance distances' separating fragments or populations (a quantification of linkages among fragments), and a current map that shows how much 'current' (or movement) flows through each location in the landscape (McRae et al. 2008). This allows us to identify 'pinch points' or locations that are critical for maintaining connectivity across the landscape (McRae et al. 2008). Validation of least-cost path models and circuit theory models, using empirical data on animal movement, or estimated actual connectivity, has shown circuit-based models to generally be better able to predict animal movement (McRae et al. 2008; McClure et al. 2016).

For both cost-distance and resistance-distance modelling, a key step is quantifying 'costs' or 'resistances' of different matrix types. Rayfield et al. (2010) show through simulation models that cost-distance models are highly sensitive to the assigned cost values. Zeller et al. (2012) reviewed connectivity models and found that most studies used expert opinion to assign resistance values to matrix types, a practice also undertaken for tiger connectivity modelling (e.g. Areendran et al. 2012; Rathore et al. 2012). Expert opinion, however, may be misleading, especially when it relates to a process such as dispersal, which is infrequent, occurs at large scales and is often hard to detect. A number of connectivity studies also use habitat suitability as a proxy for species connectivity requirements, which may provide very misleading results in heterogeneous landscapes (Zeller et al. 2012; Vasudev and Fletcher 2015; Abrahms et al. 2016). Quantifying resistance values of the matrix is thus a vital step for tiger connectivity models. This can be done using observed tiger movement behaviour, data from radiotelemetry or calibrations of matrix resistance derived from estimates of actual connectivity (i.e. from genetic assignment tests, multi-state capture-recapture models or dynamic occupancy models, e.g. Ramesh et al. 2016).

Spatially explicit individual-based models (IBMs) can be used to model animal connectivity. These models follow simulated individuals that make decisions based on predetermined rules, and track population- or landscape-scale consequences of the behaviour of these individuals (DeAngelis and Gross 1992). IBMs are powerful and flexible and can incorporate current theoretical advances in movement ecology (e.g. Revilla and Wiegand 2008). However, IBMs are data hungry and often sensitive to model parameters that sometimes are difficult to reliably estimate. Kanagaraj et al. (2013) assess connectivity across tiger populations in the Terai Arc Landscape in India and Nepal, using a spatially explicit IBM. They use habitat suitability as a proxy for matrix resistance and a range of plausible tiger movement parameters. Their simulations indicated that fragments in this landscape are currently isolated but pointed to potential for restoring connectivity (Kanagaraj et al. 2013). Carroll and Miquelle (2006) use an IBM to model population viability of tigers in the Russian Far East, within which was included the modelling of 'floaters', or dispersing tigers. Dispersal was modelled using a random walk, and results of the model included the identification of corridors connecting patches in the landscape (Carroll and Miquelle 2006). Both the above models suffered from a paucity of movement-related information. Kanagaraj et al. (2013) in fact also found that their model was highly sensitive to the specification of movement parameters, emphasizing the importance of obtaining such information through observations of tiger movement, which would greatly enhance the utility of such simulation models.

Stochastic dynamic optimization also provides a modelling approach to develop optimal movement decisions while incorporating animal perceptive abilities, landscape characteristics, internal states and other relevant factors (Williams and Nichols 1984). Such formulations would be an extension of the least-cost modelling approach but with relaxation of many of its assumptions. In this approach, dispersal can be viewed as a series of decisions taken by an individual, based on its perceived environment, current internal state and perceived rewards. A stochastic dynamic optimization approach allows us to formulate movement decisions based on simple mathematical rules of maximizing perceived rewards while minimizing perceived costs and/or subject to certain constraints, based on the current state of the individual. The state of the individual could comprise its internal motivation (sensu Nathan et al. 2008), fatigue and current state of knowledge of the landscape. Rewards may include the presence, accessibility and quality—in terms of habitat structure, resources, or mates—of potential destination fragments, as well as resources in the matrix, depending on the state of the individual. Costs may include actual physiological costs of moving, predation risk, mortality costs associated with moving through the matrix—for instance, road mortality—or moving through unfamiliar areas and/or missed opportunity costs (Zeller et al. 2012; Vasudev et al. 2015). Optimization models can also take into account stochastic effects, and uncertainty in rewards and costs. However, such models may still be data hungry, requiring information on step-wise rewards, costs and state of dispersers; the sensitivity of models to these parameters is unknown. While these optimization approaches have been applied to animal seasonal migration (Bauer and Klaassen 2013), they have not been developed to any significant extent for dispersal and provide a valuable avenue of exploration.

13.4.7 Using Remotely Sensed Data to Depict Heterogeneous Landscapes

Remotely sensed data and derived products have largely been used to model potential connectivity in combination with approaches such as least-cost paths (Adriaensen et al. 2003) and circuit theory (McRae et al. 2008). All these approaches rely on a cost/resistance layer, which is an estimate of how landscape features or environmental parameters affect animal movement within the landscape (Zeller et al. 2012). To evaluate potential tiger connectivity, studies have used spatially explicit information on land-use, land-cover, linear intrusions, human settlements and human populations to derive cost layers (Joshi et al. 2013; Sharma et al. 2013a; Kanagaraj et al. 2013; Ramesh et al. 2016). The benefits of remotely sensed data lie in their being widely available at fine resolutions for large landscapes; on-ground observations on such large spatial scales would be practically impossible.

While there is no standard method for deriving resistance layers, the overall framework is to assign the highest resistance to landscape features that impede movement (Zeller et al. 2012). Further, multiple layers can be combined to obtain a single layer that reflects the cumulative resistance of all landscape features. One of the biggest challenges in developing resistance layers is that landscape features/patterns do not directly reflect ecological costs (e.g. travel time, mortality risk, habitat suitability, ability to cross physical barriers) which affect animal movement, and converting remotely sensed products or spatially explicit data to a meaningful resistance layer needs to be carefully thought through (Vasudev et al. 2015).

A review of 96 research papers by Zeller et al. (2012) shows that broadly five different sources of biological data and three different analytical approaches have been used to convert remotely sensed products into resistance values. While remotely sensed products are now easily available, a key consideration before using remotely sensed/GIS data is to examine the spatial resolution at which the data are available and how this relates to the way the animal perceives its environment (grain size), and any potential mismatches need to be addressed (Zeller et al. 2012). It is also important to consider the implications of classification errors, and their effects on derived connectivity maps and subsequent management/conservation decisions. The same holds true while using remotely sensed indices (e.g. Normalized Difference Vegetation Index, NDVI), where issues of saturation and non-linear relationships with ecological processes need to be carefully examined.

Irrespective of the source of information, and the analytical approach adopted to assign resistance values to landscape features/pattern, it is important that the estimated resistances are validated against independently collected information on actual movement from telemetry studies, studies on marked animals (e.g. photographic capture-recapture) or genetic data (Revilla and Wiegand 2008). However, most often such information is likely to be missing. In this case, it may be preferable to use a range of resistance values, and the resulting connectivity maps should be examined by experts who are familiar with the target species and the ecological condition of the landscape. Such an assessment would help in determining the optimal resistance values and sensitivity to resistance values and also in minimizing any uncertainty in the resulting outputs. It must be kept in mind, however, that our ability to make meaningful inference from analyses lacking key movement-related information is limited.

While remotely sensed products have been used widely to identify potential connectivity, they have also been applied in understanding dispersal using multi-state capture-recapture models (see Sect. 4.2. for details on the model). The model parameterization incorporates both survival at a given location and movement between locations (Hestbeck et al. 1991). Remotely sensed information characterizing the quality of location (e.g. metrics on compactness of tiger habitat) can be used to understand the role of landscape features on survival. Similarly, landscape metrics such as density of linear features surrounding the location, mesh size and inter-patch distances describing landscape features between two locations, determined along least-cost/current flow/random walks, can be incorporated to empirically understand which landscape features promote dispersal. The possibility of linking environmental and landscape features to understand dispersal using multi-state models needs to be further explored, and this opens a new line of ecological studies which can greatly contribute to our understanding of tiger dispersal.

13.4.8 Spatial Conservation Prioritization

Conservation planning and prioritization exercises can determine where and on what to invest limited conservation resources (Moilanen et al. 2009). Such exercises bring increased transparency and efficiency to conservation programmes and ultimately enhance the probability of successful conservation. Connectivity is inherently spatial, and hence, it is vital to include spatial components in prioritization exercises for connectivity conservation. Basically, these methods provide a way to use existing knowledge emanating from various observational and inferential approaches, to formally inform conservation decision-making (Fig. 13.2). Circuit theory can provide information on conservation bottlenecks or locations in the landscape that are vital for the maintenance of landscape connectivity (McRae et al. 2008). Individual-based models can provide information on potential corridors, or of impacts of dispersal mortality on species. For tigers, least-cost path modelling has been used to identify corridors (Carroll and Miquelle 2006; Rathore et al. 2012)

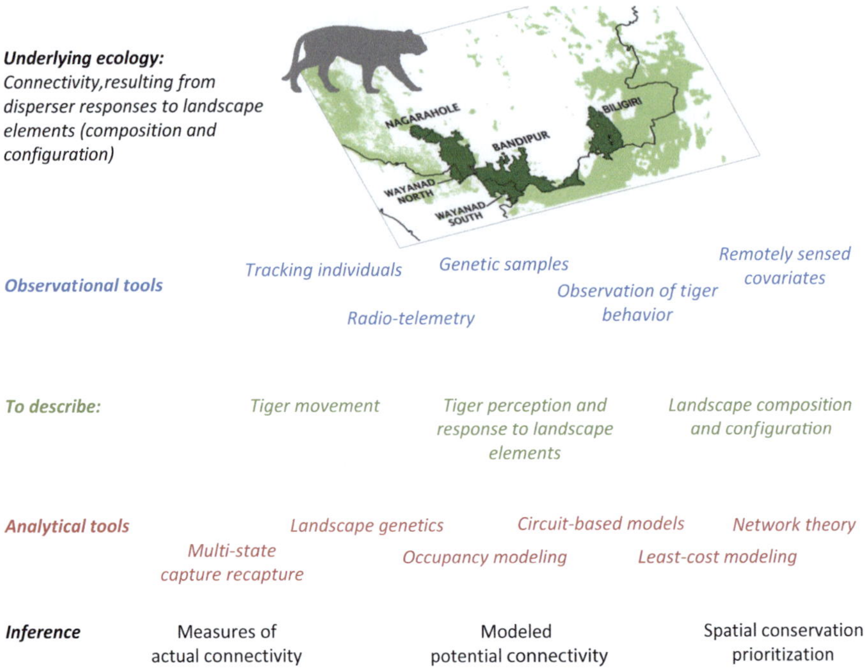

Plate 13.2 An overview of methods and approaches available for studying, predicting and making inferences on tiger connectivity. Here, for illustrative purposes, a landscape in Western Ghats is shown, but methods are generally applicable across landscapes. Map sourced from WCS India, tiger image credited to Arjun Srivathsa

and the impact of developmental projects on tiger connectivity (Areendran et al. 2012); both studies, however, suffered from a lack of data on matrix resistances. Occupancy modelling approaches have been used to identify optimal locations for road-crossing structures in highways (Rayan and Linkie 2015).

Network theory has gained much attention in the past few years due to its utility in depicting sets of interconnected populations, and its strong theoretical foundations (Minor and Urban 2007). Networks are composed of individual habitat fragments, termed ‘nodes’, interconnected by linkages, termed ‘edges’. Nodes can be characterized through their size, habitat quality, population size or any other characteristic, while linkages can be characterized by their strength or degree of flow. Networks can be symmetrical, such that the flow of individuals or genes from fragment A to B is identical to the flow from B to A, or it can be asymmetrical, where this assumption is relaxed. Quantification of linkages can be through data on actual connectivity using mark-recapture data (e.g. Fletcher et al. 2013) or through potential connectivity using circuit theory (e.g. Vasudev and Fletcher 2015). While network approaches have been hailed due to the minimal data requirements (Minor and Urban 2007), it is highly recommended that linkages between patches be quantified realistically to obtain meaningful and useful inferences.

With information on habitat fragments and nodes, there are a number of metrics that provide analyses of the importance of each node (habitat patch) or link for overall landscape connectivity (Minor and Urban 2007). Isolation metrics determine clusters of populations and fragments that are genetically and demographically unconnected to their neighbours (Minor and Urban 2007). A number of these metrics are similar to those used in metapopulation theory. Recent extensions of network modularity have been used to identify critical scales or thresholds in natural landscapes, which can be used to inform conservation activities and delineate species conservation landscapes (Fletcher et al. 2013). A number of prioritization models also aim to combine information on the feasibility of conserving certain areas, with their conservation value or impact on species viability (Moilanen et al. 2009).

Multilocation population projection matrices provide an explicit modelling framework that includes both within- and between-population dynamics. Asymptotic population growth rate and sensitivity/elasticity of growth rate to population-specific vital rates can be computed directly. The multilocation extension of Fisher's reproductive value provides a metric reflecting the relative importance of an animal in each potential age/stage to ultimate population size and is thus a natural metric by which to place value on different subpopulations (e.g. Rousset 1999). Empirical approaches to this theory include multistate capture-recapture modelling, which can be used to directly estimate the proportional contribution of animals in any subpopulation to growth of the entire metapopulation system, an explicit accounting of subpopulation 'value' (see Sect. 4.3; also Nichols 2016).

Optimization theory can also be used on networks to achieve conservation objectives (e.g. maximize species presence or landscape connectivity), under certain conservation/management constraints, feasibility and vulnerability (Chadès et al. 2011). These methods are yet to be used for tigers but hold great potential for informing future conservation of the species.

13.5 Tiger Connectivity Conservation: Challenges and Opportunities

13.5.1 Existing Evidence of Population Isolation

Several landscape genetic studies explore connectivity and how land-use types impact resistance to tiger movement. These studies have mostly focused on the Central Indian landscape (see Joshi et al. 2013; Sharma et al. 2013a; Thatte et al. in review). A single study in Northwestern India (Reddy et al. 2012) establishes connectivity but does not estimate resistance. Overall, these studies indicate that tiger connectivity is high, despite fragmentation; this may perhaps be due to the relatively recent time frame of fragmentation in these landscapes or to the ability of tigers to traverse large distances in human-inhabited lands.

Multiple approaches and genetic datasets from different studies (e.g. Thatte et al., in review; Joshi et al. 2013) suggest that human settlements and linear barriers such as roads are detrimental to tiger connectivity. Because many tiger landscapes have not been studied (e.g. Western Ghats, Terai Arc, Northeast India), we are uncertain about how generalizable these results are across landscapes and land-use types. Future studies should focus on trying to investigate connectivity in multiple landscapes.

13.5.2 Conservation Implications of Small Isolated Populations

Small isolated populations are generally extinction prone for multiple reasons. Small populations are subject to demographic stochasticity, such that they may be impacted by random variation in birth rates, death rates or resulting sex ratios. They are also subject to genetic stochasticity, that is, inbreeding depression, loss of potentially adaptive genetic diversity and the accumulation of mutations (O'Grady et al. 2006). Small, isolated populations tend to have closely related individuals and low genetic variation. Genetic variability is the raw material for evolution, a necessity for species to adapt to a changing environment due to habitat loss, fragmentation and climate change. Low genetic variability reduces individual survival and fitness, and decreases population persistence (Lacy 1997; Frankham 2005).

There are few studies that have isolated the impact of inbreeding on extinction probability. However, O'Grady et al. (2006), using existing evidence of inbreeding in the wild, showed through simulation models that inbreeding is likely to have an impact on small isolated populations of many endangered species. Populations with low heterozygosity, a consequence of inbreeding, have impaired sperm quality that impacts reproductive success (Ralls et al. 1988). Ralls et al. (1988) demonstrated significant effects of inbreeding on fitness in 38 zoo species and concluded that even a 0.25 level of inbreeding results in a 33% decline in juvenile survival. Several studies have subsequently reviewed inbreeding effects in wild populations (e.g. Keller and Waller 2002). Demonstrated inbreeding effects in wild felids include poor sperm quality, low testosterone levels and chryptorchidism (undescended testes) in Florida panthers, all of which impact population reproductive rates. Further, these effects were reversed when genetically dissimilar individuals were introduced into the population. Inbred populations of the Florida panther showed improved reproductive output following genetic restoration (Hostetler et al. 2013).

Thus far, inbreeding depression has not been demonstrated in wild tiger populations, although foetal and juvenile mortality have been observed in inbred captive tigers (Xu et al. 2007). It is also evident that most populations of the species are much smaller than what is recommended for long-term viability (Traill et al. 2007; Walston et al. 2010). As an apex predator, tigers exist at relatively low densities (Karanth et al. 2004). In source sites, population sizes range between 14 and 300 individuals, while Traill et al. (2007) recommend population sizes of over 4000 individuals for long-term viability. Smith and McDougal (1991) estimate that the effective population size for tigers is little less than half the adult population (i.e. the largest tiger population would have an effective population size of <150 individuals).

Through a spatially explicit individual-based simulation model, Kenney et al. (2014) show that in small (12 female breeding tigers) and medium sized (24 female breeding tigers) populations, even one immigrant male tiger per generation can drastically reduce the probability that the population will go extinct in the next 100 years. Currently, tigers appear to be able to traverse human-dominated landscapes across large distances (Joshi et al. 2013; Ramesh et al. 2016). The only wild tiger population that appears to be isolated based on genetic data is a population at the westernmost edge of the range, Ranthambore (U Ramakrishnan, unpublished data). While this population does not yet show signs of inbreeding depression, relatedness between individuals is high and genetic variation is low (Natesh et al., in review).

Small populations may also by 'rescued' by immigration, ameliorating negative impacts of demographic stochasticity (in addition to increasing genetic variability). Joshi et al. (2013) show movement of individuals from high-density sites such as Kanha to low-density tiger populations such as Nagarjunasagar-Srisailam Tiger Reserve (NSTR). Tigers are also now moving into Kawal Tiger Reserve in Telangana from source sites, leading to population recovery (I Siddiqui, unpublished data). Maintaining such connectivity is critical given ongoing efforts to increase protection of NSTR to aid tiger population recovery.

13.5.3 Current Connectivity Conservation

By far the most common conservation strategy used to enhance connectivity is the demarcation of corridors, or narrow stretches of habitat that connect fragments. Corridors are mentioned, for instance, in the Wildlife Protection Act of India (1972), while proposing a clear definition and legal framework for protection in future amendments. A number of corridors have been demarcated for tigers, some examples being the Rajaji-Corbett and Kanha-Pench corridors in India, the Central Annamite corridor in Vietnam and the Tennaserim Hills in Myanmar. Katarniaghat-Bardia and Dudhwa-Basanta transboundary corridors link forests of India and Nepal, while the recently declared Sredne-Ussuriiskii refuge links tiger habitat in Russia to China.

Tiger corridors have largely been demarcated through structural connectivity measures, with little spatial information on where tigers move, or their importance for maintaining connectivity. Nonetheless, there is evidence that tigers use some of these forested corridors (Karanth et al. 2011; Rayan and Linkie 2015) and that they house populations of tiger prey (Rayan and Linkie 2015) and are likely of conservation value for other biodiversity as well.

Corridors rarely benefit from the same level of protection, or conservation effort, as Protected Areas (PAs). Corridors, even when government-owned, may be managed as multiple-use lands, or may not be prioritized for conservation. Corridors are therefore especially susceptible to forest loss or degradation. Corridors, being by nature narrow stretches of habitat, are also susceptible to edge effects and associated consequences for habitat quality. Kanagaraj et al. (2013) also showed through a simulation model that narrow corridors may not serve well to link habitat fragments. Prey may be differentially available in corridors (Rayan and Linkie 2015). Dispersing tigers that do not encounter prey may turn to domestic cattle for food, leading to human-tiger conflict. Human settlements and other developmental projects may restrict the use of corridors (Joshi et al. 2013; Rayan and Linkie 2015).

Roads and other linear developmental projects have a disproportionate impact on tiger connectivity (Joshi et al. 2013). Areendran et al. (2012) assessed the impact of developmental projects on connectivity in the Nilgiri landscape, southwestern India, through use of cost-distance modelling. Even at conservative estimates of their impacts, roads drastically increased dispersal cost and negatively impacted landscape connectivity (Areendran et al. 2012).

Dispersal is a risky phase, during which animals often face high mortality risks. The inability of tigers to successfully traverse a human-dominated landscape is a serious issue that can have impacts on landscape connectivity and population demographic parameters. Marked dispersing tigers in the Russian Far East suffered 100% mortality while traversing the matrix (Goodrich et al. 2008). Animals do not only move through forested corridors, and those dispersing through human-dominated areas face difficulties of prey availability (Karanth et al. 2011; Rayan and Linkie 2015), decreased hunting efficiency (Kerley et al. 2002) and heightened poaching (Goodrich et al. 2008). Dispersing tigers are also more conflict prone, as (a) they are more likely to encounter people, (b) they are often moving through areas of low prey availability and hence may resort to killing cattle and (c) due to high levels of perceived risk and unfamiliar environments faced during dispersal, tigers may display behaviours that aggravate conflict. Indeed, movement conduits face high levels of human-tiger conflict (e.g. Ahmed et al. 2012). Such conflict, whether it leads to tiger mortality or removal to captivity, ultimately has the impact of limiting successful dispersal. Roads and highways in the matrix might also cause increased dispersal mortality. Lastly, tigers may be unable to locate suitable habitat due to the configuration of the matrix, leading to dispersal culs-de-sac or dead ends (Kanagaraj et al. 2013). To ensure that current rates of connectivity persist into the future, these factors need to be addressed within tiger conservation programmes.

13.5.4 A Prospectus for Tiger Connectivity Conservation: Corridors and Beyond

Connectivity conservation globally has largely focussed on the demarcation and maintenance of corridors (Worboys et al. 2010). Corridors are extremely important where present, as they represent areas of natural vegetation and form habitat for a number of species. Further, corridors, in addition to aiding dispersal, may also serve as secondary habitat for species. Structural connectivity or habitat contiguity is thus critical as it not only reinforces linkages in the landscape for multiple species and ecological processes but also serves to increase the overall natural vegetation cover. Thus, forest conversion to inhospitable matrix in areas that may serve to enhance structural connectivity needs to be strictly avoided, while habitat restoration or afforestation efforts should be carefully planned using spatial prioritization exercises such that these efforts lead to increased linkages between fragmented populations.

However, corridors, while valuable, are typically difficult to manage and often face threats of habitat loss, poaching, encroachment and edge effects. Corridors may not always be within the jurisdiction of the government, and hence imposing the legal framework that applies to PAs may be problematic. Lastly, there is evidence that while animals use corridors for dispersal, having a hospitable matrix that can occasionally be used adds to path redundancy and the overall resilience of connectivity towards land-use change. Thus, managing the matrix through incentive measures, participatory programmes or legal initiatives is vital for the overall health of the landscape.

Animals use not only corridors for dispersal but also move through the matrix. Tigers are known to move through coffee plantations, sugarcane fields and even human settlements (Ramesh et al. 2016). A more meaningful perspective towards connectivity conservation is a mechanistic approach, whereby the main *threats* to successful dispersal and connectivity are addressed (Vasudev et al. 2015).

Corridors as well as different matrix types vary in their utility due to certain factors, such as lack of cover, limited food resources or human presence (e.g. Rayan and Linkie 2015). Roads and other linear developmental projects are hugely detrimental to connectivity. Ideally, ecological costs of connectivity loss should be included in planning stages of these developmental projects such that they minimally impact wildlife. Alternatively, overpasses and viaducts have been recommended to allow wildlife crossing of highways (Rayan and Linkie 2015). These methods are often costly, and it is imperative to design them based on scientific information on tiger movement such that they are maximally effective. Further, monitoring of overpasses and viaducts is critical to ensure conservation success.

The lack of food resources in corridors and the matrix often leads to tigers preying on livestock. Increasingly, the response to such incidents is the removal of tigers either through killing or moving into captivity. Such removal directly impacts connectivity and, under extreme cases, completely precludes dispersal. Fostering an environment where removals immediately follow the sighting of tigers in human-use lands will impact connectivity and population viability (see Goswami and Vasudev 2017). Fencing tigers and other wildlife within forests is also not a solution that is viable in the long term. Context-specific participatory solutions that disincentivize practices that are harmful to tiger connectivity and encourage wildlife-friendly measures are required within tiger conservation landscapes.

Lastly, while all these approaches are focussed on enhancing connectivity in the landscape, it is crucial to remember that connectivity may also have negative impacts on species (Fletcher et al. 2016), such as loss of local adaptions (Marshall and Spalton 2000) or spread of diseases (McCallum and Dobson 2002). To date, these impacts have not been detected in tigers. Theoretical models suggest that 'one migrant per generation' is an appropriate level of dispersal to be maintained among populations (Wang 2004), but this rule is based on assumptions that are unlikely to hold in natural populations (Wang 2004), and how much connectivity should be maintained in a landscape is still undetermined. Connectivity requirements (and limitations) for inbreeding avoidance, demographic rescue or habitat recolonization are also bound to be drastically different, with demographic contributions requiring far greater movement than genetic contributions.

Overall there is increased need for incorporation of species dispersal biology into connectivity conservation (Sawyer et al. 2011; Zeller et al. 2012; Vasudev et al. 2015). Tiger conservation investments into corridors and the matrix, unless informed by science, may not be successful in increasing landscape connectivity. In the face of rapid habitat and environmental changes, the ability of conservation scientists to predict the response of tiger populations to such changes is critical. Without increasing our understanding of tiger movement behaviour, and incorporating this knowledge into connectivity conservation planning, tigers may only persist in small, isolated populations in the future.

References

Abrahms B, Sawyer SC, Jordan NR et al (2016) Does wildlife resource selection accurately inform corridor conservation? J Appl Ecol. doi:10.1111/1365-2664.12714

Adriaensen F, Chardon JP, De Blust G et al (2003) The application of "least-cost" modelling as a functional landscape model. Landsc Urban Plan 64:233–247. doi:10.1016/S0169-2046(02)00242-6

Ahmed RA, Prusty K, Jena J et al (2012) Prevailing human carnivore conflict in Kanha-Achanakmar corridor, Central India. J Zool 7:158–164. doi:10.5829/idosi.wjz.2012.7.2.6335

Areendran G, Raj M, Raj K (2012) Modeling impact of economic development projects on Tiger conservation landscape–a case study from Nilgiris, India. Asian J Geoinformatics 12(1)

Arnason AN (1972) Parameter estimates from mark-recapture experiments on two populations subject to migration and death. Res Popul Ecol (Kyoto) 13:97–113. doi:10.1007/BF02521971

Bauer S, Klaassen M (2013) Mechanistic models of animal migration behaviour – their diversity, structure and use. J Anim Ecol 82:498–508. doi:10.1111/1365-2656.12054

Brown JH, Kodric-Brown A (1977) Turnover rates in insular biogeography: effect of immigration on extinction. Ecology 58:445–449

Carroll C, Miquelle DG (2006) Spatial viability analysis of Amur tiger *Panthera tigris altaica* in the Russian Far East: the role of protected areas and landscape matrix in population persistence. J Appl Ecol 43:1056–1068. doi:10.1111/j.1365-2664.2006.01237.x

Chadès I, Martin TG, Nicol S et al (2011) General rules for managing and surveying networks of pests, diseases, and endangered species. Proc Natl Acad Sci U S A 108:8323–8328. doi:10.1073/pnas.1016846108

Chundawat RS, Sharma K, Gogate N et al (2016) Size matters: scale mismatch between space use patterns of tigers and protected area size in a tropical dry Forest. Biol Conserv 197:146–153

Clobert J, Baguette M, Benton TG et al (2012) Dispersal ecology and evolution. Oxford University Press, Oxford

Coffman CJ, Nichols JD, Pollock KH (2001) Population dynamics of *Microtus pennsylvanicus* in corridor-linked patches. Oikos 93:3–21. doi:10.1034/j.1600-0706.2001.930101.x

Cushman SA, Elliot NB, Macdonald DW et al (2016) A multi-scale assessment of population connectivity in African lions (*Panthera leo*) in response to landscape change. Landsc Ecol 31(6):1337–1353. doi:10.1007/s10980-015-0292-3

DeAngelis DL, Gross LJ (1992) Individual-based models and approaches in ecology. Chapman and Hall, London

Doerr VAJ, Barrett T, Doerr ED (2011) Connectivity, dispersal behaviour and conservation under climate change: a response to Hodgson et al. J Appl Ecol 48:143–147. doi:10.1111/j.1365-2664.2010.01899.x

Eaton MJ, Hughes PT, Hines JE, Nichols JD (2014) Testing metapopulation concepts: effects of patch characteristics and neighborhood occupancy on the dynamics of an endangered lagomorph. Oikos 123:662–676. doi:10.1111/oik.01008

Fletcher RJ, Revell A, Reichert BE et al (2013) Network modularity reveals critical scales for connectivity in ecology and evolution. Nat Commun. doi:10.1038/ncomms3572

Fletcher RJ, Burrell NS, Reichert BE et al (2016) Divergent perspectives on landscape connectivity reveal consistent effects from genes to communities. Curr Landsc Ecol Rep 1:67–79. doi:10.1007/s40823-016-0009-6

Frankham R (2005) Genetics and extinction. Biol Conserv 126:131–140. doi:http://dx.doi.org/10.1016/j.biocon.2005.05.002

Goodrich JM, Kerley LL, Smirnov EN et al (2008) Survival rates and causes of mortality of Amur tigers on and near the Sikhote-Alin biosphere Zapovednik. J Zool 276:323–329. doi:10.1111/j.1469-7998.2008.00458.x

Goswami VR, Vasudev D (2017) Triage of conservation needs: the juxtaposition of conflict mitigation and connectivity considerations in heterogeneous, human-dominated landscapes. Front Ecol Evol. doi:10.3389/fevo.2016.00144

Hall LA, Beissinger SR (2014) A practical toolbox for design and analysis of landscape genetics studies. Landsc Ecol 29:1487–1504. doi:10.1007/s10980-014-0082-3

Hanski I (1998) Metapopulation dynamics. Nature 396:41–49. doi:10.1016/0169-5347(89)90061-X

Hestbeck JB, Nichols JD, Malecki RA (1991) Estimates of movement and site fidelity using mark-resight data of wintering Canada geese. Ecology 72:523–533. doi:10.2307/2937193

Hostetler JA, Onorato DP, Jansen D, Oli MK (2013) A cat's tale: the impact of genetic restoration on Florida Panther population dynamics and persistence. J Anim Ecol 82:608–620. doi:10.1111/1365-2656.12033

Jhala YV, Qureshi Q, Gopal R (2015) The status of tigers in India 2014. National Tiger Conservation Authority, New Delhi & The Wildlife Institute of India, Dehradun, India

Joshi A, Vaidyanathan S, Mondol S et al (2013) Connectivity of tiger (*Panthera tigris*) populations in the human-influenced forest mosaic of central India. PLoS One 8:e77980. doi:10.1371/journal.pone.0077980

Kanagaraj R, Wiegand T, Kramer-Schadt S, Goyal SP (2013) Using individual-based movement models to assess inter-patch connectivity for large carnivores in fragmented landscapes. Biol Conserv 167:298–309. doi:10.1016/j.biocon.2013.08.030

Karanth KU, Nichols JD, Kumar NS et al (2004) Tigers and their prey: predicting carnivore densities from prey abundance. Proc Natl Acad Sci U S A 101:4854–4858. doi:10.1073/pnas.0306210101

Karanth KU, Nichols JD, Kumar NS, Hines JE (2006) Assessing tiger population dynamics using photographic capture – recapture sampling. Ecology 87:2925–2937

Karanth KU, Gopalaswamy AM, Kumar NS et al (2011) Monitoring carnivore populations at the landscape scale: occupancy modelling of tigers from sign surveys. J Appl Ecol 48:1048–1056. doi:10.1111/j.1365-2664.2011.02002.x

Keller LF, Waller DM (2002) Inbreeding effects in wild populations. Trends Ecol Evol 17:230–241. doi: 10.1016/S0169-5347(02)02489-8

Kenney J, Allendorf FW, Mcdougal C, Smith JLD (2014) How much gene flow is needed to avoid inbreeding depression in wild tiger populations? Proc R Soc Lond B Biol Sci 281:20133337

Kerley LL, Goodrich JM, Miquelle DG et al (2002) Effects of roads and human disturbance on Amur tigers. Conserv Biol 16:97–108. doi:10.1046/j.1523-1739.2002.99290.x

Kramer-Schadt S, Revilla E, Wiegand T et al (2004) Fragmented landscapes, road mortality and patch connectivity: modelling influences on the dispersal of Eurasian lynx. J Appl Ecol 41:711–723. doi:10.1111/j.0021-8901.2004.00933.x

Lacy RC (1997) Importance of genetic variation to the viability of mammalian populations. J Mammal 78:320–335
Langrock R, King R, Matthiopoulos J et al (2012) Flexible and practical modeling of animal telemetry data: hidden Markov models and extensions. Ecology 93:2336–2342. doi:10.2307/23436257
Lebreton JD, Nichols JD, Barker RJ et al (2009) Modeling individual animal histories with multistate capture-recapture models. In: Caswell H (ed) Advances in ecological research. Academic, San Diego, pp 87–173
Lindenmayer DB, Fischer J (2007) Tackling the habitat fragmentation panchreston. Trends Ecol Evol 22:127–132. doi: 10.1016/j.tree.2006.11.006
MacArthur RH, Wilson EO (1967) The theory of island biogeography. Princeton University Press, Princeton
MacKenzie DI, Nichols JD, Royle JA et al (2006) Occupancy estimation and modeling: inferring patterns and dynamics of species occurrence, Amsterdam, Academic
Maehr DS, Land ED, Shindle DB et al (2002) Florida panther dispersal and conservation. Biol Conserv 106:187–197. doi:10.1016/S0006-3207(01)00245-2
Malviya M, Ramesh K (2015) Human-felid conflict in corridor habitat: implications for tiger and leopard conservation in Terai Arc Landscape, India. Hum Wildl Interac 9:48–57
Manel S, Holderegger R (2013) Ten years of landscape genetics. Trends Ecol Evol 28:614–621. doi:10.1016/j.tree.2013.05.012
Marshall TC, Spalton JA (2000) Simultaneous inbreeding and outbreeding depression in reintroduced Arabian oryx. Anim Conserv 3:241–248
McCallum H, Dobson A (2002) Disease, habitat fragmentation and conservation. Proc R Soc Lond B Biol Sci 269:2041–2049
McClure ML, Hansen AJ, Inman RM (2016) Connecting models to movements: testing connectivity model predictions against empirical migration and dispersal data. Landsc Ecol. doi:10.1007/s10980-016-0347-0
McRae BH, Dickson BG, Keitt TH, Shah VB (2008) Using circuit theory to model connectivity in ecology, evolution, and conservation. Ecology 89:2712–2724. doi:10.1890/07-1861.1
Minor ES, Urban DL (2007) Graph theory as a proxy for spatially explicit population models in conservation planning. Ecol Appl 17:1771–1782
Moilanen A, Wilson KA, Possingham HP (2009) Spatial conservation prioritization: quantitative methods and computational tools. Oxford University Press, Oxford
Morales JM, Haydon DT, Frair J et al (2004) Extracting more out of relocation data: building movement models as mixtures of random walks. Ecology 85:2436–2445. doi:10.1890/03-0269
Nathan R, Getz WM, Revilla E et al (2008) A movement ecology paradigm for unifying organismal movement research. Proc Natl Acad Sci U S A 105:19052–19059
Nichols JD (2016) And the first one now will later be last: time-reversal in Cormack-Jolly-Seber models. Stat Sci 31:175–190. doi:10.1214/16-STS546
Nichols JD, Kendall WL (1995) The use of multi-state capture-recapture models to address questions in evolutionary ecology. J Appl Stat 22:835–846. doi:10.1080/02664769524658
Nichols JD, Pollock KH (1990) Estimation of recruitment from immigration versus in situ reproduction using Pollock's robust design. Ecology 71:21–26. doi:10.2307/1940243
Nichols JD, Sauer JR, Pollock KH, Hestbeck JB (1992) Estimating transition probabilities for stage-based population projection matrices using capture-recapture data. Ecology 73:306–312. doi:10.2307/1938741
O'Grady JJ, Brook BW, Reed DH et al (2006) Realistic levels of inbreeding depression strongly affect extinction risk in wild populations. Biol Conserv 133:42–51. doi:10.1016/j.biocon.2006.05.016
Patterson T, Thomas L, Wilcox C et al (2008) State–space models of individual animal movement. Trends Ecol Evol 23:87–94. doi:10.1016/j.tree.2007.10.009
Piry S, Alapetite A, Cornuet J-M et al (2004) GENECLASS2: a software for genetic assignment and first-generation migrant detection. J Hered 95:536–539. doi:10.1093/jhered/esh074

Pritchard JK, Stephens M, Donnelly P (2000) Inference of population structure using multilocus genotype data. Genetics 155:945–959

Ralls K, Ballou JD, Templeton A (1988) Estimates of lethal equivalents and the cost of inbreeding in mammals. Conserv Biol 2:185–193. doi:10.1111/j.1523-1739.1988.tb00169.x

Ramesh K, Cushman SA, Sarkar MS et al (2016) Multi-scale prediction of landscape resistance for tiger dispersal in central India. Landsc Ecol. doi:10.1007/s10980-016-0363-0

Rathore CS, Dubey Y, Shrivastava A et al (2012) Opportunities of habitat connectivity for tiger (*Panthera tigris*) between Kanha and Pench national parks in Madhya Pradesh, India. PLoS One. doi:10.1371/journal.pone.0039996

Rayan DM, Linkie M (2015) Conserving tigers in Malaysia: a science-driven approach for eliciting conservation policy change. Biol Conserv 184:18–26. doi:10.1016/j.biocon.2014.12.024

Rayfield B, Fortin MJ, Fall A (2010) The sensitivity of least-cost habitat graphs to relative cost surface values. Landsc Ecol 25:519–532. doi:10.1007/s10980-009-9436-7

Reddy PA, Gour DS, Bhavanishankar M et al (2012) Genetic evidence of tiger population structure and migration within an isolated and fragmented landscape in northwest India. PLoS One 7:e29827. doi:10.1371/journal.pone.0029827

Revilla E, Wiegand T (2008) Individual movement behavior, matrix heterogeneity, and the dynamics of spatially structured populations. Proc Natl Acad Sci 105:19120–19125. doi:10.1073/pnas.0801725105

Riley SPD, Pollinger JP, Sauvajot RM et al (2006) A southern California freeway is a physical and social barrier to gene flow in carnivores. Mol Ecol 15:1733–1741. doi:10.1111/j.1365-294X.2006.02907.x

Rousset F (1999) Reproductive value vs sources and sinks. Oikos 86:591. doi:10.2307/3546664

Runge JP, Runge MC, Nichols JD (2006) The role of local populations within a landscape context: defining and classifying sources and sinks. Am Nat 167:925–938. doi:10.1086/503531

Sanderson WEW, Forrest J, Loucks C et al (2010) Setting priorities for tiger conservation: 2005–2015. In: Tilson R, Nyhus PJ (eds) Tigers of the world, pp 143–161

Sawyer SC, Epps CW, Brashares JS (2011) Placing linkages among fragmented habitats: do least-cost models reflect how animals use landscapes? J Appl Ecol 48:668–678. doi:10.1111/j.1365-2664.2011.01970.x

Schwarz CJ, Schweigert JF, Arnason AN (1993) Estimating migration rates using tag-recovery data. Biometrics 49:177–193. doi:10.2307/2532612

Sharma S, Dutta T, Maldonado JE et al (2013a) Forest corridors maintain historical gene flow in a tiger metapopulation in the highlands of central India. Proc R Soc B Biol Sci 280:20131506. doi:10.1098/rspb.2013.1506

Sharma S, Dutta T, Maldonado JE et al (2013b) Spatial genetic analysis reveals high connectivity of tiger (*Panthera tigris*) populations in the Satpura-Maikal landscape of Central India. Ecol Evol 3:48–60. doi:10.1002/ece3.432

Skvarla JL, Nichols JD, Hines JE, Waser PM (2004) Modeling interpopulation dispersal by banner-tailed kangaroo rats. Ecology 85:2737–2746. doi:10.1890/03-0599

Slatkin M (1987) Gene flow and the geographic structure of natural populations. Science 236:787–792. doi:10.1126/science.3576198

Smith JLD (1993) The role of dispersal in structuring the Chitwan tiger population. Behaviour 124:165–195

Smith JLD, Mcdougal C (1991) The contribution of variance in lifetime reproduction to effective population size in tigers. Conserv Biol 5:484–490. doi:10.1111/j.1523-1739.1991.tb00355.x

Taylor PD, Fahrig L, Henein K, Merriam G (1993) Connectivity is a vital element of landscape structure. Oikos 68:571–573. doi:10.2307/3544927

Traill LW, Bradshaw CJA, Brook BW (2007) Minimum viable population size: a meta-analysis of 30 years of published estimates. Biol Conserv 139:159–166. doi:http://dx.doi.org/10.1016/j.biocon.2007.06.011

Treves A, Karanth KU (2003) Human-carnivore conflict and perspectives on carnivore management worldwide. Conserv Biol 17:1491–1499. doi:10.1111/j.1523-1739.2003.00059.x

Turner MG, Gardner RH (2001) Landscape ecology in theory and practice. Springer-Verlag, New York
Vasudev D, Fletcher RJ (2015) Incorporating movement behavior into conservation prioritization in fragmented landscapes: an example of western hoolock gibbons in Garo Hills, India. Biol Conserv 181:124–132. http://dx.doi.org/10.1016/j.biocon.2014.11.021
Vasudev D, Fletcher RJ, Goswami VR, Krishnadas M (2015) From dispersal constraints to landscape connectivity: lessons from species distribution modeling. Ecography 38:967–978. doi:10.1111/ecog.01306
Walston J, Robinson JG, Bennett EL et al (2010) Bringing the tiger back from the brink-the six percent solution. PLoS Biol 8:6–9. doi:10.1371/journal.pbio.1000485
Wang J (2004) Application of the one-migrant-per-generation rule to conservation and management. Conserv Biol 18:332–343. doi:10.1111/j.1523-1739.2004.00440.x
Wikramanayake ED, Dinerstein E, Robinson JG et al (1998) An ecology-based method for defining priorities for large mammal conservation: the tiger as case study. Conserv Biol 12:865–878. doi:10.1046/j.1523-1739.1998.96428.x
Williams BK, Nichols JD (1984) Optimal timing in biological processes. Am Nat 123:1–19
Worboys GL, Francis WL, Lockwood M (2010) Connectivity conservation management: a global guide. Earthscan, New York
Williams BK, Nichols JD, Conroy MJ (2002) Analysis and management of animal populations. Academic, San Diego
Xu YC, Fang SG, Li ZK (2007) Sustainability of the South China tiger: implications of inbreeding depression and introgression. Conserv Genet 8:1199–1207. doi:10.1007/s10592-006-9276-9
Yackulic CB, Reid J, Davis R et al (2012) Neighborhood and habitat effects on vital rates: expansion of the barred owl in the Oregon Coast Ranges. Ecology 93:1953–1966. doi:10.1890/11-1709.1
Yackulic CB, Reid J, Nichols JD et al (2014) The roles of competition and habitat in the dynamics of populations and species distributions. Ecology 95:265–279. doi:10.1890/13-0012.1
Zeller KA, McGarigal K, Whiteley AR (2012) Estimating landscape resistance to movement: a review. Landsc Ecol 27:777–797. doi:10.1007/s10980-012-9737-0

Informed Decision Processes for Tiger Conservation: A Vision for the Future

14

James D. Nichols, K. Ullas Karanth, Arjun M. Gopalaswamy, G. Viswanatha Reddy, John M. Goodrich, and Dale G. Miquelle

14.1 Introduction

Humans have been involved in efforts to manage wildlife populations for centuries. Modern wildlife management and conservation are often traced back to the publication in 1933 of *Game Management* by Aldo Leopold. In this seminal volume, Leopold described both rudimentary population models and animal "census methods," recognizing the important roles these methods would play in active

J.D. Nichols (✉)
Crofton, MD, USA
e-mail: jamesdnichols2@gmail.com

K.U. Karanth
Wildlife Conservation Society (WCS), New York, NY, USA

Centre for Wildlife Studies, Bengaluru, India

Wildlife Conservation Society, India Program, Bengaluru, India

National Centre for Biological Sciences-TIFR, Bengaluru, India
e-mail: ukaranth@wcs.org; https://www.wcs.org/; http://cwsindia.org/; http://wcsindia.org/home/; https://www.ncbs.res.in/

A.M. Gopalaswamy
Statistics and Mathematics Unit, Indian Statistical Institute, Bengaluru, India

Department of Zoology, University of Oxford, Oxford, UK
e-mail: arjungswamy@gmail.com; https://www.isibang.ac.in; https://www.zoo.ox.ac.uk/

G.V. Reddy
Indian Forest Service, Government of Rajasthan, Jaipur, India
e-mail: gvreddy.rajforests@gmail.com

J.M. Goodrich
Panthera, New York, NY, USA
e-mail: jgoodrich@panthera.org; https://www.panthera.org/

K.U. Karanth, J.D. Nichols (eds.), *Methods For Monitoring Tiger And Prey Populations*, DOI 10.1007/978-981-10-5436-5_14

management. It is doubtful that a member of the wildlife profession from Leopold's time would even recognize the methods currently used for population modeling and inference, as these methods have progressed dramatically in formalism and associated rigor. In stark contrast to the dramatic improvements in these classes of methodology, our approach to decision-making has evolved very little since the time of Leopold. Instead, most current approaches to decision-making still entail experts looking at models and data and then rendering a judgment about what action(s) is most appropriate. Such decisions based on the thought processes of the decision-maker are not necessarily bad, but they are neither transparent nor objective, and they will usually be suboptimal. It is our view that the future of the wildlife and conservation professions should move to adopt modern, decision-theoretic approaches to making decisions in ways that are transparent, objective, defensible, and scientific.

As we saw in Chap. 1, tiger conservation is attracting substantial interest within tiger range states as well as from the international community. In Chaps. 2, 3, 4, 5, 6, 7, 8, 9, 10, 11, 12, and 13 of this volume, we examined rigorous methods for modeling and making inferences about tiger and prey populations. Given all these resources and powerful monitoring methods that are tailored well to fit modern decision-theoretic approaches, we believe that programs of tiger conservation should strongly consider adopting these approaches for making conservation decisions. In this somewhat theoretical and speculative chapter, we explore how such a tiger conservation agenda can be advanced.

We begin by briefly considering the basic components of structured decision processes, as these are required for virtually any informed decision. We then consider the special case of recurrent decision problems characterized by uncertainty, outlining an adaptive management approach to these problems. Finally, we move to specific examples associated with conservation of tigers and prey and briefly discuss application of decision-theoretic approaches.

14.2 Elements of an Informed Decision Process

14.2.1 Objectives

Management and conservation entail actions designed to meet specified objectives. Indeed, the effectiveness of any action is measured only by the degree to which it furthers stated objectives. In the absence of objectives, there is no basis for selecting one management action over another or deciding whether or not specific management actions have been successful. Thus, objectives are not optional but rather are essential components of management processes.

D.G. Miquelle
Wildlife Conservation Society (WCS), New York, NY, USA

Department of Ecology, Far Eastern Federal University, Vladivostok, Russia
e-mail: dmiquelle@wcs.org

Objectives are based on human values, specifically on the values of decision-makers and other stakeholders with interests in the managed system. Realistic objectives can be arrived at only by reconciling varied, and sometimes conflicting, demands of different segments of human society as well as data on tiger ecology. Therefore, the process of developing a conservation program should incorporate special efforts to include all potential stakeholders in order to arrive at objectives that reflect all relevant opinions and to instill a sense of "buy-in" to the program by all stakeholders. The usual case in conservation problems is for multiple objective components that compete with one another, in the sense that achievement of one objective component may be deleterious to another. Costs associated with management actions are often included as a component of objectives as well, because resources available for conservation are nearly always limited. Inclusion of multiple objectives and corresponding trade-offs in a decision process may require a common currency, use of threshold constraints, or even an explicit weighting scheme. Above all, it is important to devote adequate time and effort to developing objectives, as they drive the entire process.

14.2.2 Actions

The terms "conservation" and "management" imply taking actions that move the system in the direction of achieving objectives. Following the development of objectives, it is important to identify potential actions that can be used to attain objectives. Conceptually, actions may be continuous variables (e.g., harvest rate, fraction of land that is burned) or discrete (e.g., select one of three fixed harvest rates or fractions of lands to be burned). The optimization methods currently used to solve structured decision problems are better adapted to deal with selecting from a set of discrete actions, although this methodological constraint will likely be overcome in the near future. The primary consideration in selecting a set of potential actions is the nature of the objective(s) and the potential for considered actions to influence the system in ways relevant to those objectives.

14.2.3 Model(s)

Models provide a basis for making predictions about the consequences of different actions, thus providing a basis for selecting the action that is most likely to perform best with respect to objectives. Models are viewed by some managers as entities devised by academics that have no place in real-world management. We believe that this perspective reflects a misunderstanding of conservation and management. If a decision-maker really has no basis for projecting consequences of different management actions, then he/she cannot claim to be engaged in informed management. Instead, the decision of which action to implement at any decision point might as well be decided by a throw of dice. Models need not be mathematical constructs (although these are frequently the most useful) and may instead be formulations found in the mind of the decision-maker. The important point is that models for projecting system responses to management actions are not optional in informed management. They are required.

Uncertainty about managed systems often leads to use of multiple competing models of predicted system response. That is, we are not certain exactly how the system responds to our management actions and thus develop multiple models of possible responses. Development of a decision based on multiple models typically requires that the models be weighted by our relative degrees of confidence in their predictive abilities, based on past predictions. Adaptive resource management (ARM) is designed to make decisions in the face of such uncertainty, while simultaneously reducing it (e.g., Nichols and Williams 2013, see below).

14.2.4 Monitoring

Tiger monitoring programs (the primary subject of this book) provide estimates of system state variables and vital rates for multiple purposes. For a single population at a specific site, the key state variable might be tiger abundance and the associated vital rates of survival, recruitment, and movement. For a metapopulation of a single species across multiple sites, the relevant state variable may be the fraction of sites occupied, with corresponding vital rates of local extinction and colonization.

In many informed decision processes, the optimal decision depends on the current state of the system. So we take very different actions if our population is very small than if it is large, for example. Estimation of system state for making state-dependent decisions is a primary function of monitoring. Another function of monitoring is assessment of progress toward management objectives, which frequently include state variables themselves. Still another function is initial and updated parameter estimation for management models.

A fourth important function of monitoring data is for use in learning. When uncertainty about system response leads to multiple models, management is improved when this uncertainty is reduced. Reduction in uncertainty occurs via the insertion of a scientific step into the management process (see discussion of ARM below). This scientific step compares model-based predictions against monitoring results in order to learn which model predicts best. More specifically, informed decisions based on multiple models require use of model "weights" ($w_{i,t}$ is the weight associated with model i at time t) that reflect our relative degrees of confidence in the different models. These weights are simply numbers between 0 and 1 that sum to 1 for all of the models in the model set (denote number of models as m).

If we are managing a focal species, such as the tiger, our monitoring program (see below) provides an estimate of population size or density (system state) at time t. Using the model weights (and thus the relative model influences) at time t, we decide on a specific action to take. Based on this information available at time t (i.e., estimated population size, x_t, and action taken, a_t), each model makes a prediction of the distribution of population sizes expected at time $t + 1$. $\Pr_i(x_{t+1} | x_t, a_t)$ denotes the probability under model i of the estimated population size at $t + 1$ (x_{t+1}), given the population size at time t and action taken at time t. Model weights are updated using Bayes formula:

$$w_{i,t+1} = \frac{w_{i,t}\mathrm{Pr}_i\left(x_{t+1}|x_t, a_t\right)}{\sum_{i=1}^{m} w_{i,t}\mathrm{Pr}_i\left(x_{t+1}|x_t, a_t\right)} \tag{14.1}$$

The numerator of the above expression is the product of the prior weight (the relative degree of confidence in model i accrued up until time t) and the probability of the new monitoring data (x_{t+1}) having arisen, given that model i is a good approximating model. The denominator sums this product of prior weight and likelihood of new estimates for all of the models in the model set.

Repeated use of expression (14.1) hopefully results in one of the members of the model set attaining weights that approach 1, with weights of other models becoming very small. This evolution of model weights results in increased confidence in the model that makes the best predictions. This increased confidence is accompanied by increased influence on decisions and hopefully in better management.

14.2.5 Decision Algorithm

At each decision point, information from the preceding four elements must be combined to produce an informed decision about which action should be taken. Frequently in natural resource management and conservation, such decisions come from the thought processes of decision-makers. Decisions can be made in this way, but they will seldom be optimal.

Some conservation programs use formal optimization approaches to arrive at decisions. One-step optimization approaches are based only on immediate *returns* (the values of variables directly relevant to the objective function) and are appropriate for one-time decisions, such as land acquisitions or drawing up park boundaries, perhaps. However, many ongoing conservation programs are better characterized as dynamic decision processes (Fig. 14.1), such as patrolling to reduce poaching, fire suppression, or removal of "problem" tigers. In such dynamic processes, the action taken at any decision point, t, produces returns and drives the system to a new state. Because of state-dependent decisions, the new state determines what actions are viable the next time step. This combination of system state dynamics and state-dependent decisions requires a dynamic optimization approach that considers not just single-step returns, but returns from subsequent decision points into the future. Stochastic dynamic optimization approaches to achieve such optimization include stochastic dynamic programming (e.g., Bellman 1957; Williams et al. 2002) and other approaches such as genetic algorithms.

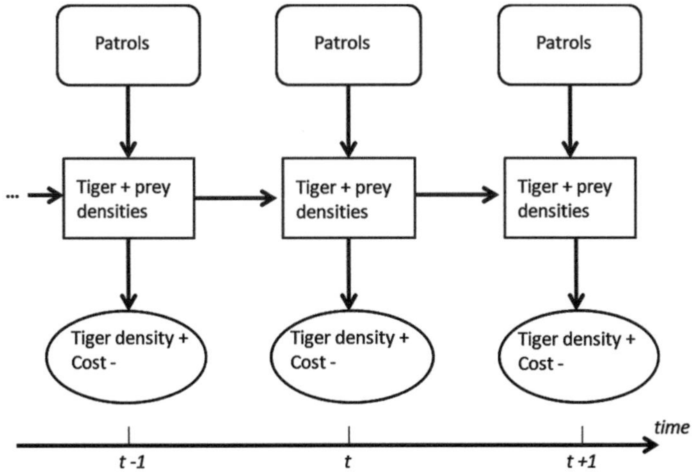

Plate 14.1 Diagrammatic representation of tiger reserve management as a recurrent decision problem. At each decision point (t), decisions are made about whether and where to deploy patrols. These patrol decisions might be made every year, for example. The decisions are based on objectives (e.g., tiger numbers exceeding some objective threshold at minimal cost) and on current tiger numbers, as estimated via monitoring. The selected patrol actions lead to changes in numbers of tigers and prey species, and these changes are assessed via monitoring before the next decision point. Achievement of tiger densities that exceed the threshold is viewed as a positive outcome of management, whereas patrol costs are viewed as negative outcomes. These outcomes combine to represent the management *returns*, by which the success of management is judged. The new estimates of tiger densities are then used to make the state-dependent decision about patrols the next time step, and the process proceeds

14.3 Adaptive Resource Management

14.3.1 General Description

ARM is an informed decision process developed to deal with uncertainty, specifically with uncertainty about system responses to management actions (Walters 1986; Williams et al. 2007; Nichols and Williams 2013). ARM provides an approach for making decisions in the face of uncertainty and for reducing that uncertainty in order to improve future management. ARM was developed for recurrent decision processes with multiple decision points through time (Fig. 14.1). However, ARM can also be applied to sequences of "one-time" decisions (e.g., land purchases, establishing boundaries of a tiger reserve) that are sufficiently similar that lessons learned from one decision may be useful to subsequent decisions elsewhere.

Although not yet widely used in tiger conservation, ARM has been used successfully in North America in the management of waterfowl harvest (Johnson et al. 1993; Nichols et al. 2007; 2015; U.S. Fish and Wildlife Service 2013; Cooch et al. 2014), horseshoe crab harvest for both crab and shorebird objectives (Smith et al. 2013; McGowan et al. 2015), and hiking in Denali National Park with golden eagle disturbance objectives (Martin et al. 2011; Fackler et al. 2014). ARM programs typically include setup, iterative, and possibly double-loop phases.

14.3.2 Setup Phase

The initial phase of any ARM program entails assembling and developing all of the above components of an informed decision process, i.e., objectives, actions, models, monitoring, and decision algorithms. The articulation of objectives and development of a formal objective function is an extremely important step that requires full stakeholder involvement. Not only the decision-maker, associated management personnel, and scientists knowledgeable of the system should be involved in establishing objectives, but also other conservation groups, individuals whose businesses or livelihoods are somewhat dependent on the system, and members of the general public should be invited to participate. Objectives in conservation typically entail trade-offs, such that ultimate objective functions nearly always represent compromises of stakeholders with competing demands. Broad stakeholder involvement is also useful in deciding on what actions should be considered as part of the decision process, as sometimes actions that are potentially effective (e.g., predator control, resettlement of villages from inside protected areas) may not be equally acceptable to all stakeholders. Necessarily, such stakeholders involved in setting objectives may include reserve managers; conservationists; political, social, and community leaders; as well as some scientists.

In contrast to the establishment of objectives and potential actions, the development of system models and monitoring programs is primarily the task of wildlife managers and scientists. Interested stakeholders should be invited to participate in these processes, but expertise in modeling and monitoring is typically found in conservation and scientific professionals. Model development should seek to incorporate existing uncertainty about system response to management actions. In particular, all stakeholders should be able to see in the model set at least one model that reflects their understanding of how the system works. Monitoring should then be focused on state variables that are found in the objectives and that are needed for state-dependent decisions, as well as other variables related to objectives. Learning is sometimes best achieved via a focus on key rate parameters governing system behavior, requiring the monitoring of key system vital rates as well.

Selection of the decision algorithm is also best handled by professionals familiar with dynamic optimization (Bellman 1957; Puterman 1994; Williams and Nichols 2014). However, this expertise is rare. If it is not available, decisions can still be made by key individuals in the absence of optimization algorithms, as long as there is recognition that such subjective decisions are likely to be suboptimal. ARM processes that proceed without use of optimization to select actions can still represent major improvements over widely prevalent forms of *ad hoc* management and still permit learning.

14.3.3 Iterative Phase

Full implementation of ARM can immediately follow the development of decision components in the setup phase. This iterative or operational phase of ARM proceeds as follows. At each decision point, an action is selected based on all of the decision process elements: objectives, potential actions, estimated state of the system, models (and associated weights) of system response to management actions, and the decision algorithm that selects the optimal action.

The selected action is implemented, yielding returns and driving the system to a new state, which is identified by the monitoring program. Learning occurs as the new estimate of system state obtained via monitoring is compared against the predictions of each model using Eq. 14.1. Weights increase for models that predict well [larger $\Pr_i(x_{t+1}|x_t, a_t)$] and decrease for models that predict poorly. The next decision is then made, with most process components remaining the same (objectives, actions, models), but with an updated estimate of system state and updated model weights. The entire process iterates through time in this manner, with model weights hopefully becoming large (approaching 1 eventually) for one of the models in the set and becoming small for the others.

14.3.4 Double-Loop Phase

Whenever there is doubt or dissatisfaction with any of the decision process elements, it is possible to revisit the setup phase and reconsider any process component. For example, changing attitudes and values could bring about a change in program objectives. New technologies or ideas could lead to new potential actions to consider. If none of the models seems to predict adequately, then one or more could be modified using new ideas about system dynamics. New technologies and/or inference methods could lead to new monitoring methods, and new software for solving dynamic optimization problems could be implemented as well (e.g., Chades et al. 2012; Fackler and Pacifici 2014; Fackler et al. 2014). Double-loop reconsideration of components is always possible, although the expectation is that such considerations will usually occur over longer time steps than the iterative process itself.

14.3.5 ARM Advantages

ARM has some important advantages over the kinds of processes typically used to manage natural resources. Many conservation decision processes exist in the absence of clearly specified and agreed upon objectives. Without clear objectives, the terms "management" and "conservation" are not really applicable, as there is never a means to decide whether actions have been useful or not. Such ill-defined processes represent little more than "random walks in action space." As a formal decision process, ARM requires objectives, and these objectives drive the entire process, as appropriate.

Related to objectives is the ability of ARM processes to readily accommodate competing perspectives of multiple stakeholders. Objective functions frequently represent compromises among stakeholders with very different values. In addition, different values are often accompanied by different ideas about how the focal system "works" and can be managed. Within ARM, these different ideas can be incorporated directly into the process via competing models of system response to management actions. And the model weights that dictate the relative influence of these models in conservation decisions result not from political pressure, but from relative abilities of the different models to predict system change. This direct incorporation of varying stakeholder values and ideas about system dynamics facilitates stakeholder buy-in and should lead to a general sense of fair play.

Despite pleas for evidence-based conservation (e.g., Sutherland et al. 2004), conservation frequently employs actions about which there is little evidence of effectiveness. ARM places a premium on learning (accruing evidence) and accomplishes this by inserting a scientific step into the decision process. This inclusion of science satisfies current demands for both evidence-based decisions and integration of science and management (Nichols et al. 2015). The nature of this scientific step also insures that science is directed explicitly at management, as opposed to peripheral issues.

Another advantage of many ARM processes is that management decisions are optimal. In some management settings, this may not be so important. However, for very contentious issues, especially those that entail litigation, it may prove very useful to be able to make the following kind of statement: "Given the management objectives to which we all agreed, the set of actions that was available to us, and our understanding of the system state and dynamics as provided by our monitoring and encoded in our model set; we made the smartest decision that we could have made, and we can prove this to you!"

A final advantage worthy of emphasis is the transparency of the process. From the setup phase, with its inclusion of relevant stakeholders, to the development of the model set based on different ideas of the stakeholders, there is an open attempt to include multiple perspectives. In addition, whenever dynamic optimization is used in the process, there is no suspicion that the decision of which action to take is based on an overemphasis on the values and ideas of the decision-maker. These values and ideas are transparently found in the objective function and the model set, and, conditional on them, the optimal action is not an opinion but a provable fact.

14.4 Structured Decisions and Tiger Conservation

14.4.1 Land Purchase

Many tiger conservation efforts focus on various types of protection, including habitat, prey populations, and tigers themselves. Decisions about habitat protection sometimes entail purchase or acquisition of land parcels as new tiger reserves, as additions to existing reserves or as corridors connecting reserves or other source populations. Objectives will usually be framed as making adequate purchases with minimal cost. Although such land acquisitions are one-time decisions, they may permit the learning that characterizes recurrent decisions if multiple purchases are made that share common features. We believe that structured decision processes can be useful for all land acquisition decisions and that ARM and consequent learning can be applied sometimes to sequences of related acquisition decisions.

14.4.2 Voluntary Resettlement

Another kind of one-time decision for existing tiger reserve lands entails encouraging voluntary relocation of families of local people residing in reserve inholdings (Karanth 2007). Such relocation efforts reduce poaching and conflict, and consolidate habitats of tigers and prey species (Karanth et al. 2004). Objectives will again include efforts to minimize monetary costs and potential social costs, while maximizing gains to tiger and prey species populations. As with decisions about land acquisitions, resettlement projects may be viewed as one-time or recurrent decisions, depending on the consideration of similar actions in other locations.

14.4.3 Livestock Grazing Control

Grazing control is a common approach to improving and maintaining habitat within tiger reserves and other protected areas of India. The focus of this management action is on reducing the number of livestock grazing in a reserve, or at least restricting it to a limited number of buffer zones. Such control efforts require ongoing efforts to detect illegal grazing and enforce restrictions. System state variables that can be monitored to draw inferences about control effectiveness include numbers and distribution of livestock, status (e.g., richness and/or spatial extent) of native plant communities, status (e.g., abundance or density) of wild ungulate prey species, and finally, tiger densities themselves. Allocation of ranger effort to grazing detection and control will most typically be viewed as a recurrent decision problem.

14.4.4 Patrols and Law Enforcement

For existing reserves that focus on tiger conservation, protection usually takes the form of patrols by staff to discourage poaching of tigers and prey animals. Decisions can be viewed as recurrent, as patrol effort, timing, and locations are all decision components that can be varied over time depending on perceived poaching threats (Fig. 14.1). Objectives would include costs associated with different patrol efforts, and benefits would include recovery and well-being of prey and tiger populations.

System models for projecting consequences of patrol efforts could focus directly on response of tiger populations to patrol effort. Another approach would be to construct models that projected responses of prey populations to patrol effort (and reduced poaching) and then responses of tiger populations to both patrol effort and changes in prey populations. This latter approach would include both key prey species and tigers as state variables. Still a third approach would project responses of poaching activity to patrol effort, treating poaching activity itself as a system state variable. Prey populations would then be modeled as functions of poaching activity, and tiger populations as a function of both prey populations and tiger poaching activity. These modeling approaches are successively more mechanistic, and selection of the appropriate modeling for any situation will be tied directly to the associated monitoring program(s).

Regardless of which of the above three modeling approaches is selected, monitoring efforts supporting decision processes based on patrols as a primary management action would include information on tiger abundance or density (for specific reserves) or tiger distribution (for landscapes including multiple reserves), depending on the scale at which management decisions are being made. Tiger monitoring for specific reserves would likely entail spatially explicit capture-recapture methods using data from camera traps or perhaps from scat collection with subsequent DNA analysis for individual identification (Gopalaswamy et al. 2012; Royle et al. 2009a, b, Chap. 9). Monitoring of tiger distribution would likely be based on sign survey data analyzed using occupancy modeling (Karanth et al. 2011, Chap. 4).

The second and third modeling approaches listed above would require monitoring of selected prey species. In most cases, this would entail distance sampling along line transects (Buckland et al. 2001; Chap. 6). However, other possibilities exist, and spatially explicit capture-recapture models could be used for prey species with individual marking patterns (e.g., chital, see Chap. 9). Other possible inference methods for abundance and density exist as well (Chap. 8). If the decision process is to focus on a larger landscape scale, occupancy methods may be adequate for some purposes, although prey density estimates will be more useful for assessing models and making decisions than simple occupancy estimates.

Finally, the most mechanistic approach listed above incorporated poaching activity as a model state variable. Data relevant to this variable might include patrol encounters with poachers or poacher sign (kill sites), arrest records, local media reports, and even "market surveys" of local shops. Of course incomplete detection characterizes these data sources, just as they do animal monitoring data. That is, certainly some poaching activity will go undetected. However, capture-recapture and occupancy thinking can be used to draw inferences about illegal activities in the face of nondetection (e.g., Barber-Meyer 2010; Sharma et al. 2014).

The presentation here of three possible approaches to modeling and monitoring decision processes that focus on patrols as a primary management action can be used to make several general points about such processes. First, there is no single "right way" to develop models and monitoring for any process. The approach of directly modeling tiger dynamics as a function of patrol effort requires simpler models and less monitoring effort than the other two approaches. The focus is directly on the response of tigers to the management action, with no need to model or monitor state variables associated with the underlying mechanisms of tiger response. One disadvantage of this approach may be the expected time lag between management action and tiger response. In situations where tigers are at carrying capacity relative to existing prey numbers, patrol efforts are expected to first influence prey populations, and tiger numbers may then respond to changes in prey numbers. However, where tiger numbers are below carrying capacity, frequently due to direct poaching of tigers, a response in tiger numbers could occur without an increase in prey. Modeling and monitoring that incorporate prey populations and poaching activity should reduce time lags in system responses to management, with changes in poaching activity expected to shortly follow changes in patrol effort. In fact, there is rarely good reason not to monitor poaching activity, because the cost is largely covered by existing patrol efforts and tiger and prey monitoring efforts.

Another potential disadvantage to focusing directly on the relationship between tiger numbers and patrol effort is that it may be more difficult to diagnose problems with models, should they exist. For example, multiple possible explanations might underlie a drop in tiger numbers or a failure of tiger numbers to respond to patrol efforts. Perhaps the patrol efforts were not effective at influencing poaching activities. Or perhaps poaching was reduced, but not to the extent needed to increase tiger numbers. Or maybe tiger numbers did not respond to increased prey numbers as anticipated. Diagnoses of problems with model predictions should be easier for programs that include intermediate state variables such as prey abundance and poaching activity.

14.4.5 Summary

There are multiple ways to model and monitor system dynamics in order to support decision processes. Perhaps the key point is the need to integrate all components of the decision process. The entire process is driven by objectives. So when these objectives include tiger populations, for example, then management actions should be selected based on their potential to influence tiger numbers. Similarly, models should project consequences of potential actions on tiger numbers, and monitoring should be focused on tiger dynamics. If more mechanistic models that include responses of prey species and even illegal activities are thought to be useful, then monitoring should be tailored to these models and should include state variables in addition to tigers. Good, and even optimal, decisions can be made using any of these described approaches, and all permit learning for better management in the future.

14.5 Conclusions

This chapter has not offered a critique of current approaches to conservation, either in general or for tigers in particular. Many tiger conservation efforts have been very effective and should be applauded. Instead, we began the chapter by simply noting that some classes of methods integral to conservation (specifically, inference and modeling) have evolved tremendously over the past century, whereas the standard approach to conservation decision-making has changed little. In this chapter, we have briefly described some modern approaches to decision-making, believing that they deserve consideration in tiger conservation.

Structured decision processes can be viewed as providing a formal structure for making logical, common-sense decisions. Separate specification and discussion of the elements of informed decision processes are useful not only in insuring consideration of all relevant elements, but also as a means of minimizing obfuscatory shifts from one element to another in arguments advocating specific actions. Discussing each element separately and then eventually combining them is a very useful approach to keeping discussions properly focused. Structured decision processes are typically transparent and defensible, characteristics that not only engender trust and buy-in, but that are also useful in cases of litigation.

ARM is a special case of structured decision-making developed for decisions that are recurrent (as opposed to one-time decisions) and characterized by uncertainty. The recurrent nature of these decision problems admits the possibility of learning about the predictive abilities of one or more system models by comparing model-based predictions against estimates of system state variables and vital rates obtained via monitoring. This insertion of a scientific step into the decision process effectively integrates science and management, leading to better decisions in the future as uncertainty is reduced.

Tiger conservation nearly always requires difficult decisions. Limited tiger numbers, limited tiger habitat, limited potential management actions, and limited resources available for tiger conservation all place pressure on those involved in conservation to make wise decisions. We believe that structured decision-making and ARM should be given serious consideration as means to facilitate such decisions for existing and future tiger conservation programs.

References

Barber-Meyer SM (2010) Dealing with the clandestine nature of wildlife-trade market surveys. Conserv Biol 24:918–923

Bellman R (1957) Dynamic programming. Princeton University Press, Princeton

Buckland ST, Anderson DR, Burnham KP, Laake JL, Borchers DL, Thomas L (2001) Introduction to distance sampling: estimating abundance of biological populations. Oxford University Press, Oxford

Chades I, Carwardine J, Martin TG, Nicol S, Sabbadin R (2012) MOMDPs: a solution for modelling adaptive management problems. Twenty-sixth AAAI Conference of the Association for the Advancement of Artificial Intelligence (AAAI-12) July 22–26, Toronto, pp 267–273

Cooch EG, Guillemain M, Boomer GS, Lebreton J-D, Nichols JD (2014) The effects of harvest on waterfowl populations. Wildfowl Spec Issue 4:220–276

Fackler PL, Pacifici K (2014) Addressing structural and observational uncertainty in resource management. J Environ Manag 133:27–36

Fackler PL, Pacifici K, Martin J, McIntyre C (2014) Efficient use of information in adaptive management with an application to managing recreation near Golden Eagle nesting sites. PLoS One 9(8):e102434. doi:10.1371/journal.pone.0102434

Gopalaswamy AM, Royle JA, Delampady M, Nichols JD, Karanth KU, Macdonald DW (2012) Density estimation in tiger populations: combining information for strong inference on tiger density. Ecology 93:1741–1751

Johnson FA, Williams BK, Nichols JD, Hines JE, Kendall WL, Smith GW, Caithamer DF (1993) Developing an adaptive management strategy for harvesting waterfowl in North America. Transactions of the North American Wildlife and Natural Resources Conference 58:565–583

Karanth KK (2007) Making resettlement work: the case of India's Bhadra Wildlife Sanctuary. Biol Conserv 139:315–324

Karanth KU, Nichols JD, Kumar NS, Link WA, Hines JE (2004) Tigers and their prey: predicting carnivore densities from prey abundance. Proc Natl Acad Sci 101:4854–4858

Karanth KU, Gopalaswamy AM, Kumar NS, Vaidyanathan S, Nichols JD, MacKenzie DI (2011) Monitoring carnivore populations at landscape scales: occupancy modeling of tigers from sign surveys. J Appl Ecol 48:1048–1056

Martin J, Fackler PL, Nichols JD, Runge MC, McIntyre C, Lubow BL, McCluskie MG, Schmutz JA (2011) An adaptive management framework for optimal control of recreational activities in Denali National Park. Conserv Biol 25:316–323

McGowan CP, Smith DR, Nichols JD, Lyons JE, Sweka J, Kalasz K, Niles LJ, Wong R, Brust J, Davis M, Spear B (2015) Implementation of a framework for multi-species, multi-objective adaptive management in Delaware Bay. Biol Conserv 191:759–769

Nichols JD, Runge MC, Johnson FA, Williams BK (2007) Adaptive harvest management of North American waterfowl populations: a brief history and future prospects. J. Ornithol 148(Suppl 2):S343–S349

Nichols JD, Williams BK (2013) Adaptive management. In: Encyclopedia of environmetrics, 2nd edn. Wiley, New York, pp 1–6

Nichols JD, Johnson FA, Williams BK, Boomer GS (2015) On formally integrating science and policy: walking the walk. J Appl Ecol 52:539–543

Puterman ML (1994) Markov decision processes: discrete stochastic dynamic programming. Wiley, New York

Royle JA, Nichols JD, Karanth KU, Gopalaswamy AM (2009a) A hierarchical model for estimating density in camera trap studies. J Appl Ecol 46:118–127

Royle JA, Karanth KU, Gopalaswamy AM, Kumar NS (2009b) Bayesian inference in camera trapping studies for a class of spatial capture-recapture models. Ecology 90:3233–3244

Sharma K, Wright B, Joseph T, Desai N (2014) Tiger poaching and trafficking in India: estimating rates of occurrence and detection over four decades. Biol Conserv 179:33–39

Smith DR, McGowan CP, Daily JP, Nichols JD, Sweka JA, Lyons JE (2013) Evaluation of a multi-species adaptive management framework: must uncertainty impede effective decision-making? J Appl Ecol 50:1431–1440

Sutherland WJ, Pullin AS, Dolman PM, Knight TM (2004) The need for evidence-based conservation. Trends Ecol Evol 19:305–308

US Fish and Wildlife Service (2013) Adaptive harvest management: 2013 duck hunting season. U.S. Department of Interior, Washington, DC

Walters CJ (1986) Adaptive management of renewable resources. Macmillan Publishing Company, New York

Williams BK, Nichols JD (2014) Optimization in natural resources conservation. In: Guntenspergen G (ed) Application of threshold concepts in natural resource decision making. Wiley, New York, pp 45–65

Williams BK, Nichols JD, Conroy MJ (2002) Analysis and management of animal populations. Academic, San Diego

Williams BK, Szaro RC, Shapiro CD (2007) Adaptive management: the U.S. department of the interior technical guide. U.S. Department of the Interior, Washington, DC

MIX
Papier aus verantwortungsvollen Quellen
Paper from responsible sources
FSC® C105338

If you have any concerns about our products,
you can contact us on
ProductSafety@springernature.com

In case Publisher is established outside the EU,
the EU authorized representative is:
Springer Nature Customer Service Center GmbH
Europaplatz 3, 69115 Heidelberg, Germany

Printed by Libri Plureos GmbH
in Hamburg, Germany